W9-BCL-189

MONOGRAPHS ON
STATISTICS AND APPLIED PROBABILITY

Probability, Statistics and Time
M.S. Bartlett

The Statistical Analysis of Spatial Pattern
M.S. Bartlett

Stochastic Population Models in Ecology and Epidemiology
M.S. Bartlett

Risk Theory
R.E. Beard, T. Pentikäinen and E. Pesonen

Bandit Problems: Sequential Allocation of Experiments
D.A. Berry and B. Fristedt

Residuals and Influence in Regression
R.D. Cook and S. Weisberg

Point Processes
D.R. Cox and V. Isham

Analysis of Binary Data
D.R. Cox

The Statistical Analysis of Series of Events
D.R. Cox and P.A.W. Lewis

Analysis of Survival Data
D.R. Cox and D. Oakes

Queues
D.R. Cox and W.L. Smith

Stochastic Modelling and Control
M.H.A. Davis and R. Vinter

Stochastic Abundance Models
S. Engen

The Analysis of Contingency Tables
B.S. Everitt

An Introduction to Latent Variable Models
B.S. Everitt

Finite Mixture Distributions
B.S. Everitt and D.J. Hand

Population Genetics
W.J. Ewens

Classification
A.D. Gordon

Monte Carlo Methods
J.M. Hammersley and D.C. Handscomb

Identification of Outliers
D.M. Hawkins

Generalized Linear Models
P. McCullagh and J.A. Nelder

Distribution-free Statistical Methods
J.S. Maritz

Multivariate Analysis in Behavioural Research
A.E. Maxwell

Applications of Queueing Theory
G.F. Newell

Some Basic Theory for Statistical Inference
E.J.G. Pitman

Density Estimation for Statistics and Data Analysis
B.W. Silverman

Statistical Inference
S.D. Silvey

Models in Regression and Related Topics
P. Sprent

Regression Analysis with Applications
G.B. Wetherill

Sequential Methods in Statistics
G.B. Wetherill and K.D. Glazebrook

(Full details concerning this series are available from the Publishers)

The Statistical Analysis of Compositional Data

J. AITCHISON
Professor of Statistics
University of Hong Kong

LONDON NEW YORK
CHAPMAN AND HALL

First published in 1986 by
Chapman and Hall Ltd
11 New Fetter Lane, London EC4P 4EE
Published in the USA by
Chapman and Hall
29 West 35th Street, New York, NY 10001

Printed in Great Britain by
J.W. Arrowsmith Ltd, Bristol

ISBN 0 412 28060 4

British Library Cataloguing in Publication Data

Aitchison, J.
The statistical analysis of compositional data.
——(Monographs on statistics and applied probability)
1. Mathematical statistics
I. Title II. Series
519.5 QA276

ISBN 0-412-28060-4

Library of Congress Cataloging-in-Publication Data

Aitchison, J. (John), 1926-
The statistical analysis of compositional data.

(Monographs on statistics and applied probability)
Includes bibliographies and indexes.
1. Multivariate analysis. 2. Correlation (Statistics)
I. Title. II. Title: Compositional data. III. Series.
QZ278.A37 1986 519.5'35 86-2685
ISBN 0-412-28060-4 (U.S.)

To M,
the constant
among many variables

Contents

Preface

As long ago as 1897 Karl Pearson, in a now classic paper on spurious correlation, first pointed out dangers that may befall the analyst who attempts to interpret correlations between ratios whose numerators and denominators contain common parts. He thus implied that the analysis of compositional data, with its concentration on relationships between proportions of some whole, is likely to be fraught with difficulty. History has proved him correct: over the succeeding years and indeed right up to the present day, there has been no other form of data analysis where more confusion has reigned and where more improper and inadequate statistical methods have been applied.

The special and intrinsic feature of compositional data is that the proportions of a composition are naturally subject to a unit-sum constraint. For other forms of constrained data, in particular for directional data where there is a unit-length constraint on each direction vector, scientist and statistician alike have readily appreciated that new statistical methods, appropriate to the special nature of the data, are required; and there now exists an extensive literature on the successful statistical analysis of directional data. It is paradoxical that for compositional data, subject to an apparently simpler constraint, such an appreciation and development have been slower to emerge. In applications the unit-sum constraint has been widely ignored or wished away and inappropriate 'standard' statistical methods, devised for and successfully applied to unconstrained data, have been used with disastrous consequences. The special difficulties of compositional data have been recognized by a number of scientists and statisticians; their considerable activity, however, seems to have been divided between warning their colleagues that for compositional data standard concepts of correlation are uninterpretable, and illustrating in greater and greater detail the variety of ways in which standard statistical procedures go wrong, rather than the more positive approach of creating a meaningful new methodology.

The purpose of this monograph is to present a clear and unified account of such a new statistical methodology devised in recent years specifically for compositional data. The style of presentation is applied mathematical, letting practical problems of compositional data analysis from a wide variety of disciplines motivate the introduction of statistical concepts, principles and procedures, which are in turn illustrated by applications to compositional data sets. Emphasis is on the provision of useful and usable results rather than on elaborate proofs of their derivation. An essential role, however, of any first text in a new and substantial area of statistics is to make the reader familiar enough with the tools of the statistical trade and sufficiently confident in their use to allow their ready adaptation to unfamiliar compositional data problems. As a help towards instilling this confidence, each chapter ends with a set of problems for the reader.

Compositional data analysts will appreciate that the multivariate form of their data demands that the statistical methods of analysis must also be multivariate. Thus for an understanding of this text the reader will require not only to be familiar with the basics of statistical inference but also to have had some exposure, however minimal, to ideas of standard multivariate analysis. Any analyst, who in the past has been able to apply, however mistakenly, standard multivariate procedures to compositional data, will have no difficulty with the methods advocated in this monograph. The great beauty of the new methodology appropriate to unit-sum contrained data is its essential simplicity.

A user-friendly, interactive microcomputer statistical package CODA has been produced to allow easy implementation of the many special methods required for efficient compositional data analysis. Information on the technical details and availability of the CODA manual and diskette are provided in Appendix C.

This monograph owes much to many people. To colleagues at the University of Hong Kong, some of whom have collaborated in research into aspects of compositional data analysis which have made this text possible. To discussants at conferences and seminars in Australia, Canada, New Zealand, the People's Republic of China, the UK, USA, and in particular to the discussants of a paper read to the Royal Statistical Society in London in January 1982. To postgraduate students in classes at Princeton University during the Fall semester 1981 and at the University of Hong Kong during the 1984–85 academic session for their challenging questions. To Mrs Lucile Lo for her efficient technical word-processing and production of a perfect typescript, with patience and good humour in spite of the many hindrances placed in her way by the author. To my wife Muriel for her ready acceptance of the loss of vacations during the long hot Hong Kong working summers of the final draft.

Reference system

The numbering system for definitions, tables, figures, properties (propositions), equations, questions and problems is by decimal point within chapter. Typical examples with obvious abbreviations are Definition 2.1, Table 1.6, Fig. 6.2, Property 5.3, Question 1.2.4. Equations are numbered on the right and are referred to simply by bracketed number; for example, (4.2) is the second numbered equation in Chapter 4.

A number of compositional data sets are used repeatedly in the text to motivate the introduction of concepts, to develop methodology, to illustrate applications and to present new problems to the reader. For ease of reference these data sets are collected in Appendix D at the end of the monograph and referred to in the text as Data 1, Data 2,....

To avoid interruption to the exposition, references to published work are mainly confined to bibliographic notes at the end of each chapter. References in the text to the Bibliography are by author and year of publication.

Properties of elementary matrices are listed in Appendix A and referred to in the text as F1, F2, G1, H1, etc.

CHAPTER 1

Compositional data: some challenging problems

It is sometimes considered a paradox that the answer depends not only on the observations but on the question: it should be a platitude.

Harold Jeffreys: *The Theory of Probability*

1.1 Introduction

Any vector **x** with non-negative elements $x_1, \ldots, x_D$ representing proportions of some whole is subject to the obvious constraint

$$x_1 + \cdots + x_D = 1. \tag{1.1}$$

Compositional data, consisting of such vectors of proportions, play an important role in many disciplines and often display appreciable variability from vector to vector. Because of this variability some form of statistical analysis is essential for the adequate investigation and interpretation of the data. All too often, however, the unit-sum constraint (1.1) is either ignored or improperly incorporated into the statistical modelling and there results an inadequate or irrelevant analysis with a doubtful or distorted inference. The sole purpose of this monograph is to develop a simple and appropriate statistical methodology for the analysis of compositional data and to demonstrate its use in applications.

We start by presenting in this chapter a variety of problems of compositional data analysis from a number of different disciplines. These serve not only to illustrate the breadth of the subject but also to motivate the concepts introduced and the methods developed in subsequent chapters. A first step in this motivation and development is to identify and formulate questions appropriate to each particular situation.

In some areas of application the original data sets are necessarily large and so, in order to provide the reader with some forms of problem, we have been forced to reduce the data sets to manageable size within the constraints of textbook illustration. In so doing we have taken care to preserve the essential nature of each problem and to present the full methodology required for its analysis. Since reference to a particular data set may take place at many points throughout the text, all data sets are collected in Appendix D at the end of the monograph to allow easier location. These data sets are numbered sequentially and referred to briefly as Data 1, Data 2,....

1.2 Geochemical compositions of rocks

The statistical analysis of geochemical compositions of rocks is fundamental to petrology. Commonly such compositions are expressed in terms of percentages by weight of ten or more major oxides. For example, the compositions of Table 1.1 consist of the percentages by weight of ten major oxides for six specimens, three each of Permian and post-Permian rocks, selected from a set of 102 specimens investigated by Carr (1981). The variation, not only between specimens of different types but also between specimens of the same type, clearly indicates that some form of statistical analysis is required for a full appreciation of the pattern of major-oxide compositions of Permian and post-Permian rocks.

In order to study the nature of such problems with smaller data sets we present in Data 1 and 2 mineral compositions of 50 rock specimens, 25 of the type hongite (Data 1) and 25 of the type kongite (Data 2). Each composition consists of the percentages by weight of five minerals, albite, blandite, cornite, daubite, endite, which we conveniently abbreviate to A, B, C, D, E in future discussion.

Table 1.1. *Some typical major-oxide compositions of Permian and post-Permian rocks*

	Percentage compositions of major oxides by weight									
Type	SiO_2	TiO_2	Al_2O_3	TotFe	MnO	MgO	CaO	Na_2O	K_2O	P_2O_5
Permian	60.54	1.32	15.22	6.95	0.21	2.33	3.18	4.81	4.84	0.60
	54.30	1.24	16.67	8.70	0.07	4.24	8.34	3.41	2.52	0.49
	52.17	0.82	20.05	8.38	0.10	2.28	9.29	3.22	2.99	0.69
Post-	55.95	1.26	18.54	7.24	0.28	1.20	3.30	6.14	5.67	0.45
Permian	45.40	1.34	20.14	8.00	0.06	9.29	9.59	3.89	1.38	0.90
	46.59	1.06	15.99	11.20	0.30	10.50	10.45	2.03	1.45	0.43

Each of these data sets displays the characteristic features of a compositional data set.

1. Each row of the data array corresponds to a single rock specimen, more generally to a *replicate*, a single experimental or observational unit.
2. Each column corresponds to a single mineral, more generally to a specific ingredient or part of each composition.
3. Each entry is non-negative.
4. The sum of the entries in each row is 1, or equivalently 100 per cent.

Thus the entry x_{ri} in the rth row and ith column records the proportion or component of the ith ingredient or part in the rth replicate composition. For the 25 replicates of 5-part compositions in each of Data 1 and 2, r runs through $1,\ldots,25$ and i through $1,\ldots,5$. When we consider compositional data arrays more generally we shall suppose that there are N replicates of D-part compositions, so that the unit-sum property of rows or replicates imposes the constraints

$$\sum_{i=1}^{D} x_{ri} = 1 \qquad (r = 1,\ldots,N). \tag{1.2}$$

If we express the complete data array as the $N \times D$ matrix $\mathbf{X}$ then constraints (1.2) can be expressed in matrix form:

$$\mathbf{X}\mathbf{j}_D = \mathbf{j}_D, \tag{1.3}$$

where $\mathbf{j}_D$ denotes a $D \times 1$ vector with each of its D entries equal to 1.

A first step towards the development of sensible statistical methods of analysing the hongite and kongite data sets is to formulate some typical questions.

Questions 1.2

1. How can we satisfactorily describe the pattern of variability of hongite compositions?

2. For a new rock specimen with (A, B, C, D, E) composition (44.0, 20.4, 13.9, 9.1, 12.6) and claimed to be hongite, can we say whether it is fairly typical of hongite in composition? If not, can we place some measure on its atypicality?

3. For a particular rock type, can we define a covariance or correlation structure which will allow us to pose and test meaningful hypotheses about that structure?

4. What forms of statistical independence are possible within the unit-sum constraint?

5. To what extent can we obtain insight into the pattern of variability of the compositions by concentrating on the relative values of the components in a subset, say (A, B, C), of each composition? How can we quantify what, if anything, is being lost in this subcompositional approach to compositional data analysis?

6. Are the patterns of variability of hongite and kongite different? If so, can a convenient form of classification be devised on the basis of the composition?

7. Can we investigate whether a rule for classification based on only a selection of the components would be as effective as use of the full composition?

Similar questions can be posed about Data 3 and 4 which give the percentages by weight of minerals A, B, C, D, E in 25 rock specimens of types boxite and coxite, respectively. These data sets also contain extra information in the recorded depth of location of each specimen and, additionally for each of the coxite specimens, a measure of porosity. We shall pose questions involving this concomitant information later; meanwhile we concentrate on one noticeable feature of the compositions, namely that the boxite and coxite compositions are much less variable than the hongite and kongite compositions.

Questions 1.2 (continued)

8. Can we somehow quantify the variability of a composition by defining some measure of total variability?

1.3 Sediments at different depths

In sedimentology, specimens of sediments are traditionally separated into three mutually exclusive and exhaustive constituents, sand, silt and clay, and the proportions of these parts by weight are quoted as (sand, silt, clay) compositions. Data 5 records the (sand, silt, clay) compositions of 39 sediment samples at different water depths in an Arctic lake. Of obvious interest here are the following questions.

Questions 1.3

1. Is sediment composition dependent on water depth?

2. If so, how can we quantify the extent of the dependence?

For the compositional data set of this problem $N = 39$ and $D = 3$, and the compositional data matrix $\mathbf{X}$ is of order 39×3. The additional feature here is the introduction of a concomitant variable or *covariate*, water depth, which may account for some of the variation in the compositions. In familiar statistical terminology we are faced with a multivariate regression problem with sediment composition as regressand and water depth as regressor. The unfamiliar aspect of the problem is the unit-sum constraint on the components of the regressand vector.

Similar questions can clearly be posed about the possible dependence of the boxite compositions on depth for the 25×5 compositional data array of Data 3.

1.4 Ternary diagrams

A convenient way of displaying the variability of 3-part compositions such as the (sand, silt, clay) compositions of Data 5 is in what is variously termed a *ternary diagram*, a *reference triangle*, or *barycentric*

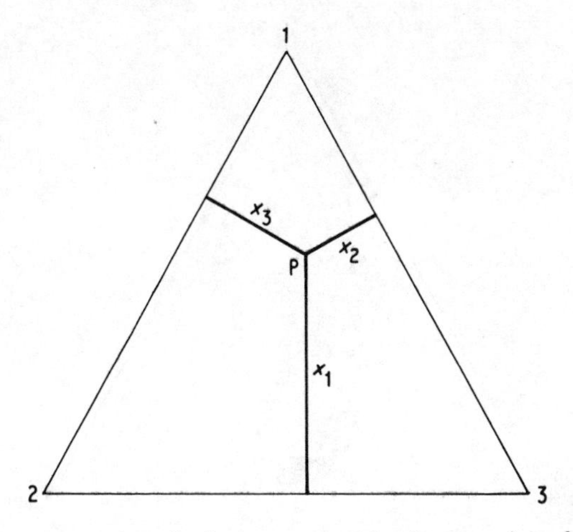

Fig. 1.1. *Representation of a 3-part composition (x_1, x_2, x_3) in the reference triangle 123.*

coordinate space. Such ternary diagrams are extremely popular in some disciplines, particularly in the geological sciences. We here illustrate the use of a ternary diagram for the sediment compositions just discussed.

The triangle of Fig. 1.1 with vertices 1, 2, 3 is equilateral and has unit altitude. For any point P in triangle 123 the perpendiculars x_1, x_2, x_3 from P to the sides opposite 1, 2, 3 satisfy

$$x_i \geqslant 0 \quad (i = 1, 2, 3), \qquad x_1 + x_2 + x_3 = 1. \tag{1.4}$$

Moreover, corresponding to any vector (x_1, x_2, x_3) satisfying (1.4), there is a unique point in triangle 123 with perpendicular values x_1, x_2, x_3. There is therefore a one-to-one correspondence between 3-part compositions and points in triangle 123, and so we have a simple means of representing 3-part compositions. In such a representation we may note that the three inequalities in (1.4) are strict if and only if the representative point lies in the interior of triangle 123. Also, the larger a component x_i is, the further the representative point is away from the side opposite the vertex i, and thus, roughly speaking, the nearer the point is to the vertex i. Moreover, 3-part compositions with two components, say x_2 and x_3, in constant ratio are represented by points on a straight line through the complementary vertex 1. Conveniently scaled triangular coordinate paper is commercially available for the construction of ternary diagrams.

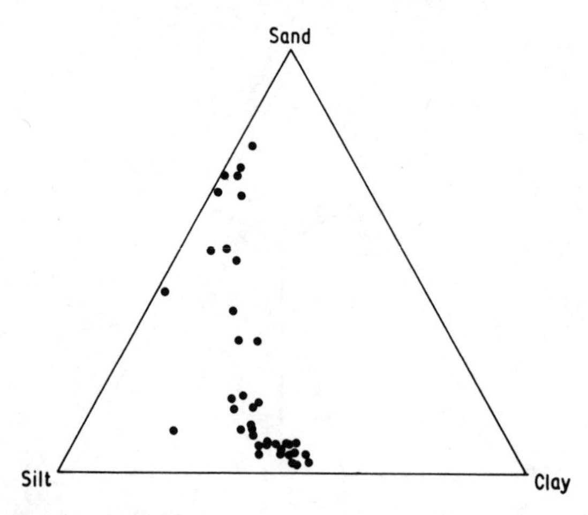

Fig. 1.2. *Ternary diagram for (sand, silt, clay) compositions of 39 sediment samples from an Arctic lake.*

With this form of representation, each (sand, silt, clay) composition of Data 5, with percentages such as (77.5, 19.5, 3.0) expressed in their decimal fraction forms (0.775, 0.195, 0.030), can be plotted as a point in the ternary diagram of Fig. 1.2. The diagram shows clearly that there is a substantial cluster of compositions with low proportions of sand accompanied by roughly equal proportions of silt and clay. We can also judge the extent of the variability of the ratio of two components from the sweep of straight lines through the complementary vertex required to cover all the representative points. For example, for the sediments the sand/clay ratio is much more variable than the silt/clay ratio.

The ternary diagram obviously cannot in itself reveal any possible dependence of sediment composition on water depth. For this purpose we require to construct a 3-dimensional display such as the triangular bar in Fig. 1.3, where the depth axis is along the length of the bar. We could then fully represent Data 5 by placing each composition in the ternary diagram produced by the cross-section of the bar at the appropriate depth for that composition. We would then have to try to identify any systematic trend in the representational points along the length of the bar. Although such 3-dimensional displays have a useful expository role in describing the structure of a problem, they have clearly only a limited part to play in data analysis. Moreover, it is impossible to extend this form of visual display if the

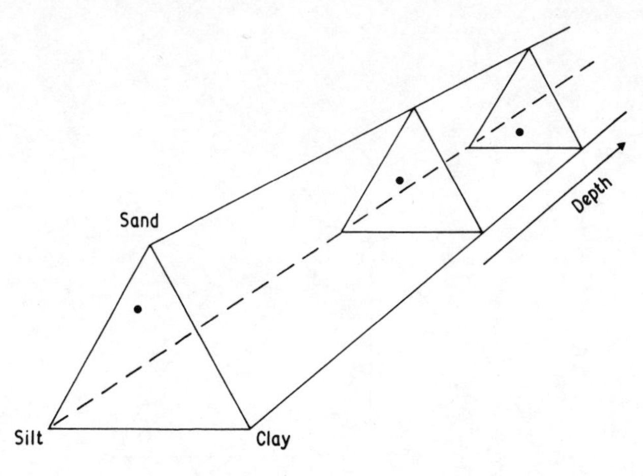

Fig. 1.3. *Triangular bar diagram for representation of sediment compositions and depths.*

composition has more than three parts or if there are more than two explanatory factors under consideration.

1.5 Partial analyses and subcompositions

Before we proceed to pose questions about further compositional data sets, we first use ternary diagrams to make more specific the broad issues raised in Question 1.2.5 about certain forms of partial analysis of the hongite and kongite compositions of Data 1 and 2.

We have seen in Section 1.4 how to obtain a useful visual representation of the variability of 3-part compositions. For 2-part compositions we have an even simpler representation on a line joining points 1 and 2 of unit length, in the manner illustrated in Fig. 1.4, a representative point P being near the end-point or vertex i if the corresponding component x_i is large.

For 4-part compositions we have to move into 3-dimensional space to obtain a picture of the variability, with a regular tetrahedron 1234 of unit altitude taking the place of the triangle 123 used as a frame of reference for 3-part compositions. In Fig. 1.5 the component x_i

Fig. 1.4. *Representation of a 2-part composition* (x_1, x_2) *on the line joining points 1 and 2.*

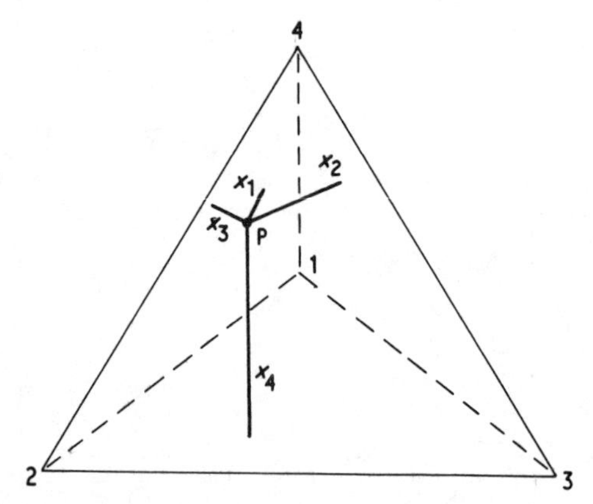

Fig. 1.5. *Representation of a 4-part composition* (x_1, x_2, x_3, x_4) *in the reference tetrahedron 1234.*

corresponds to the perpendicular from the representative point P to the triangular face opposite the vertex i. For compositions with more than four parts there is no satisfactory way of obtaining a visual representation of the variability and so there has been a tendency to be content with only partial analyses which focus attention on some subcompositions formed from the composition.

Corresponding to any subset of mineral constituents of hongite, say the subset A, B, C, we can form a subcomposition simply by rescaling the original proportions of these selected minerals so that the scaled proportions sum to 1. For example, the first hongite specimen H1 of Data 1 yields an ABC subcomposition expressed in percentages as follows:

$$100 \times (48.8, 31.7, 3.8)/(48.8 + 31.7 + 3.8)$$
$$= (57.9, 37.6, 4.5).$$

The complete set of 25 ABC subcompositions for the hongite specimens is given in Table 1.2. Since each such subcomposition is essentially a 3-part composition, it can be represented in a ternary diagram ABC, as in Fig. 1.6. In a similar way, the ADE

Table 1.2. *ABC subcompositions for the 25 specimens of hongite of Data 1*

Specimen no.	*Percentages* A	B	C	*Specimen no.*	*Percentages* A	B	C
H1	57.9	37.6	4.5	H16	40.1	9.1	50.8
H2	59.5	29.4	11.1	H17	48.8	50.1	1.1
H3	46.1	11.3	42.6	H18	58.3	38.0	3.7
H4	62.1	29.1	8.8	H19	53.9	20.1	26.0
H5	51.8	44.8	3.4	H20	55.8	14.4	29.8
H6	63.2	31.7	5.1	H21	57.8	21.0	21.2
H7	54.3	40.1	5.6	H22	61.1	30.6	8.3
H8	41.8	6.3	51.9	H23	57.1	40.0	2.9
H9	51.8	14.7	33.5	H24	57.1	20.8	22.1
H10	47.4	51.8	0.8	H25	57.0	30.9	12.0
H11	61.8	24.1	14.1				
H12	53.0	43.8	3.2				
H13	40.8	10.6	48.6				
H14	53.3	16.6	30.1				
H15	58.1	22.0	19.9				

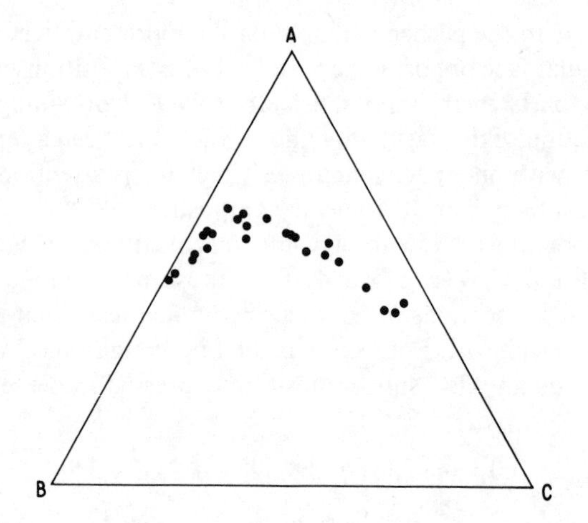

Fig. 1.6. *ABC subcompositions for 25 hongite specimens.*

subcompositions for hongite can be computed and represented in the ternary diagram ADE of Fig. 1.7.

There is a marked contrast in the patterns of variability of ABC and ADE subcompositions. The first appears to have widely scattered replicates in a banana- or boomerang-shaped configuration with extensive variation in the ratio of B to C compared with the other two ratios; the second is much more compact, possibly pear-shaped and

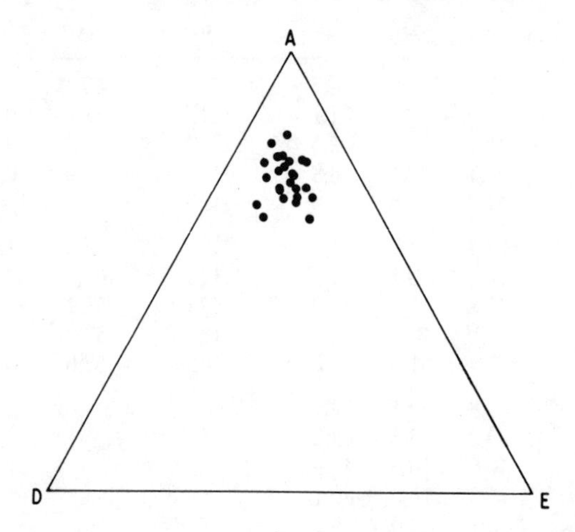

Fig. 1.7. *ADE subcompositions for 25 hongite specimens.*

with no great variation in any of the ratios of its mineral components A, D, E.

A whole sequence of awkward questions now springs to mind.

Questions 1.5

1. Should we interpret the obvious curvature in Fig. 1.6 as indicative of some kind of trend in the hongite ABC subcompositions?

2. In what sense may the ABC subcomposition be capturing the essence of the pattern of variability of the full composition?

3. Which of the subcompositions, ABC or ADE, is better in this role? If we conclude that ABC retains most of the information in the whole composition, is it possible that one of the 2-part subcompositions, AB, AC, BC, would be just as satisfactory?

4. Can we detect any differences between the hongite ABC subcompositional pattern of Fig. 1.6 and that for kongite in Fig. 1.8? If not, how can we choose a 3-part subcomposition which somehow captures the essence of the hongite and kongite patterns individually and yet emphasizes the differences between the patterns?

5. If we find none of the ten possible 3-part subcompositions satisfactory in this respect, must we conclude that mineral composition serves no useful purpose in distinguishing between hongite and

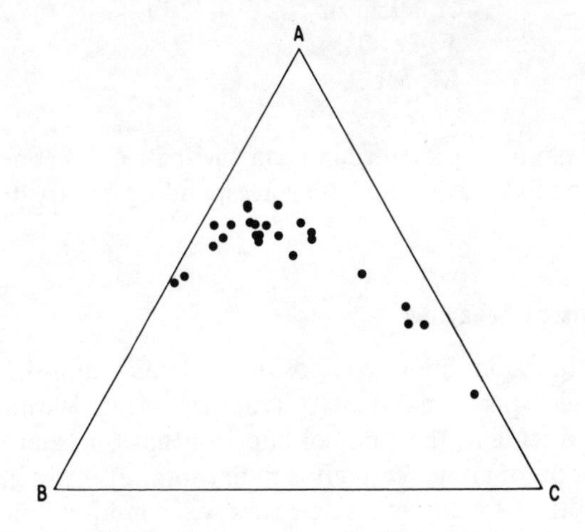

Fig. 1.8. *ABC subcompositions for 25 kongite specimens.*

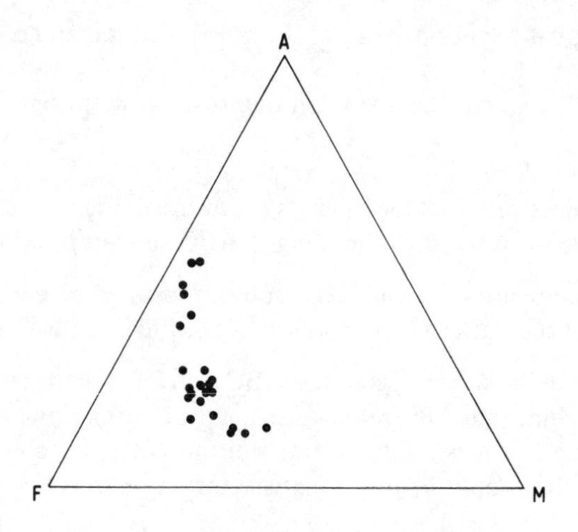

Fig. 1.9. *AFM ternary diagram for 23 aphyric Skye lavas.*

kongite? More generally, are there any other graphic representations of compositional data which may prove more satisfactory tools than ternary diagrams for 3-part subcompositions?

A popular version of such partial analyses in geochemistry is the AFM diagram, formed from the relative proportions of

A: alkali or $Na_2O + K_2O$,
F: Fe_2O_3,
M: MgO.

A typical example is presented in Data 6 with its AFM compositions of 23 aphyric Skye lavas and with corresponding ternary diagram in Fig. 1.9.

1.6 Supervisory behaviour

So far our questions about compositional data analysis have arisen from geological problems about rocks and sediments. We now turn to a different discipline, the study of human behaviour, and present a simple version of an increasingly popular form of investigation.

The results of a study of a single supervisor in his relationship to three supervisees are recorded in Data 7. Instruction in a technical subject took place in sessions of one hour and with one supervisee at a time. Each supervisee attended six sessions, once every two weeks over a twelve-week period. All of the eighteen sessions were recorded and for each session the 'statements' of the supervisor were classified into four categories:

C: commanding, giving a specific instruction to the supervisee;
D: demanding, posing a specific question to the supervisee;
E: expository, providing the supervisee with an explanation;
F: faulting, pointing out faulty technique to the supervisee.

Thus for each session the proportions of statements in the four categories form a (C, D, E, F) composition and the compositions are set out in Data 7 in a two-way format according to fortnight and supervisee.

There are two main questions of interest.

Questions 1.6

1. Does the supervisor's behaviour change over time?

2. Is the supervisor's behaviour consistent with respect to his three supervisees?

The problem here is similar to that of Section 1.3, where we were concerned about the manner in which sediment composition might be dependent on a single covariate, water depth. Our current interest is to discover the extent to which the compositions in our 18×3 data matrix are dependent on two factors, fortnight and supervisee. We can anticipate, therefore, that any resolutions of the two problems will be closely related.

1.7 Household budget surveys

An important aspect of the study of consumer demand is the analysis of household budget surveys, in which attention often focuses on the expenditures of a sample of households on a number of mutually exclusive and exhaustive commodity groups and their relations to total expenditure, income, type of housing, household composition, and so on. In the investigation of such data the pattern or composition of expenditures, the proportions of total expenditures allocated to the commodity groups, can be shown to play a central role in a form of budget-share approach to the analysis. Assurances of confidentiality and limitations of space preclude the publication of individual budgets from an actual survey, but we can present a reduced version of the problem which retains its key characteristics.

In a sample survey of single persons living alone in rented accommodation, twenty men and twenty women were randomly selected and asked to record over a period of one month their expenditures on the following four mutually exclusive and exhaustive commodity groups:

1. Housing, including fuel and light.
2. Foodstuffs, including alcohol and tobacco.
3. Other goods, including clothing, footwear and durable goods.
4. Services, including transport and vehicles.

The results are recorded in Data 8.

Interesting questions are readily formulated.

Questions 1.7

1. To what extent does the pattern or budget share of expenditures for men depend on the total amount spent?

2. Are there differences between men and women in their expenditure patterns?

3. Are there some commodity groups which are given priority in the allocation of expenditure?

For a satisfactory answer to such questions we clearly require a flexible form of statistical analysis of compositions. An idea of the nature of the first question can be provided in diagrammatic form if we consider a division of total expenditure into only three commodity groups. Then for each household the data of pattern of expenditure and total expenditure can be represented as a point within a triangular bar as in Fig. 1.3. Distance along the bar now represents total expenditure; budget pattern determines the compositional point within the ternary diagram cross-section at that distance. The first question posed then requires us to detect, if we can, any trend in the location of the data points as we move along the bar.

1.8 Steroid metabolite patterns in adults and children

The daily excretions of certain steroid metabolites in the urine have been found useful in distinguishing between the various forms of a rare syndrome. The statistical diagnostic system has been devised from records of adult excretions and is not directly applicable to children, whose excretion quantities are appreciably smaller than those of adults. To explore the possibility of modifying the diagnostic system for application to children it is therefore necessary to investigate whether the excretion patterns of children differ from those of adults and to obtain some quantitative picture of what these differences, if any, may be. The details of the diagnostic problem need not concern us here, except to provide motivation for the compositional analysis of the urinary excretions of adults and children.

Data 9 shows the urinary excretions (mg/24 hours) of 37 normal adults and 30 normal children of

1. total cortisol metabolites, w_1;
2. total corticosterone metabolites, w_2;
3. total pregnanetriol and Δ-5-pregnentriol, w_3.

For each of these vectors (w_1, w_2, w_3) we can then construct the corresponding composition or relative pattern

$$(x_1, x_2, x_3) = (w_1, w_2, w_3)/(w_1 + w_2 + w_3)$$

and the size

$$t = w_1 + w_2 + w_3$$

of the excretion. Questions that then arise are the following.

Questions 1.8

1. Does the relative pattern depend on the size of the excretion in adults, in children?

2. Are there differences between adults and children in the relative patterns of their excretions?

1.9 Activity patterns of a statistician

The activities of an academic statistician were divided into the following six categories:

1. Teaching
2. Consultation
3. Administration
4. Research
5. Other wakeful activities
6. Sleep

Data 10 shows the proportions of the 24 hours devoted to each activity, recorded on each of 20 days, selected randomly from working days in alternate weeks, so as to avoid any possible carry-over effects, such as a short-sleep day being compensated by make-up sleep on the succeeding day.

The six activities may be divided in two categories: 'work' comprising activities 1, 2, 3, 4; and 'leisure' comprising activities 5, 6. Our analysis may then be directed towards the work pattern consisting of the relative times spent on the four work activities, the leisure pattern, and the division of the day into work time and leisure time.

Questions 1.9

1. To what extent, if any, do the patterns of work and of leisure depend on the times allocated to these major divisions of the day?

2. Is the ratio of sleep to other wakeful activities dependent on the times spent in the various work activities?

1.10 Calibration of white-cell compositions

A cytologist is interested in the possibility of introducing into his laboratory a new method of determining the white-cell composition of a blood sample, that is the proportions of the three kinds of white cells,

G: granulocytes,
L: lymphocytes,
M: monocytes,

among the total of white cells observed. The current method involves time-consuming, microscopic inspection and is known to be accurate, whereas the proposed method is a quick automatic image analysis whose accuracy is still largely undetermined. In an experiment to assess the effectiveness of the proposed method, each of 30 blood samples was halved, one half being assigned randomly to one method, the other half to the other method. The resulting 30 pairs of 3-part compositions are recorded in Data 11. It is fairly obvious that the two methods produce different compositions but all that may be necessary is some form of calibration of the image analysis compositions to obtain substantial conformity with the microscopic inspection compositions.

The main questions here are thus the following.

Questions 1.10

1. Can the image analysis method safely replace the microscopic inspection method?

2. How can such a calibration of one composition with another be achieved?

An important feature of this problem is that, although we have apparently 60, 3-part compositions, these are essentially 30 pairs of related compositions. A statistical question we thus have to face is how we can develop models which will allow description of the pattern of joint or conditional variability of two compositions.

1.11 Fruit evaluation

The yatquat tree produces each season a single large fruit whose quality is assessed in terms of the relative proportions by volume of flesh, skin and stone. In an experiment to investigate whether a certain hormone influences quality, an arboriculturist uses 40 yatquat trees, randomly allocates 20 trees to the hormone treatment and leaves untreated the remaining 20 trees. Data 12 provides, in addition to the fruit compositions of the present season, the compositions of the fruit of the same 40 trees for the preceding season when none of the trees was treated.

Questions 1.11

1. Has the hormone treatment any effect on yatquat quality?

2. If so, what is the nature of this effect?

As in the white-cell problem of Section 1.10, we have here two compositions, past and present fruit compositions, for each experimental unit or yatquat tree, but with an additional factor of either treatment or no treatment.

1.12 Firework mixtures

The properties of many substances or objects, such as gasolines, metal alloys and cakes, depend on the particular mixture, or composition, of their ingredients. The purpose of experiments with different mixtures is to obtain some understanding of the nature and extent of the dependence of the properties on the composition. In the analysis of such experiments the composition is confined to the role of a covariate. An example from the fireworks industry will allow us to concentrate on more specific questions concerning this form of compositional data analysis. Data 13 shows two measured properties, brilliance and vorticity, of 81 girandoles composed of different mixtures of five ingredients (1, 2, 3, 4, 5). Of these ingredients, 1 and 2 are regarded as the primary light-producing ingredients, 3 as the principal propellant, and 4 and 5 as binding agents for 3. An overall objective of the experimentation is to investigate these supposed roles of the ingredients and to produce as simple an explanation as possible of how the properties relate to the ingredients. Questions of some importance are obviously the following.

Questions 1.12

1. Does the composition of the firework mixture have any effect at all on brilliance or vorticity?

2. If so, does brilliance depend solely on the light-producing ingredients 1 and 2? If not, in what way do the other ingredients contribute to brilliance?

3. If vorticity depends on the composition, is this effect due entirely to the principal propellant 3 or are the other ingredients involved in determining this response to the mixture?

1.13 Clam ecology

From the many colonies of clams in East Bay, 20 were selected at random and from each a sample of clams was taken. Each sample was sieved into three size ranges, large, medium, small; then each size range was sorted by shell colour, dark, light. For each colony the proportions of clams in each colour–size combination was estimated and the corresponding compositions are recorded in Data 14. Each composition is of the form

	Large	*Medium*	*Small*
Dark	x_{11}	x_{12}	x_{13}
Light	x_{21}	x_{22}	x_{23}

A similar study was conducted in West Bay and the resulting 20 colour–size compositions are given in Data 15. The marine biologist suspects that there may be a tendency for shell colour to darken as the clam grows. We can pose some relevant questions for investigation.

Questions 1.13

1. Is there any association between size and colour in the clam colonies of East Bay; of West Bay?

2. Are there any differences between East and West Bay clams in their colour–size variation?

A new feature in this problem is that the parts of the compositions are most readily described by the two factors, colour and size, and so each composition may be set out most naturally in a two-way, or more specifically a 2×3, array.

1.14 Bibliographic notes

Perusal of almost any geological journal, such as the *Journal of Geology*, *Journal of Petrology*, *Journal of Sedimentary Petrology*, *Contributions to Mineral Petrology*, will quickly uncover many compositional data sets; and in some series there is a high probability of finding a ternary diagram within ten randomly selected pages. The concept of the ternary diagram has its origin in the nineteenth century and can be found in Ferrers (1866). Its introduction into geological analysis dates back to Becke (1897) and so is contemporaneous with the warning of Pearson (1897) on the difficulties of interpretation of correlations between components of a composition.

An early example of compositional data in biology is to be found in the extensive pigmentation survey of almost half a million Scottish schoolchildren reported by Tocher (1908). As part of this study Tocher presented for each of 33 counties of Scotland four compositions, namely the proportions of boys and of girls in each of five hair-colour and in each of four eye-colour categories. A more recent and plentiful supply of biological compositional data is to be found in pollen and foraminiferal diagrams, which present proportions of different fossils at various levels within core samples. Typical examples of such palaeoecological data can be found in Birks and West (1973) and Poore and Berggren (1975).

The first attempt at a budget-share approach to household budget studies appears to be that of Working (1943). Although experiments with mixtures are clearly of ancient origin, the first formal attempts at statistical modelling are by Quenouille (1953, 1959). An account of development since then is given by Cornell (1981).

Problems

Each of the following problems presents a new situation involving a compositional data set. Attempt to formulate as precisely as possible any questions which you regard as of interest and which you judge may be answered by an analysis of the data. Draw any diagrams that you think may be helpful in indicating answers to your questions.

1.1 Differential diagnosis between two diseases A and B is currently expensive and time-consuming. The feasibility of using 4-part serum protein compositions easily determined from blood samples for this purpose is under investigation. Data 16 records the proportions of the four serum proteins from blood samples of 30 patients, of whom 14 and 16 are known to have diseases A and B, respectively. The investigating clinician is interested in the possibility of constructing a differential diagnostic system from this experience for the assessment of disease type from the serum protein compositions of new patients, such as the six cases Cl–C6 also recorded in Data 16.

1.2 In the study of subjective performance in inferential tasks the subject is faced with a finite set of mutually exclusive and exhaustive hypotheses, and on the basis of specific information presented to him is required to divide the available unit of probability among these hypotheses. In one such study the task is presented as a problem of differential diagnosis of three mutually exclusive and exhaustive diseases of students, known under the generic title of ‘newmath syndrome’,

A: algebritis,
B: bilateral paralexia,
C: calculus deficiency.

The subject, playing the role of diagnostician, is informed that the three disease types are equally common and is shown the results of 10 diagnostic tests on 60 previous cases of known diagnosis, 20 of each type. The subject is then shown the results of the 10 tests for a new undiagnosed case and asked to assign diagnostic probabilities to the three possible disease types.

Data 17 shows the subjective assessments of 15 clinicians and 15 statisticians for the same case. For this case the objective diagnostic probabilities are known to be (0.08, 0.05, 0.87).

1.3 In a pebble analysis of glacial tills, the total number of pebbles in each of 92 samples was counted and the pebbles were sorted into four categories,

A: red sandstone,
B: gray sandstone,
C: crystalline,
D: miscellaneous.

The percentages of these four categories and the total pebble counts are recorded in Data 18. The glaciologist is interested in describing the pattern of variability of his data and whether the compositions are in any way related to abundance.

1.4 The Hong Kong Pogo-Jump Championship is similar to the triple jump except that the competitor is mounted on a pogo-stick. After a *pogo-up* towards a starting board the total *jump* distance achieved in three consecutive *bounces*, known as the *yat*, *yee* and *sam*, is recorded. Data 19 shows the yat, yee and sam measurements for the four jumps of the seven finalists in the 1985 Championship.

1.5 Data 20 records the (sand, silt, clay) compositions of 17 sediments, 10 of which (A1–A10) are identified as *offshore* and 7(B1–B7) as *nearshore*. Four new samples Cl–C4, all from the same site and hence of the same type, have been analysed and the general problem is to assess, if possible, this unknown type.

1.6 In the development of bayesite, a new fibreboard, experiments were conducted to obtain some insight into the nature of the relationship of its permeability to the mix of its four ingredients,

A: short fibres,
B: medium fibres,
C: long fibres,
D: binder,

and the pressure at which these are bonded. The results of 21 such experiments are reported in Data 21. It is required to investigate the dependence of permeability on mixture and bonding pressure.

1.7 A study of the activities of 20 machine operators during their eight-hour shifts has been conducted, and the proportions of time spent in the following categories:

A: high-quality production,
B: low-quality production,
C: machine setting,
D: machine repair,

are recorded in Data 22. Of particular interest are any insights which such data may give of relationships between productive and non-productive parts of such shifts.

1.8 In a comparative study of four different methods of assessing leucocyte composition of a blood sample, aliquots of blood samples from ten patients were assessed by the four methods. Data 23 shows the percentages in the 40 analyses of:

P: polymorphonuclear leucocytes,
S: small lymphocytes,
L: large mononuclears.

1.9 As part of a study of seventeenth-century English skulls three angles of a triangle in the cranium,

N: nasial angle,
A: alveolar angle,
B: basilar angle,

were measured for 22 female and 29 male skulls. These, together with similar measurements on 22 female and 29 male skulls of the Naqada race, are presented in Data 24. The general objective is to investigate possible sex and race differences in skull shape.

1.10 In a regional ecological study plots of land of equal area were inspected and the parts of each plot which were thick or thin in vegetation and dense or sparse in animals were identified. From this field work the areal proportions of each plot were calculated for the four mutually exclusive and exhaustive categories (thick, dense), (thick, sparse), (thin, dense), (thin, sparse). These sets of proportions are recorded in Data 25 for 50 plots from each of two different regions, A and B. Questions of general interest are whether animals and vegetation behave independently of each other in their determination of habitat, and whether there are differences between the ecological patterns in regions A and B.

CHAPTER 2

The simplex as sample space

——he hath strange places cramm'd
With observations, the which he vents
In mangled forms.

William Shakespeare: *As You Like It*

2.1 Choice of sample space

The first task of the statistician when faced with modelling a new observational or experimental situation is to devise a suitable sample space. Such a sample space is simply a set of convenient symbols which can be identified through a one-to-one correspondence with the possible outcomes of the observational or experimental process. Often the task is simple. If we are simply recording on a Geiger counter the number of particles emitted by a radioactive source in a given interval of time, a natural sample space is the set of non-negative integers. If the process records the systolic and diastolic blood pressures of a patient at a given point of time, then a convenient sample space is the first, or positive, quadrant of two-dimensional real space. When our interest is in the direction of a star relative to the Earth at a particular moment of time, we can effectively adopt a unit sphere as sample space with a point on it representing the observed direction.

Sometimes the sample space turns out to be less familiar than the set of non-negative integers, the positive quadrant or the unit sphere. This is so for the problems of Chapter 1. We shall see that a restricted part of real space termed the *simplex* and operations on its elements play a fundamental role in problems of compositional data either as the whole or as an important part of the sample space. Our aim in this chapter is therefore to introduce the simplex and these operations as simply as possible and to establish a clear notation and terminology for later development of general methods of compositional data analysis.

2.2 Compositions and simplexes

Compositions

In our informal discussion of compositional data in Chapter 1 we have used the terms *parts* and *components* in relation to a *composition*. For example, the first row of Table 1.1 is the composition of a Permian rock specimen with its ten parts the major oxides

$$SiO_2, TiO_2, Al_2O_3, \text{TotFe}, MnO, MgO, CaO, Na_2O, K_2O, P_2O_5$$

and its ten components the percentages

$$60.54, 1.32, 15.22, 6.95, 0.21, 2.33, 3.18, 4.81, 4.84, 0.60.$$

Thus parts are labels describing or identifying the constituents into which the whole has been divided, whereas components are numerical proportions in which individual parts occur. When discussing a composition in more general terms we shall use the integers

$$1, 2, \ldots, D$$

to denote the parts, and subscripted letters

$$x_1, x_2, \ldots, x_D$$

to denote the components. The components of any D-part composition $(x_1, \ldots, x_D)$ must satisfy the obvious requirements that each component is non-negative:

$$x_1 \geqslant 0, \ldots, x_D \geqslant 0, \tag{2.1}$$

and that the sum of all the components is 1:

$$x_1 + \cdots + x_D = 1. \tag{2.2}$$

It will greatly simplify our exposition throughout Chapters 2–10 if we exclude from consideration any compositional data sets with zero components. The reason for this is that the general statistical methods we develop require special adaptation when such zeros are present. We shall take up the problem of zeros in Chapter 11 but until then we replace the natural non-negativity condition (2.1) by the stronger assumption of strict positivity:

$$x_1 > 0, \ldots, x_D > 0. \tag{2.3}$$

We now incorporate these ideas into a formal definition.

Definition 2.1 A *composition* $\mathbf{x}$ of D *parts* is a $D \times 1$ vector with positive *components* $x_1, \ldots, x_D$ whose sum is 1.

The unit-sum constraint (2.2) places a much more fundamental restriction on the freedom of the components of the composition $(x_1, \ldots, x_D)$ than the positivity of the components. First we note that the composition is completely specified by the components of a d-part subvector such as $(x_1, \ldots, x_d)$, where

$$d = D - 1, \tag{2.4}$$

since, by (2.2), the remaining component has a 'fill-up' value

$$x_D = 1 - x_1 - \cdots - x_d. \tag{2.5}$$

This means that a D-part composition is essentially a d-dimensional vector and so can be represented in some convenient d-dimensional set. We can emphasize this d-dimensional aspect of a composition $(x_1, \ldots, x_D)$, and anticipate a crucial later argument, by pointing out that specification of a d-part subvector is not the only convenient way of determining the composition. For example, if we specify the d ratios,

$$r_i = x_i/x_D \qquad (i = 1, \ldots, d), \tag{2.6}$$

then the composition is completely determined as follows:

$$\begin{aligned} x_i &= r_i/(r_1 + \cdots + r_d + 1) \qquad (i = 1, \ldots, d), \\ x_D &= 1/(r_1 + \cdots + r_d + 1). \end{aligned} \tag{2.7}$$

Simplexes

For any experimental or observational situation where we record a D-part composition $(x_1, \ldots, x_D)$ we have a choice between two routes to natural sample spaces. If we are concerned with mathematical problems, such as specifying a density function on the sample space, then we have to emphasize the dimensionality of the composition and define the sample space in terms of a determining subvector such as $(x_1, \ldots, x_d)$. This leads us to the following definition of a simplex in terms of standard set notation.

Definition 2.2 The d-dimensional *simplex* is the set defined by

$$\mathcal{S}^d = \{(x_1, \ldots, x_d): x_1 > 0, \ldots, x_d > 0; x_1 + \cdots + x_d < 1\}.$$

Since in most applications the component x_D given by (2.5) is as important as any of the other components $x_1, \ldots, x_d$, we would naturally prefer, in the absence of any mathematical prerequisite, a more symmetrical approach. In this approach we may adopt the following definition.

Definition 2.3 The d-dimensional *simplex embedded in D-dimensional real space* is the set defined by

$$\mathcal{S}^d = \{(x_1, \ldots, x_D): x_1 > 0, \ldots, x_D > 0; x_1 + \cdots + x_D = 1\}.$$

The $\mathcal{S}^d$ of Definitions 2.2 and 2.3 are equivalent as sample spaces for D-part compositions: by variation of the components in either version, subject only to the conditions expressed within the braces, we arrive at the same set of all possible D-part compositions. The difference between the two versions is in their geometric representation. Figure 2.1 shows the diagrammatic representations of $\mathcal{S}^1$ and $\mathcal{S}^2$ according to the two definitions. Definition 2.2 clearly recognizes the correct dimensionality by depicting $\mathcal{S}^1$ as a subset of the one-dimensional real space $\mathcal{R}^1$, the real line, and $\mathcal{S}^2$ as a subset of two-dimensional real space $\mathcal{R}^2$, the real plane. To gain the advantage of symmetry, Definition 2.3 has to resort to the device of embedding the d-dimensional sample space $\mathcal{S}^d$ in a real space of higher dimension, namely $\mathcal{R}^D$. Thus $\mathcal{S}^1$ is a line segment, and therefore a one-dimensional subset, within $\mathcal{R}^2$; and $\mathcal{S}^2$ is an equilateral triangle, a two-dimensional subset, within $\mathcal{R}^3$. This equilateral triangle is clearly identified as the useful ternary diagram introduced in Section 1.4. We shall find it easy to switch freely between the two versions of $\mathcal{S}^d$, our sample space for recording D-part compositions, and refer to both as the d-dimensional simplex.

If we now review the various compositions in Data 1–10 we can quickly identify their simplex sample spaces with $d = 4, 4, 4, 4, 2, 2, 3, 3, 2, 5$; for Data 14, $d = 5$. For the problem of fireworks mixtures of Data 13 the 5-part compositions play an explanatory or covariate role and so the simplex $\mathcal{S}^4$ acts as a factor or design space. In the calibration problem of white-cell compositions of Data 11 each experimental unit, a blood sample, gives rise to two 3-part compositions, say (x_1, x_2, x_3) and (X_1, X_2, X_3). The appropriate sample

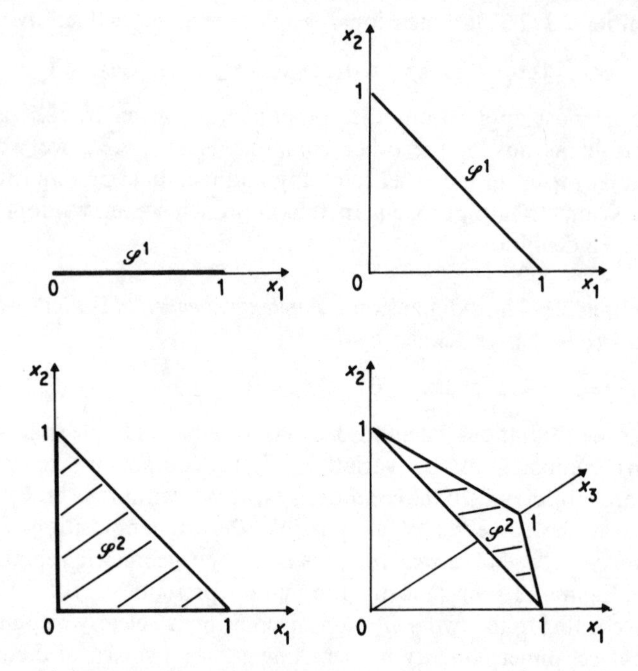

Fig. 2.1. *Different geometric representations of $\mathscr{S}^1$ and $\mathscr{S}^2$.*

space is then the following set:

$$\left\{(x_1, x_2, x_3, X_1, X_2, X_3): x_i > 0\,(i = 1, 2, 3), X_i > 0\,(i = 1, 2, 3),\right.$$

$$\left.\sum_{i=1}^{3} x_i = 1,\ \sum_{i=1}^{3} X_i = 1\right\}, \tag{2.8}$$

in mathematical terms the product set $\mathscr{S}^2 \times \mathscr{S}^2$. This product set will similarly serve as a sample space for the past and present fruit compositions of Data 12.

2.3 Spaces, vectors, matrices

Spaces

We have already used $\mathscr{R}^d$ and $\mathscr{S}^d$ to denote d-dimensional real space and the d-dimensional simplex. For some applications we also require

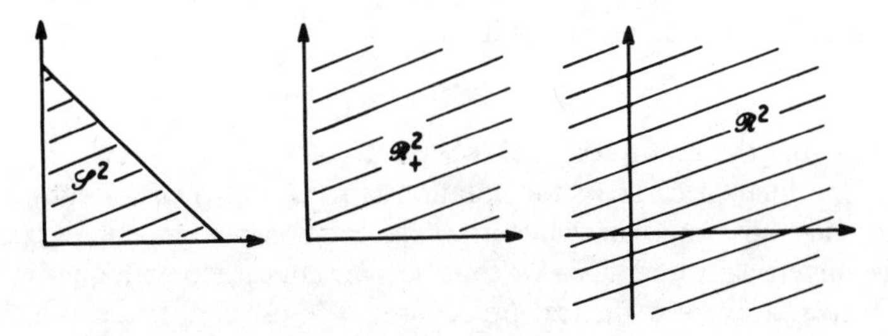

Fig. 2.2. *The relationship of* $\mathscr{S}^2$, $\mathscr{R}^2_+$ *and* $\mathscr{R}^2$.

to consider d-dimensional positive space $\mathscr{R}^d_+$, the positive orthant of $\mathscr{R}^d$, defined by

$$\mathscr{R}^d_+ = \{(w_1, \ldots, w_d): w_1 > 0, \ldots, w_d > 0\}. \tag{2.9}$$

With $\mathscr{S}^d$ as in Definition 2.2 we have the subset relationship

$$\mathscr{S}^d \subset \mathscr{R}^d_+ \subset \mathscr{R}^d \tag{2.10}$$

and Fig. 2.2 illustrates this relationship in diagrammatic form for $d = 2$. In order to clarify discussion of relationships between vectors in these three spaces, we restrict the use of the letters x, w, y to vectors and their elements in $\mathscr{S}^d$, $\mathscr{R}^d_+$, $\mathscr{R}^d$, respectively.

Vectors

We shall use the space-saving convention that a column vector may be specified by its components enclosed in round brackets: thus we have already used $(x_1, \ldots, x_D)$ to denote a $D \times 1$ or column vector or composition $\mathbf{x}$. Square brackets then denote a row vector: thus $[x_1, \ldots, x_D]$ is the transpose of $\mathbf{x}$.

When there is no chance of confusion about the dimension of the composition $(x_1, \ldots, x_D)$ we write it compactly as the vector $\mathbf{x}$. When it is necessary to distinguish between the full vector $(x_1, \ldots, x_D)$ and its diminished form $(x_1, \ldots, x_d)$, as used in Definition 2.2, we write

$$\mathbf{x}^{(D)} = (x_1, \ldots, x_d, x_D), \qquad \mathbf{x}^{(d)} = (x_1, \ldots, x_d).$$

The superscript thus indicates the number of elements of the vector and so also provides a useful notation for leading subvectors of $\mathbf{x}^{(D)}$; for example

$$\mathbf{x}^{(c)} = (x_1, \ldots, x_c). \tag{2.11}$$

In some applications we also have to use trailing subvectors of $\mathbf{x}^{(D)}$

and so we introduce the notation

$$\mathbf{x}_{(c)} = (x_{c+1}, \ldots, x_D) \tag{2.12}$$

to denote the vector $\mathbf{x}^{(D)}$ with $\mathbf{x}^{(c)}$ deleted.

One final piece of vector notation is required when we consider certain transformations which treat the components of a composition asymmetrically and when we thus become concerned with questions of invariance of statistical procedures. We use $\mathbf{x}_{-j}$ to denote the vector $\mathbf{x} = (x_1, \ldots, x_D)$ with x_j omitted. For example, if $D = 6$ then $\mathbf{x}_{-4} = (x_1, x_2, x_3, x_5, x_6)$. Note also that $\mathbf{x}_{-D} = \mathbf{x}^{(d)}$ and $\mathbf{x}_{-1} = \mathbf{x}_{(1)}$.

Matrices

We use $\mathbf{I}_d$ to denote the identity matrix of order d, $\mathbf{J}_d$ the square matrix of order d with every element 1, and $\mathbf{j}_d$ the $d \times 1$ vector with each element 1. The transpose of any matrix $\mathbf{A}$ is denoted by $\mathbf{A}'$ and the determinant, inverse (when it exists) and trace of a square matrix $\mathbf{A}$ by $|\mathbf{A}|$, $\mathbf{A}^{-1}$ and $\text{tr}(\mathbf{A})$. For any square matrix $\mathbf{A}$ we use $\text{diag}(\mathbf{A})$ to denote the diagonal matrix formed from the diagonal elements of $\mathbf{A}$. The notation $\text{diag}(a_1, \ldots, a_d)$ is also used without ambiguity to denote the diagonal matrix of order d with a_i in the ith diagonal position. Other matrices and matrix operations will be introduced as they arise naturally in the development of the study of compositional data.

We adopt the usual subscript convention for the elements of a matrix, with the first and second subscripts denoting row and column, respectively. Instead of specifying matrices *in extenso* we can then use typical-element specifications. For example, the compositional data matrix of Section 1.2, including its order, is specified as

$$\mathbf{X} = [x_{ri}: r = 1, \ldots, N; i = 1, \ldots, D],$$

and the covariance matrix of the crude components $x_1, \ldots, x_D$ of a composition as

$$[\text{cov}(x_i, x_j): i, j = 1, \ldots, D].$$

Notation All vectors and matrices will be represented by bold characters throughout the monograph.

2.4 Bases and compositions

Each single-person budget of Data 8 consists of four quantities, say w_1, w_2, w_3, w_4, recorded on the same measurement scale, namely

Hong Kong dollars. The budget pattern corresponding to $\mathbf{w} = (w_1, w_2, w_3, w_4)$ is the set of proportions $\mathbf{x} = (x_1, x_2, x_3, x_4)$ of total expenditure $t = w_1 + w_2 + w_3 + w_4$ assigned to each of the commodity groups and determined by

$$x_i = w_i/(w_1 + w_2 + w_3 + w_4) \qquad (i = 1, 2, 3, 4). \qquad (2.13)$$

We have already seen in Section 1.8 a similar relationship between a typical vector $\mathbf{w} = (w_1, w_2, w_3)$ of quantities of steroid metabolites excreted, each recorded on the same measurement scale of mg/24 hr, and the total excretion $t = w_1 + w_2 + w_3$ and relative excretion pattern $\mathbf{x} = (x_1, x_2, x_3)$ determined by

$$x_i = w_i/t \qquad (i = 1, 2, 3). \qquad (2.14)$$

In each of these examples we are forming a D-part composition $\mathbf{x}$ from a vector $\mathbf{w} \in \mathscr{R}_+^D$ of quantitative measurements simply by scaling the components of $\mathbf{w}$ so that they have unit sum.

We now embody these ideas into some useful definitions and properties. First we assign a terminology to the vector $\mathbf{w}$.

Definition 2.4 A *basis* $\mathbf{w}$ of D parts is a $D \times 1$ vector of positive components $(w_1, \ldots, w_D)$ all recorded on the same measurement scale.

Next we formalize the scaling process of (2.13) and (2.14).

Definition 2.5 The *constraining operator* $\mathscr{C}$ transforms each vector $\mathbf{w}$ of D positive components into the unit-sum vector $\mathbf{w}/\mathbf{j}'\mathbf{w}$. In other words the constraining operator defines a transformation $\mathscr{C}: \mathscr{R}_+^D \to \mathscr{S}^d$ where $d = D - 1$.

Finally we present an obvious terminology for t and $\mathbf{x}$ obtained from any basis $\mathbf{w}$.

Definition 2.6 Every basis $\mathbf{w} \in \mathscr{R}_+^D$ has a unique *size* $t = w_1 + \cdots + w_D = \mathbf{j}'\mathbf{w}$ and *composition* $\mathbf{x} = \mathscr{C}(\mathbf{w}) = \mathbf{w}/t$.

We now record and comment on a number of simple properties of bases.

Property 2.1

(a) A basis $\mathbf{w}$ completely determines its size t and its composition $\mathbf{x}$.

(b) A basis $\mathbf{w}$ is completely determined by its size t and composition $\mathbf{x}$.

Note that (a) simply reiterates Definition 2.6, and (b) follows immediately from the relationship $\mathbf{w} = t\mathbf{x}$.

Property 2.2 There is a one-to-one correspondence between $\mathbf{w}$ and $(t, \mathbf{x})$ defined by

$$\mathscr{R}_+^D \to \mathscr{R}_+^1 \times \mathscr{S}^d : \mathbf{w} \to (t, \mathbf{x}) = (\mathbf{w}'\mathbf{j}, \mathscr{C}(\mathbf{w}))$$
$$\mathscr{R}_+^1 \times \mathscr{S}^d \to \mathscr{R}_+^D : (t, \mathbf{x}) \to \mathbf{w} = t\mathbf{x},$$

and having jacobian

$$\operatorname{jac}(\mathbf{w} \mid t, \mathbf{x}) = t^d.$$

The one-to-one correspondence is a restatement of Property 2.1 and the jacobian property is easily verified.

It is important to realize that the constraining transformation $\mathscr{C}$: $\mathscr{R}_+^D \to \mathscr{S}^d$ is many–one. The relationship of the many bases corresponding to a given composition is set out in the next property.

Property 2.3 The set of bases with composition $\mathbf{x}$ is

$$\mathscr{B}(\mathbf{x}) = \{t\mathbf{x} : t > 0\}.$$

This many–one feature has a simple geometric representation. Figure 2.3 shows $\mathscr{R}_+^3$ as the space of 3-part bases and embedded within it the simplex $\mathscr{S}^2$ of 3-part compositions. To identify the unique composition $\mathbf{x}$ of a basis $\mathbf{w}$ we take the point of intersection of the simplex and the ray from the origin through $\mathbf{w}$. If, however, we start with a composition $\mathbf{x}$ we see that it could have arisen from any basis on the ray from the origin through $\mathbf{x}$ and so $\mathscr{B}(\mathbf{x})$ of Property 2.3 is represented in the diagram by this ray.

Thus a D-part composition determines a basis only up to a multiplicative factor and the basis sets $\mathscr{B}(\cdot)$ are essentially equivalence classes within $\mathscr{R}_+^D$ under the relationship of possessing a common composition. This is an important aspect, which has been overlooked in some forms of compositional data analysis: information about compositions provides information about underlying bases only up to the order of an equivalence class. In statistical terms there must be limitations in the inference process from compositions to bases. We

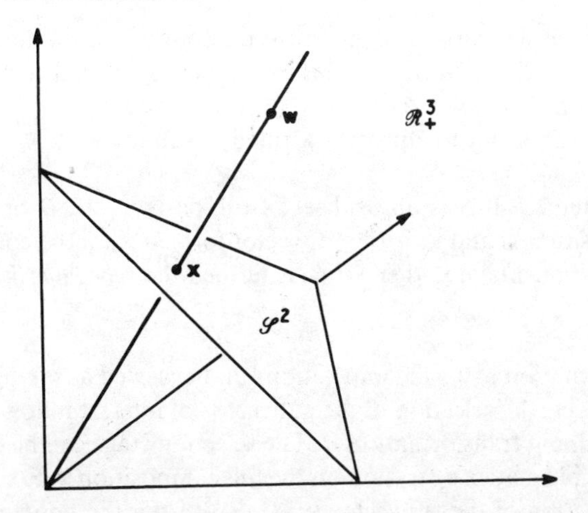

Fig. 2.3. *Diagrammatic illustration of the many-to-one nature of the relationship of bases to compositions.*

shall be able to put this into more quantitative terms as we progress through the arguments of later chapters.

2.5 Subcompositions

Given a composition $\mathbf{x}$ in $\mathscr{S}^d$ we may wish to focus attention on the relative magnitudes of a subset of the components. We have already seen an example of such a process in Section 1.5 in the formation of the ABC and ADE ternary diagrams for the hongite compositions of Data 1. Formally, for example, the components of the parts A, D, E are x_1, x_4, x_5 and the subcomposition or point (s_1, s_2, s_3) in the ADE ternary diagram is obtained by

$$(s_1, s_2, s_3) = (x_1, x_4, x_5)/(x_1 + x_4 + x_5). \tag{2.15}$$

Since x_1, x_4, x_5 are positive numbers $(x_1, x_4, x_5) \in \mathscr{R}^3_+$ and so the transformation (2.15) is identical in form to the constraining transformation of (2.13), (2.14) and Definition 2.5. There is thus no need to introduce a new operator notation for the formation of such a subcomposition as ADE. We simply write

$$(s_1, s_2, s_3) = \mathscr{C}(x_1, x_4, x_5) \tag{2.16}$$

as a shortened version of (2.15). Note that the subcomposition

(s_1, s_2, s_3) is a composition in $\mathscr{S}^2$ so that our process of forming this subcomposition is a transformation from the original sample space $\mathscr{S}^4$ to a new simplex $\mathscr{S}^2$.

We can set out this process formally as follows.

Definition 2.7 If S is any subset of the parts $1, \ldots, D$ of a D-part composition $\mathbf{x}$, and $\mathbf{x}_S$ is the subvector formed from the corresponding components of $\mathbf{x}$, then $\mathscr{C}(\mathbf{x}_S)$ is termed the *subcomposition* of the parts S.

The formation of a subcomposition can be viewed as taking place in two stages, the selection of the subvector of interest followed by the constraining transformation $\mathscr{C}$. The selection stage can be expressed as a simple matrix operation on the full composition $\mathbf{x}$. For example, the selection of the subvector (x_1, x_4, x_5) from the hongite composition $(x_1, \ldots, x_5)$ can be expressed in the following way:

$$\mathbf{x}_S = \begin{bmatrix} x_1 \\ x_4 \\ x_5 \end{bmatrix} = \begin{bmatrix} 1 & 0 & 0 & 0 & 0 \\ 0 & 0 & 0 & 1 & 0 \\ 0 & 0 & 0 & 0 & 1 \end{bmatrix} \begin{bmatrix} x_1 \\ x_2 \\ x_3 \\ x_4 \\ x_5 \end{bmatrix}.$$

Definition 2.8 A selecting matrix $\mathbf{S}$ is any matrix of order $C \times D$ $(C < D)$, with C elements equal to 1, one in each row and at most one in each column, and the remaining $C(D-1)$ elements 0.

We collect here some elementary properties of subcompositions.

Property 2.4 For a selecting matrix $\mathbf{S}$ as in Definition 2.7, $\mathscr{C}(\mathbf{Sx})$ is a subcomposition of C parts and dimension $c = C - 1$. In other words, formation of a subcomposition may be considered as the transformation

$$\mathscr{C}\mathbf{S}: \mathscr{S}^d \to \mathscr{S}^c,$$

and a subcomposition can be regarded as a composition in a simplex of lower dimension than that of the full composition.

An important feature of a subcomposition is that it preserves ratio relationships.

Property 2.5 The ratio of any two components of a subcomposition is the same as the ratio of the corresponding two components in the full composition: if $\mathbf{s} = \mathscr{C}(\mathbf{x}_S)$ then

$$s_i/s_j = x_i/x_j \qquad (i, j \in S).$$

Property 2.6 For a composition $\mathbf{x}^{(D)} = (\mathbf{x}^{(c)}, \mathbf{x}_{(c)})$ let $\mathbf{s}_1$ and $\mathbf{s}_2$ be the subcompositions $\mathbf{s}_1 = \mathscr{C}(\mathbf{x}^{(c)})$ and $\mathbf{s}_2 = \mathscr{C}(\mathbf{x}_{(c)})$. Although $\mathbf{x}^{(D)}$ determines $\mathbf{s}_1$ and $\mathbf{s}_2$ uniquely, $\mathbf{x}^{(D)}$ is not determined uniquely by $\mathbf{s}_1$ and $\mathbf{s}_2$: any composition of the form

$$t\begin{bmatrix}\mathbf{s}_1\\ \mathbf{0}\end{bmatrix} + (1-t)\begin{bmatrix}\mathbf{0}\\ \mathbf{s}_2\end{bmatrix} \qquad (0 < t < 1)$$

gives rise to the subcompositions $\mathbf{s}_1, \mathbf{s}_2$.

More generally a subcomposition is not uniquely determined by subcompositions formed from two complementary subvectors.

Formation of a subcomposition from a composition has a straightforward geometrical representation as a process of linear projection. Figure 2.4 shows two simple examples. In Fig. 2.4(a) the point P representing the 3-part composition (x_1, x_2, x_3) in the simplex 123 is projected onto the line segment of subsimplex 12 from the point or trivial subsimplex 3 to give a projection P′. Since P′2/P′1 can be easily shown to be equal to x_1/x_2 we see that P′ represents $\mathscr{C}(x_1, x_2)$ within the simplex 12. In Fig. 2.4(b) the point P represents the 4-part composition (x_1, x_2, x_3, x_4) in the simplex 1234. We can then project P onto a point in the subsimplex 12 by taking the

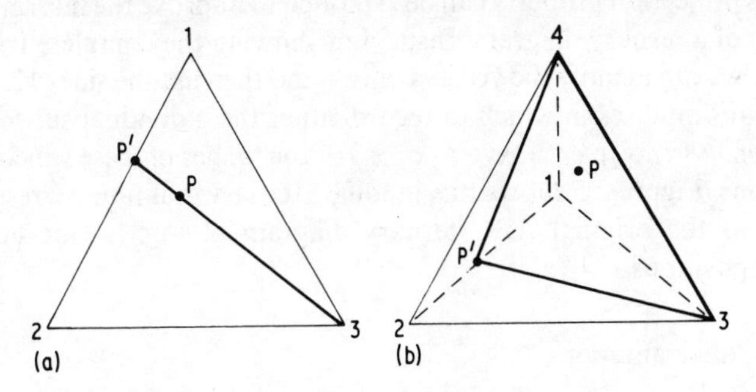

Fig. 2.4. *Formation of subcompositions represented by linear projections in the simplex.*

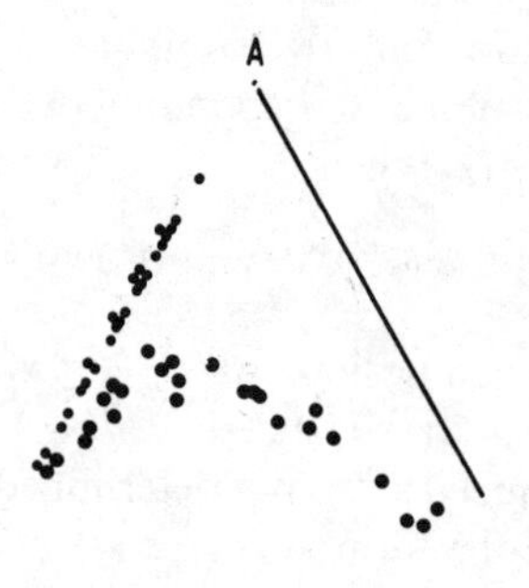

Fig. 2.5. *Modified ABC ternary diagram for hongite compositions showing AB and BC subcompositions and range of AC subcompositions.*

projection P′ to be the point of intersection of the subsimplex 12 with the plane through P and the complementary subsimplex 34. The projection P' then represents the subcomposition $\mathscr{C}(x_1, x_2)$ within the simplex 12. We collect these ideas in more general terms in the following property.

Property 2.7 The subcomposition $\mathscr{C}(\mathbf{x}^{(c)})$ is the projection on the subsimplex $12 \ldots c$ of the full composition $\mathbf{x}^{(D)}$ with the complementary subsimplex $C \ldots D$, where $C = c + 1$, acting as the source of the projecting lines.

This projection property can be exploited to improve the informativeness of a ternary diagram. Instead of showing the complete triangle 123, we can identify the vertices only – and then use the sides 12, 13, 23 as subsimplexes in which to record either the individual subcompositions $\mathscr{C}(x_1, x_2), \mathscr{C}(x_1, x_3), \mathscr{C}(x_2, x_3)$ or the ranges of these subcompositions. Figure 2.5 shows this modified ternary diagram corresponding to the original ABC ternary diagram of Fig. 1.6 for hongite compositions.

2.6 Amalgamations

For some purposes it may be of interest to amalgamate components of a composition to form a new amalgamated composition. For example, in a household expenditure enquiry we may start with a budget-share composition $(x_1, \ldots, x_9)$ of proportions of total expenditure spent on nine commodity groups:

1. Foodstuffs,
2. Housing,
3. Fuel and light,
4. Tobacco and alcohol,
5. Clothing and footwear,
6. Durable goods,
7. Miscellaneous goods,
8. Transport and vehicles,
9. Services.

If we wish, however, to consider the commodity groups defined for the single-person budgets of Data 8 then, as far as budget-share composition is concerned, interest is in the amalgamation (t_1, t_2, t_3, t_4), where

$$t_1 = x_2 + x_3,$$
$$t_2 = x_1 + x_4,$$
$$t_3 = x_5 + x_6 + x_7,$$
$$t_4 = x_8 + x_9,$$

instead of the original compositional vector $\mathbf{x}^{(9)}$.

Definition 2.9 If the parts of a D-part composition are separated into $C(\leqslant D)$ mutually exclusive and exhaustive subsets and the components within each subset are added together, the resulting C-part composition is termed an *amalgamation.*

In the example above this process of component addition to form an amalgamation can be achieved by a matrix operation $\mathbf{t}^{(4)} = \mathbf{A}\mathbf{x}^{(9)}$ with

$$\mathbf{A} = \begin{bmatrix} 0 & 1 & 1 & 0 & 0 & 0 & 0 & 0 & 0 \\ 1 & 0 & 0 & 1 & 0 & 0 & 0 & 0 & 0 \\ 0 & 0 & 0 & 0 & 1 & 1 & 1 & 0 & 0 \\ 0 & 0 & 0 & 0 & 0 & 0 & 0 & 1 & 1 \end{bmatrix}.$$

The general definition is as follows.

Definition 2.10 An *amalgamating matrix* $\mathbf{A}$ is any matrix of order $C \times D (C \leqslant D)$ with D elements equal to 1, one in each column and at least one in each row, and with the remaining $(C-1)D$ elements 0.

The following two properties of amalgamations are obvious.

Property 2.8 If a D-part composition $\mathbf{x}^{(D)}$ is premultiplied by the amalgamating matrix $\mathbf{A}$ of Definition 2.10, the resulting amalgamation $\mathbf{t}^{(C)} = \mathbf{A}\mathbf{x}^{(D)}$ is a C-part composition: in other words, the amalgamating matrix is a transformation

$$\mathbf{A}: \mathscr{S}^d \to \mathscr{S}^c,$$

where $c = C - 1$.

Property 2.9 Every $D \times D$ permutation matrix, including the identity matrix $\mathbf{I}_D$, is an amalgamating matrix.

If we suppose that the composition has been ordered in such a way that combinations are between neighbouring components, the formal general process can be set out as follows. Let the integers $a_0, \ldots, a_C$ satisfy

$$0 = a_0 < a_1 < \cdots < a_C = D \tag{2.17}$$

and define

$$t_i = x_{a_{i-1}+1} + \cdots + x_{a_i} \qquad (i = 1, \ldots, C). \tag{2.18}$$

Then $\mathbf{t}^{(C)} \in \mathscr{S}^c$ and so is a C-part and c-dimensional composition, an amalgamation of $\mathbf{x}^{(D)}$.

Note that a $C \times D$ amalgamating matrix $\mathbf{A}$ can be applied to any D-vector, such as a basis $\mathbf{w} \in \mathscr{R}_+^D$, not necessarily a composition. Amalgamating matrices, because of their simple form, have further simple properties. Since we shall always be studying amalgamations as part of a more general concept, a partition of a composition, we leave some of these properties as exercises for the reader.

2.7 Partitions

The C-part amalgamation just discussed involves a separation of the vector $\mathbf{x}^{(D)}$ into C subvectors. For example, we may have separated a 7-part composition $\mathbf{x}^{(7)}$ into three subvectors

$$(x_1, x_2 | x_3, x_4 | x_5, x_6, x_7) \tag{2.19}$$

by the two vertical bars in the formation of a 3-part amalgamation $\mathbf{t}^{(3)}$, where

$$t_1 = x_1 + x_2, \; t_2 = x_3 + x_4, \; t_3 = x_5 + x_6 + x_7. \tag{2.20}$$

When considering such an amalgamation we may be interested also in the three subcompositions associated with these subvectors. Suppose that we are studying 7-part compositions by weight of woodland loam in which parts 1 and 2 are animal, living and dead, parts 3 and 4 are vegetable, living and dead, and parts 5, 6 and 7 are mineral, coarse-, medium- and fine-grained. The amalgamation (2.20) is then a natural way of studying the overall (animal, vegetable, mineral) composition of (2.19), but we may also wish to study the ratio of living to dead animal content, that is the subcomposition $\mathscr{C}(x_1, x_2)$, and similarly the vegetable and mineral subcompositions $\mathscr{C}(x_3, x_4)$ and $\mathscr{C}(x_5, x_6, x_7)$.

The formal definition of this concept of combining the operations of subcomposition and amalgamation can be built on an extension of Definition 2.9

Definition 2.11 If the parts of a D-part composition are separated into $C(\leqslant D)$ mutually exclusive and exhaustive subsets, then the associated amalgamation together with the subcompositions formed from these subsets form a *partition* of *order* $c = C - 1$.

Note that the order of a partition is the dimension of the amalgamation. In the example above, the order is 2. When the parts of the original composition are specially arranged to suit the amalgamating process, as in the example above, the order is the number of divisions required to form the subsets.

For our woodland loam example we introduce a simpler notation for the subcompositions:

$$\begin{aligned} \mathbf{s}_1 &= (s_{11}, s_{12}) = \mathscr{C}(x_1, x_2), \\ \mathbf{s}_2 &= (s_{21}, s_{22}) = \mathscr{C}(x_3, x_4), \\ \mathbf{s}_3 &= (s_{31}, s_{32}, s_{33}) = \mathscr{C}(x_5, x_6, x_7). \end{aligned} \tag{2.21}$$

Omitting the fill-up values s_{12}, s_{22}, s_{33} and t_3, we can express the joint construction of the amalgamation and the subcompositions from the original composition by the following relations:

$$\begin{aligned} s_{11} &= x_1/(x_1 + x_2), & t_1 &= x_1 + x_2, \\ s_{21} &= x_3/(x_3 + x_4), & t_2 &= x_3 + x_4. \\ s_{31} &= x_5/(x_5 + x_6 + x_7), \\ s_{32} &= x_6/(x_5 + x_6 + x_7), \end{aligned}$$

An extremely useful feature of this joint consideration of amalgamation and subcompositions is that the transformation, here from $\mathscr{S}^6$ to $\mathscr{S}^2 \times \mathscr{S}^1 \times \mathscr{S}^1 \times \mathscr{S}^2$, is one-to-one with inverse given by

$$\begin{aligned} x_1 &= s_{11}t_1, & x_4 &= (1 - s_{21})t_2, \\ x_2 &= (1 - s_{11})t_1, & x_5 &= s_{31}t_3, \\ x_3 &= s_{21}t_2, & x_6 &= s_{32}t_3, \\ & & x_7 &= (1 - s_{31} - s_{32})t_3. \end{aligned}$$

To help with theoretical discussion of partitions we set up the notation of a general partition in the following definition.

Definition 2.12 Let

$$0 = a_0 < a_1 < \cdots < a_c < a_C = D.$$

The partition based on the separation of the D-part composition $\mathbf{x}$ into C subsets by the divisions

$$(x_{a_0+1}, \ldots, x_{a_1} | x_{a_1+1}, \ldots, x_{a_2} | \ldots | x_{a_c+1}, \ldots, x_{a_C})$$

is termed the *general partition of* $\mathbf{x}$. If

$$\begin{aligned} t_i &= x_{a_{i-1}+1} + \cdots + x_{a_i} \qquad (i = 1, \ldots, C), \\ \mathbf{s}_i &= \mathscr{C}(x_{a_{i-1}+1}, \ldots, x_{a_i}) \qquad (i = 1, \ldots, C), \end{aligned}$$

then we denote the general partition as

$$\mathscr{P}(\mathbf{x}) = (\mathbf{t}; \mathbf{s}_1, \ldots, \mathbf{s}_C).$$

The nature of the general partition as a transformation is contained in the following properties.

Property 2.10 The general partition $\mathscr{P}$ is a one-to-one transformation

$$\mathscr{P}: \mathscr{S}^d \to \mathscr{S}^c \times \prod_{i=1}^{C} \mathscr{S}^{d_i},$$

where

$$d_i = a_i - a_{i-1} - 1$$

and

$$d = c + \sum_{i=1}^{C} d_i.$$

When dealing with such a transformation we must note that $\mathscr{S}^c \times \mathscr{S}^d \neq \mathscr{S}^{c+d}$ as the simple examples of Fig. 2.6 demonstrate.

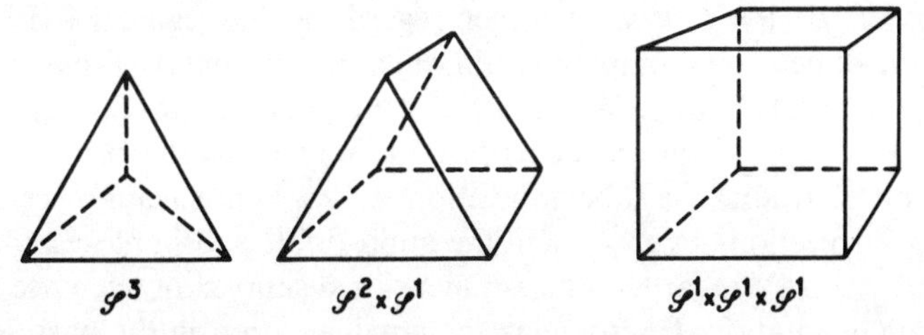

Fig. 2.6. *The different forms for spaces* $\mathcal{S}^3$, $\mathcal{S}^2 \times \mathcal{S}^1$, $\mathcal{S}^1 \times \mathcal{S}^1 \times \mathcal{S}^1$ *in their symmetric forms.*

Property 2.11 A composition is completely determined by any one of its partitions. Knowledge of $(\mathbf{t}; \mathbf{s}_1, \ldots, \mathbf{s}_C)$ completely determines $\mathbf{x}$.

Note that knowledge of the amalgamation $\mathbf{t}$ alone or of the subcompositions $\mathbf{s}_1, \ldots, \mathbf{s}_C$ alone is not sufficient to determine the composition $\mathbf{x}$.

There is a simple geometric representation of a partition of a composition $\mathbf{x}$. As an example consider the 4-part composition $\mathbf{x}$ represented in Fig. 2.7 by the point P in the tetrahedron 1234, and the partition based on the division $(x_1, x_2 | x_3, x_4)$. We recall from Property 2.7 that the subcompositions $\mathbf{s}_1$ and $\mathbf{s}_2$ of the partition are the projections P_1 and P_2 of P on the subsimplexes 12 and 34 with respect to the complementary subsimplexes 34 and 12, respectively.

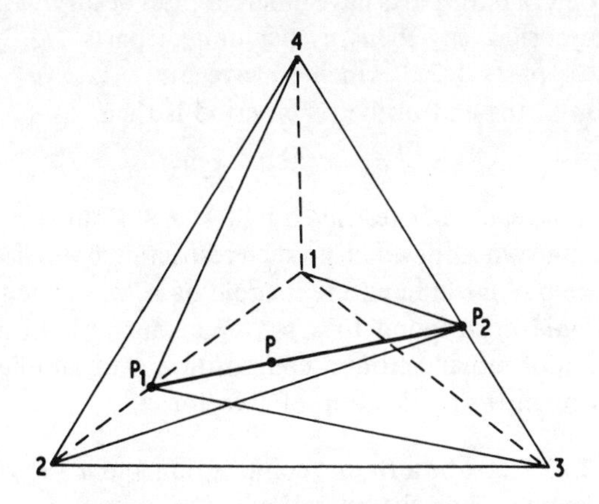

Fig. 2.7. *Geometric representation of the partition based on the division* $(x_1, x_2 | x_3, x_4)$ *of a 4-part composition.*

Since $PP_2/P_1P = t_1/t_2$ we can then regard the line segment P_1P_2 as the one-dimensional simplex within which P represents the amalgamation in the partition. We shall have no occasion to exploit this geometric representation and so describe only briefly the form for the general partition. The subcompositions $\mathbf{s}_1, \ldots, \mathbf{s}_C$ are again represented as projections $P_1, \ldots, P_C$ in the appropriate subsimplexes: then, with $P_1, \ldots, P_C$ regarded as a c-dimensional simplex of reference for the amalgamation, P represents the amalgamation in the partition.

2.8 Perturbations

For completeness we define one further operation on compositions which, as we shall later discover, plays an important role in providing a possible explanation for the patterns of observed variability in some compositional data. In this respect the operation is the counterpart of addition and multiplication in the additive and multiplicative forms of the central limit theorem leading to normal and lognormal distributions, respectively. We shall also find that the operation is useful in our study of measurement errors in compositions in Chapter 11.

We here confine ourselves to an extremely simple example as motivation for the introduction of the operation of perturbation within the simplex. Suppose that at the start of a time period we have a unit of a perishable substance consisting of three parts with composition (x_1, x_2, x_3). If the parts have different rates of survival to the end of the time period, say u_1, u_2, u_3 per unit of parts $1, 2, 3$, then the quantities of parts $1, 2, 3$ which survive are x_1u_1, x_2u_2, x_3u_3. The composition at the end of the time period is then

$$\mathbf{X} = \mathscr{C}(x_1u_1, x_2u_2, x_3u_3). \tag{2.22}$$

In this example each u_i is restricted by $0 < u_i \leqslant 1$ but clearly we can imagine circumstances in which no such restriction holds; for example if a part were biological in nature it could have the capacity to grow and this would correspond to $u_i > 1$. We can now formalize this process (2.22) of moving from a composition $\mathbf{x}$ to another composition $\mathbf{X}$ through the application of a vector $\mathbf{u}$.

Definition 2.13 Let $\mathbf{x}$ be a D-part composition and $\mathbf{u}$ a D-vector with positive elements. Then the operation

$$\mathbf{X} = \mathbf{u} \circ \mathbf{x} = \mathscr{C}(x_1u_1, \ldots, x_Du_D)$$

is termed a *perturbation* with the original composition $\mathbf{x}$ being operated on by the *perturbing vector* $\mathbf{u}$ to form a *perturbed composition* $\mathbf{X}$.

We collect here two main properties of perturbations, leaving to the reader some other simple properties as an exercise at the end of this chapter.

Property 2.12 The operation $\mathbf{u}\circ$ is a one-to-one transformation between $\mathscr{S}^d$ and $\mathscr{S}^d$. The inverse transformation is the perturbation $\mathbf{u}^{-1}\circ$, where

$$\mathbf{u}^{-1} = (1/u_1, \ldots, 1/u_D).$$

Property 2.13 Since $\mathbf{u}\circ\mathbf{x} = \mathscr{C}(\mathbf{u})\circ\mathbf{x}$, the effect of any perturbing vector $\mathbf{u}\in\mathscr{R}_+^D$ is the same as that of the perturbing vector $\mathscr{C}(\mathbf{u})\in\mathscr{S}^d$. We can therefore, if we wish, restrict consideration of perturbing vectors to the simplex $\mathscr{S}^d$ rather than $\mathscr{R}_+^D$ without any loss of generality.

2.9 Geometrical representations of compositional data

Our main aim in this chapter has been to arrive at a satisfactory sample space within which to represent compositional data and to study their variation. For an $N \times D$ compositional data matrix $\mathbf{X} = [x_{ri}]$ we would now see the d-dimensional simplex $\mathscr{S}^d$ as the natural sample space. Each row of $\mathbf{X}$ is a D-part composition, which is represented by a point in $\mathscr{S}^d$, and so the compositional data in $\mathbf{X}$ are represented by N points in $\mathscr{S}^d$.

There is, however, another geometrical representation for $\mathbf{X}$ which can prove useful. Each column of $\mathbf{X}$ is an N-vector of positive numbers and so can be represented as a point in $\mathscr{R}_+^N$, so that the compositional data in $\mathbf{X}$ are represented by D points in $\mathscr{R}_+^N$. Given those D points the data matrix $\mathbf{X}$ is completely determined.

The nature of the two representations can be illustrated by the data matrix

$$\mathbf{X} = \begin{bmatrix} 0.5 & 0.4 & 0.1 \\ 0.2 & 0.3 & 0.5 \end{bmatrix}.$$

Here $N = 2$ and $D = 3$ and Fig. 2.8(a) shows the two compositions in the rows of the data matrix $\mathbf{X}$ as points $\mathrm{R}_1, \mathrm{R}_2$ in the sample space $\mathscr{S}^2$, the ternary diagram 123. Figure 2.8(b) shows the three

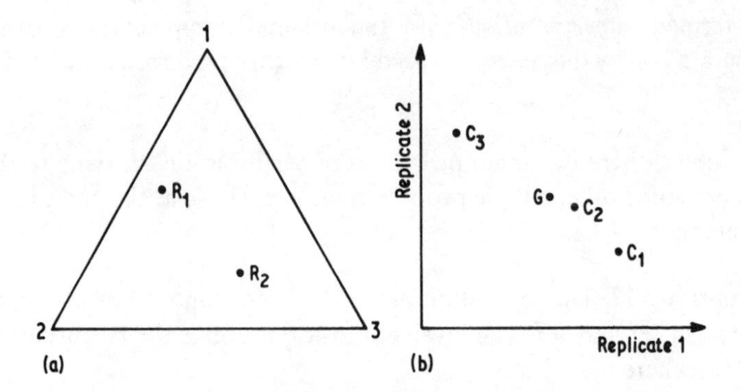

Fig. 2.8. *Sample space representation (a) and replicate space representation (b) of a compositional data array.*

columns of **X** as points C_1, C_2, C_3 in $\mathscr{R}^2_+$. In this second representation it is clear that the points may more properly be confined to the unit square in $\mathscr{R}^2_+$, more generally the hypercube in $\mathscr{R}^N_+$; we have not fussed over this refinement. The effect of the unit-sum constraint is on each row and although this places no interesting restriction on the individual points C_1, C_2, C_3, it does imply that their centre of gravity is the point $G(\frac{1}{3}, \frac{1}{3}) = \frac{1}{3}\mathbf{j}_2$.

A convenient terminology for this second form of representation is now given.

Definition 2.14 For an $N \times D$ compositional data matrix **X** the columns can be represented in the *replicate space* $\mathscr{R}^N_+$ by points $C_1, \ldots, C_D$, whose centre of gravity is $D^{-1}\mathbf{j}_N$.

We may comment on standard terminology here. In the so-called sample space the 'axes' refer essentially to the parts of the composition and each plotted point refers to a replicate; in the replicate space the axes correspond to replicate number and each plotted point refers to a part. To emphasize the duality in the roles of part and replicate there would be a case for replacing the term 'sample space' by 'part space'. We have not done so here, realizing that the standard terminology is now historically so well ingrained that any attempt at change may cause resentment and confusion.

2.10 Bibliographic notes

Although the special constrained nature of compositional data has long been recognized, there appears to have been a reluctance in the statistical analysis of compositional data to recognize explicitly the natural sample space as the set familiar to mathematicians as the regular simplex. This is perhaps surprising in view of the prolific use of the ternary diagram representation of the 2-dimensional simplex in so many applications and the familiarity of the simplex in other areas of statistics, for example as the design space for experiments with mixtures (Scheffé, 1963) and as the parameter space for multinomial experiments (Lindley, 1964).

The move from basis to composition and from composition to subcomposition by the constraining operator $\mathscr{C}$ is well established and appears as a formalized procedure in, for example, Chayes (1960, 1971) under the concept of *closure*, with bases and compositions identified with *open* and *closed* arrays. Although amalgamations and partitions have also been used in the past, their first formal treatment seems to be that of Aitchison (1982). The concept of perturbation is familiar in other areas of statistics, for example as the relationship of proportions of genotypes before and after genetic selection (Edwards, 1977, Chapter 2) and as the operation of Bayes's formula to change a prior probability assessment into a posterior probability assessment through the perturbing influence of the likelihood function; see Problem 2.13 below.

The sample space and replicate space representations of a compositional data set are simple adaptations of what are commonly termed Q and R mode representations of standard data matrices, as described for example in Jöreskog, Klovan and Reyment (1976, Chapters 4, 5).

For a geometric construction for the perturbation of a 3-part composition, see Aitchison (1974).

Problems

2.1 Identify the dimensions of the compositions in each of Problems 1.1–1.10.

2.2 Illustrate the relationship

$$\mathscr{S}^3 \subset \mathscr{R}^3_+ \subset \mathscr{R}^3$$

by sketching the three spaces.

2.3 Construct modified ternary diagrams for all four 3-part subcompositions which can be constructed from Data 7, arranging the triangles in the pattern of Fig. 2.9.

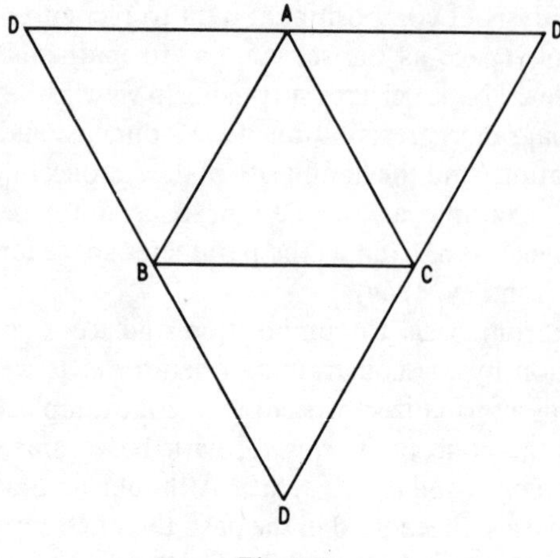

Fig. 2.9

2.4 In which of Problems 1.1–1.10 did your questions involve the relationship of a basis to its composition and size?

2.5 The size of a 3-part basis (w_1, w_2, w_3) is known to be t and the ratio w_1/w_2 is known to be r. Find a general form for the set of such bases and also for the corresponding set of compositions. Show the relationship diagrammatically.

2.6 If $\mathbf{A}_1$ and $\mathbf{A}_2$ are amalgamating matrices of orders $C_1 \times D$ and $C_2 \times C_1$ respectively, then $\mathbf{A}_2\mathbf{A}_1$ is an amalgamating matrix of order $C_2 \times D$.

2.7 For 4-part compositions represented in the tetrahedron 1234, the amalgamation (t_1, t_2) is defined by $t_1 = x_1 + x_2, t_2 = x_3 + x_4$. Show that the compositions for which $(t_1, t_2) = (0.5, 0.5)$ lie on the plane defined by the midpoints of the edges 13, 14, 23, 24.

2.8 Establish the following properties where $\mathbf{w} \in \mathscr{R}_+^D$, noting that the properties therefore also hold for $\mathbf{w} \in \mathscr{S}^d$.

(a) $\mathscr{C}(t\mathbf{w}) = \mathscr{C}(\mathbf{w})$ for any $t > 0$.
(b) $\mathscr{C}\{\mathscr{C}(\mathbf{w})\} = \mathscr{C}(\mathbf{w})$.
(c) $\mathscr{C}(\mathbf{w}) = \mathbf{w}$ if and only if $\mathbf{w} \in \mathscr{S}^d$.
(d) $\mathscr{C}(\mathbf{A}\mathbf{w}) = \mathbf{A}\mathscr{C}(\mathbf{w})$ for any amalgamating matrix $\mathbf{A}$.
(e) $\mathscr{C}(\mathbf{u} \circ \mathbf{w}) = \mathscr{C}(\mathbf{u}) \circ \mathscr{C}(\mathbf{w})$.

2.9 Suppose that the partition $(\mathbf{t}; \mathbf{s}_1, \mathbf{s}_2, \mathbf{s}_3)$ corresponding to the division $(x_1, x_2 | x_3, x_4, x_5 | x_6, x_7)$ of a 7-part composition has

$$\mathbf{t} = (0.2, 0.5, 0.3)$$
$$\mathbf{s}_1 = (0.15, 0.85)$$
$$\mathbf{s}_2 = (0.14, 0.38, 0.48)$$
$$\mathbf{s}_3 = (0.4, 0.6).$$

Determine the unique composition from which this partition arose and then construct the partition corresponding to the division

$$(x_1, x_2, x_3 | x_4, x_5 | x_6, x_7).$$

2.10 For the statistician's activity patterns of Data 10 form partitions appropriate to a division of the statistician's day into work and leisure parts.

2.11 For the compositions of eight-hour shifts of Data 22 define a suitable partition for the investigation of the productive and non-productive parts of the shifts. Compute the partition corresponding to composition O1 and attempt to provide its geometric representation in a reference tetrahedron 1234.

2.12 Establish the following properties for perturbations.

(a) $\mathbf{j}_D \circ \mathbf{x} = \mathbf{x}$ for any D-part composition $\mathbf{x}$.
(b) $\mathbf{u}_2 \circ (\mathbf{u}_1 \circ \mathbf{x}) = \mathbf{u}_1 \circ (\mathbf{u}_2 \circ \mathbf{x})$, where $\mathbf{u}_1$ and $\mathbf{u}_2$ are any perturbing vectors.
(c) $(\mathbf{j}_C, \mathbf{0}_{D-C}) \circ \mathbf{x} = \mathscr{C}(\mathbf{x}^{(C)})$, where $\mathbf{0}_{D-C}$ is a $(D - C)$-vector with all its elements 0.
(d) $\mathbf{S}(\mathbf{u} \circ \mathbf{x}) = \mathbf{Su} \circ \mathbf{Sx}$ for any selection matrix $\mathbf{S}$.
(e) If $\mathbf{x}$ is a D-part composition and $\mathbf{w} \in \mathscr{R}_+^D$ satisfies $\mathbf{u} \circ \mathbf{w} = \mathbf{u} \circ \mathbf{x}$ for some perturbing vector $\mathbf{u}$ then $\mathbf{w} \in \mathscr{B}(\mathbf{x})$.

2.13 Let $\{h_1, \ldots, h_D\}$ be a set of D mutually exclusive and exhaustive hypotheses and let $\{l_1, \ldots, l_D\}$ denote the likelihood with respect to a given event. Interpret Bayes's formula, which transforms a prior assignment of probabilities $\{p_1, \ldots, p_D\}$ to a posterior assignment $\{P_1, \ldots, P_D\}$ after the occurrence of the event by

$$P_i = p_i l_i \Big/ \sum_{j=1}^{D} p_j l_j \qquad (i = 1, \ldots, D)$$

as a simplex operation.

2.14 In relation to the pogo-jump results of Data 19, identify bases, compositions and sizes, and in terms of these consider the reformulation of any questions which you posed in Problem 1.4.

CHAPTER 3

The special difficulties of compositional data analysis

All things are difficult before they are easy.

Thomas Fuller: *Gnomologia*

3.1 Introduction

In one of his penetrating *Mathematical Contributions to the Theory of Evolution*, Pearson (1897) wrote 'On a form of spurious correlation which may arise when indices are used in the measurement of organisms'. In the following introductory excerpt we have made only small notational changes to conform with the notation of this monograph.

If $x_1 = f_1(w_1, w_3)$ and $x_2 = f_2(w_2, w_3)$ be two functions of the three variables w_1, w_2, w_3, and these variables be selected at random so that there exists no correlation between w_1, w_2 or w_1, w_3 or w_2, w_3, there will still be found to exist correlation between x_1 and x_2. Thus a real danger arises when a statistical biologist attributes the correlation between two functions like x_1 and x_2 to *organic* relationship. The particular case that is likely to occur is when x_1 and x_2 are indices with the same denominator for the correlation of indices seems at first sight a very plausible measure of organic correlation.

Pearson then derived some approximate relationships between the correlation of the indices x_1 and x_2 and the first and second moments of w_1, w_2, w_3, and demonstrated that when w_1, w_2, w_3 are pairwise uncorrelated the correlation between the simple indices $x_1 = w_1/w_3$ and $x_2 = w_2/w_3$ is approximately

$$\frac{\alpha_3^2}{\{(\alpha_1^2 + \alpha_3^2)(\alpha_2^2 + \alpha_3^2)\}^{1/2}}, \tag{3.1}$$

where $\alpha_1, \alpha_2, \alpha_3$ are the coefficients of variation of w_1, w_2, w_3. With (3.1) as an assessment of spurious correlation between indices he then proceeded to illustrate the difficulties of interpreting observed correlations between indices in a number of biological examples before drawing the following conclusion.

From the above examples it will be seen that the method which judges of the intensity of organic correlation by the reduction of all absolute measures to indices, the denominators of which are some *one* absolute measurement, is not free from obscurity; for this method would give the major portion of the observed index correlation had the parts of the animal been thrown together at random, i.e., if there were no organic correlation at all.

In notes appended to Pearson's paper both Galton and Weldon joined in the warning to biologists. Galton contributed an artificial example, 'believing that it may be useful in enabling others to realise the genesis of spurious correlation'. Weldon appreciated that his own earlier interpretation of his correlations of measurement indices for Plymouth shrimps was wrong and, after following Pearson's suggestion to form 420 spurious shrimps by random assembly of measurements from his records, reported as follows.

When the carapace lengths and post-spinous lengths of these spurious shrimps were divided by body length, and the correlation between the resulting indices was determined the value of r was found to be 0.38, the value for real shrimps being 0.81, or the correlation due to the use of indices forms 47 per cent of the observed value.

In his context of biological application, Pearson was mainly concerned with indices of the form w_1/w_3 and w_2/w_3. His remarks must surely apply with even more force to ratios such as $w_1/(w_1 + w_2 + w_3)$ and $w_2/(w_1 + w_2 + w_3)$, where not only are there common elements in the denominators of the two ratios but also elements common to numerator and denominator in each ratio. Such ratios are essentially components of a composition. Thus in the statistical analysis of compositional data we are likely to find our conclusions 'not free from obscurity' if we insist on the blind application of 'standard' methods such as evaluation and interpretation of product-moment correlations. It seems surprising that the warnings of three such eminent statistician-scientists as Pearson, Galton and Weldon should have largely gone unheeded for so long: even today uncritical applications of inappropriate statistical methods to compositional data with consequent dubious inferences are regularly reported. One of the reasons for such misguided effort is undoubtedly the slowness of the emergence of statistical methods appropriate to the special nature of the simplex sample space.

In this chapter we present a concise account of the various difficulties as they have been perceived by those analysts of compositional data who have been aware of them. It may seem strange that we should preoccupy ourselves with history and with a catalogue of 'how-not-to-do-it' methods instead of embarking immediately on the new positive approach that this monograph advocates. There is, however, much to be learned from such an approach: the solutions which we put forward are motivated by questions arising from dissatisfaction with previous attempts. To someone unaware of previous approaches the statistical methods emerging may seem esoteric, whereas they are, in historical perspective, the natural consequences of properly posed questions.

3.2 High dimensionality

Many workers, particularly geologists, have perceived the major, possibly the only, difficulty with compositional data to be the inability of the human eye to see in more than three dimensions. For example, the view of Iddings (1903), that 'since compositions involve so many components, projection into two-dimensional diagrams is needed for comparison of types and samples', has been frequently reiterated, and as recently as 1978 it is expressed in almost identical words by Barker (1978).

Three short comments on this view are worth making.

Projections as a partial analysis

No matter how many projections of the data are examined, it remains an undoubted fact that such an approach, in restricting attention to a selection of subcompositions rather than the composition as a whole, is subject to criticism as a partial analysis. Analogous partial analyses in more familiar situations are the examination of all bivariate plots of data from vectors in $\mathscr{R}^d$ in the hope of obtaining a picture of the multivariate pattern of variability, and the investigation of association in a three-way contingency table by consideration of only the two-way marginal contingency tables. We are well aware that we may miss some important feature of the variability in the first analogy and fail to detect a second-order, three-factor interaction in the second.

Graphical distortions due to the constraint

The view that graphical analysis, if it were possible, would resolve problems is particularly naive in the case of compositional data. As we have seen in Chapter 2 the unit-sum constraint confines compositional vectors to a simplex and there is no guarantee that patterns perceived in such a constrained space should necessarily have the same interpretation as in the more familiar spaces such as $\mathscr{R}^2$ in which we are accustomed to work graphically. For example, we reiterate Question 1.5.1. Does the curvature in the data of Fig. 1.6 have any significance as a trend of some kind or is this pattern just what we might expect from natural variability?

Thus even if we are dealing with a simple 3-part composition, the question of the reliability of graphical analysis with its accompanying subjective judgement has to be raised. Many writers have made this point in some detail, and such titles as 'Trends in ternary petrologic variation diagrams – fact or fantasy?' (Butler, 1979c) are not uncommon in the literature. A view expressing some despair is put forward by the President of the International Association for Mathematical Geology in his Presidential address (Reyment, 1977): '... I feel I should express a certain amount of dismay at the relatively slow rate of penetration of quantitative thinking into the rank and file of geology. Petrologists are still largely unaware of the dangers of closure (the effect of the unit-sum constraint) in their diagrams...'.

As an alternative to actual projection within the simplex, another form of graphical display has often been advocated in which d of the components have been plotted individually against the remaining Dth component, often chosen as the component with the greatest apparent variability. The Harker diagram, well known in geology, chooses SiO_2 as the component against which to compare all the other major-oxide components in the search for relationships. The dangers of interpreting such diagrams in the usual regression or correlation way, again for reasons associated with the unit-sum constraint and to be discussed in Section 3.3, have also been pointed out. For example, Chayes (1962) asserts that '... inferences based on intuitive geometrical examination or cookbook statistical testing of these diagrams will more often be wrong than right'.

Modern multivariate analysis

There have been substantial advances in multivariate statistical analysis in the past two or three decades, particularly since modern computing facilities have provided easy access to data analysis techniques which would have been prohibitive some years ago. Since it is well known that no aspect of a multivariate measurement should be discarded without good reason, it must surely now be possible to approach compositional data as a whole without resort to partial analyses. That this can be achieved for compositional data is a main theme of this monograph.

3.3 Absence of an interpretable covariance structure

The traditional and almost universally applied way of attempting to gain insight into the interdependence of the components of a D-part composition $\mathbf{x}$ has been through product-moment covariances or correlations of the crude components $x_1, \ldots, x_D$. Let $\text{var}(x_i)$, $\text{cov}(x_i, x_j)$, $\text{corr}(x_i, x_j)$ denote the variance of x_i, the covariance of x_i and x_j, the correlation between x_i and x_j, respectively, and recall the relationships

$$\text{var}(x_i) = \text{cov}(x_i, x_i) \qquad (i = 1, \ldots, D), \tag{3.2}$$

$$\text{corr}(x_i, x_j) = \text{cov}(x_i, x_j)\{\text{var}(x_i)\text{var}(x_j)\}^{-1/2} \qquad (i,j = 1, \ldots, D). \tag{3.3}$$

For ease of future reference we introduce a formal definition.

Definition 3.1 The *crude covariance structure* of a composition $\mathbf{x}$ is the set of all

$$\kappa_{ij} = \text{cov}(x_i, x_j) \qquad (i,j = 1, \ldots, D)$$

with $D \times D$ *crude covariance matrix*

$$\boldsymbol{K} = [\kappa_{ij}: i,j = 1, \ldots, D],$$

and *crude correlations*

$$\rho_{ij} = \kappa_{ij}(\kappa_{ii}\kappa_{jj})^{-1/2} \qquad (i = 1, \ldots, d; j = i+1, \ldots, D).$$

Crude investigation then either concentrates on individual values of the $\frac{1}{2}dD$ crude correlations $\text{corr}(x_i, x_j)$ or on the whole covariance structure through properties of the covariance matrix $\boldsymbol{K}$. All of these characteristics of the pattern of variability of a composition can be estimated in a standard way from a compositional data matrix

$$\mathbf{X} = [x_{ri}: r = 1, \ldots, N : i = 1, \ldots, D].$$

Writing

$$n = N - 1 \tag{3.4}$$

we have as estimate of κ_{ij},

$$\hat{\kappa}_{ij} = n^{-1} \sum_{r=1}^{N} (x_{ri} - \bar{x}_i)(x_{rj} - \bar{x}_j) \qquad (i, j = 1, \ldots, D), \tag{3.5}$$

where

$$\bar{x}_i = N^{-1} \sum_{r=1}^{N} x_{ri} \qquad (i = 1, \ldots, D). \tag{3.6}$$

For example, for the twenty-five 5-part compositions of hongite in Data 1 the estimated 5×5 crude covariance matrix is

$$10^{-3} \times \begin{bmatrix} 2.98 & 3.30 & -5.98 & -0.07 & -0.23 \\ 3.30 & 13.99 & -14.22 & -1.38 & -1.69 \\ -5.98 & -14.22 & 17.77 & 1.02 & 1.41 \\ -0.07 & -1.38 & 1.02 & 0.55 & -0.13 \\ -0.23 & -1.69 & 1.41 & -0.13 & 0.64 \end{bmatrix}. \tag{3.7}$$

The difficulties, foreseen by Pearson (1897), of interpreting such product-moment covariances and correlations between the crude components of a composition were brought forcefully to the attention of geologists by Chayes (1960, 1962), Krumbein (1962) and Sarmanov and Vistelius (1959), and of biologists by Mosimann (1962, 1963) and have continued to be a matter of concern right up to the present (Chayes, 1983). There are even complete texts (Chayes, 1971; Le Maître, 1982) describing analyses of compositions in terms of their crude covariance structure and discussing the problems of interpretation. The difficulties have been expressed in a variety of ways.

Negative bias difficulty

Let us consider the general case of a D-part composition $\mathbf{x}$, subject to the now familiar unit-sum constraint,

$$x_1 + \cdots + x_D = 1.$$

Since

$$\operatorname{cov}(x_1, x_1 + \cdots + x_D) = 0,$$

we have

$$\mathrm{cov}(x_1, x_2) + \cdots + \mathrm{cov}(x_1, x_D) = -\mathrm{var}(x_1). \tag{3.8}$$

The right-hand side of (3.8) is negative except for the trivial situation where the first component is constant. Thus at least one of the covariances on the left must be negative or, equivalently, there must be at least one negative element in the first row of the crude covariance matrix $\boldsymbol{K}$. The same negative bias must similarly occur in each of the other rows of $\boldsymbol{K}$ so that at least D of its entries must be negative. These negative properties are equally displayed by estimated covariance matrices, for example by (3.7). Hence correlations are not free to range over the usual interval $(-1, 1)$ subject only to the non-negative definiteness of the covariance or correlation matrix, and there are bound to be problems of interpretation.

The problem is at its most acute for a 2-part composition. In this case it is trivial to show that

$$\mathrm{cov}(x_1, x_2) = -\mathrm{var}(x_1) = -\mathrm{var}(x_2),$$

and that

$$\mathrm{corr}(x_1, x_2) = -1.$$

Thus, instead of the correlation having the freedom to take any value in the interval -1 to $+1$, it is a specified value. For the case $D = 3$ we would normally think of the three correlations $\rho_{12}, \rho_{13}, \rho_{23}$ as freely taking values in the range $(-1, 1)$ subject only to the non-negative definiteness condition

$$1 + 2\rho_{12}\rho_{13}\rho_{23} - \rho_{12}^2 - \rho_{13}^2 - \rho_{23}^2 \geqslant 0.$$

In particular, since $1 + 2\rho^3 - 3\rho^2 \geqslant 0$ for $0 \leqslant \rho \leqslant 1$, we could normally envisage all three correlations being equal and positive. Such a set of values for crude correlations for a 3-part composition is, however, prohibited by the negative bias property.

The obvious conclusion from this exposure of a tendency of negative bias is that whatever interpretation may be placed on a covariance such as $\mathrm{cov}(x_i, x_j)$ it must be radically different from the standard interpretation of covariances between components of a completely unrestricted vector.

Subcomposition difficulty

Since knowledge of a composition allows us to construct any subcomposition, we might expect to find some simple relationship between the crude covariance matrix of a subcomposition and that of the full composition. No such relationship exists. Moreover, the apparently erratic way in which the crude covariance associated with two specific parts can fluctuate in sign as we move from full composition to lower and lower dimensioned subcompositions has been another source of puzzlement. We can illustrate these fluctuations with Data 1 and the ABDE, ABD, ADE and BDE subcompositions of the ABCDE composition of hongite, for which the estimated covariance matrices are, respectively, $10^{-3} \times$

$$\begin{array}{cccc} A & B & D & E \\ \end{array} \qquad \begin{array}{ccc} A & B & D \end{array}$$

$$\begin{bmatrix} 2.03 & -4.31 & 1.17 & 1.11 \\ -4.31 & 10.83 & -3.29 & -3.22 \\ 1.17 & -3.29 & 1.63 & 0.49 \\ 1.11 & -3.22 & 0.49 & 1.63 \end{bmatrix}, \begin{bmatrix} 4.71 & -6.80 & 2.09 \\ -6.80 & 11.07 & -4.27 \\ 2.09 & -4.27 & 2.18 \end{bmatrix}, \tag{3.9}$$

$$\begin{array}{ccc} A & D & E \end{array} \qquad \begin{array}{ccc} B & D & E \end{array}$$

$$\begin{bmatrix} 2.54 & -1.15 & -1.39 \\ -1.15 & 1.46 & -0.31 \\ -1.39 & -0.31 & 1.70 \end{bmatrix}, \begin{bmatrix} 2.87 & -1.48 & -1.39 \\ -1.48 & 1.02 & 0.45 \\ -1.39 & 0.45 & 0.94 \end{bmatrix}.$$

From each of the five crude covariance matrices in (3.7) and (3.9), we can compute from (3.3) the correlations between the pairs of parts, and the relevant values of these are presented for each subcomposition in Table 3.1. From this simple example we see that, as we move successively from 5-part composition to 4-part and 3-part subcompositions, these correlations may change substantially and with no apparent pattern in the changes.

Table 3.1. *Fluctuations in crude correlations with subcomposition considered*

Composition or subcompostion	*Crude correlation between parts*					
	AB	AD	AE	BD	BE	DE
ABCDE	0.51	−0.05	−0.16	−0.50	−0.56	−0.22
ABDE	−0.92	0.65	0.61	−0.78	−0.79	0.30
ABD	−0.94	0.65		−0.87		
ADE		−0.60	−0.67			−0.20
BDE				−0.86	−0.85	0.46

Another difficulty frequently commented on is that variances may display different and unrelatable rank orderings as we form subcompositions. Comparison of the covariance matrices (3.7) and (3.9), for example, shows that ordering the common parts B, D, E, in descending order of their variances gives B, E, D for the full composition and B, D, E for the BDE subcomposition.

Basis difficulty

The construction of a composition $\mathbf{x} = \mathscr{C}(\mathbf{w})$ from a basis $\mathbf{w}$ is a constraining operation similar to the construction of a subcomposition from a composition. We may therefore expect the same difficulty in relating the crude covariance matrix of the composition to the covariance matrix of its basis $\mathbf{w}$ in applications where such a relationship may be of interest, as in Sections 1.7–1.8. For the steroid metabolite excretions of the 37 adults in Data 9, for example, the estimates of the crude correlations obtained from $\mathbf{w}$ and $\mathbf{x}$ are

$$\mathrm{corr}(w_1, w_2) = 0.73, \mathrm{corr}(w_1, w_3) = 0.38, \mathrm{corr}(w_2, w_3) = 0.37,$$

$$\mathrm{corr}(x_1, x_2) = -0.37, \mathrm{corr}(x_1, x_3) = -0.92, \mathrm{corr}(x_2, x_3) = -0.033.$$

Once again we confirm an absence of any relationship and therefore of any sensible interpretation of such correlations.

Null correlation difficulty

In the standard interpretation of covariances and correlations it is well known that the value zero is invariably tied to the concept of no association. Equally, ideas of independence of random variables usually lead to a zero value for a covariance or correlation. For covariances between components of a composition, with their tendency to negative bias, the use of the value zero as a criterion of independence is obviously suspect. Aware of this difficulty of defining independence within terms of a composition itself, a number of analysts, in particular Chayes and Kruskal (1966), have resorted to constructing imagined bases (the open set) from which the observed compositions (the closed set) could conceptually have arisen. The complicated technical details need not concern us here. The question then posed is what correlations between components of the compositions would have arisen if the components of the bases had been uncorrelated. The answer to this yields null correlations, not necessarily zero, as hypothetical values against which to test the estimated correlations obtained directly from the compositional data matrix $\mathbf{X}$.

The concept of null correlation here is, of course, analogous to Pearson's spurious correlation of indices discussed in Section 3.1 except that Pearson's bases were not imagined but real, so allowing direct estimation of the basis moment structure and hence, through his approximate formula, of the spurious correlation effect. No such bases exist for the compositional data problem under discussion.

The method has a number of technical difficulties which should first be listed before considering much more fundamental and conceptual objections to it.

1. The distributions of the test statistics are not known (Mosimann, 1962, p. 81; Chayes and Kruskal, 1966, p. 696) and do not fall within the framework of any standard testing approach such as generalized likelihood ratio tests.
2. The tests of null correlations are carried out separately for each pair of components. This procedure therefore is open to the same kind of criticism as the application of all $\frac{1}{2}dD$ t-tests of pairwise comparison of D treatments without a preliminary overall F-test. The theory lacks the analogue of such an overall test (Chayes and Kruskal, 1966, p. 696).
3. When the tests do detect significant non-null correlations it is by no means safe (Miesch, 1969) to conclude that the corresponding quantities in the basis are correlated. Thus, despite the fact that the battery of pairwise tests, criticized in (2), is not designed as an overall test of the hypothesis that all correlations of the basis are zero, this hypothesis is the only one which the battery effectively tests. No satisfactory analysis of the non-null case is available.

By far the greatest objection to this process is that it introduces an entity, the imaginary basis, which for truly compositional data just does not exist. It is simply a peg on which to hang an argument that has proved intransigent to other approaches. This might be acceptable, despite the approximations and other technical difficulties involved, if the concept of 'opening up' a composition into an unrestricted basis were a sound one in itself. We have already seen in Property 2.3 that corresponding to any given composition $\mathbf{x}$ there is an infinite set $\mathscr{B}(\mathbf{x})$ of bases. Such a feature immediately raises a philosophical question as to how it is possible to answer a question about an entity by asking a related question about an object which not only does not exist but also bears a many-to-one relationship to the given entity. In order to make an inference about a composition by the opening-up approach we have first to make an inference about corresponding bases, of which there are many. How can such an inference be sound? For a direct discussion of this aspect of the Chayes–Kruskal approach we refer the reader to Kork (1977). Since we shall be able to make a similar point in a setting where there is an exact relationship between the covariance structures of the composition and its imaginary basis, we will not enter into an unnecessarily complicated discussion here.

3.4 Difficulty of parametric modelling

In complex situations such as those involving compositional data it is difficult to see how analysis of the patterns of variability can ever be wholly successful in the absence of a rich enough parametric class of distributions over the appropriate sample space. For example, the multinormal class and its transformed classes such as the multivariate lognormal class have proved themselves to be flexible instruments in the analysis of data in $\mathscr{R}^d$ and $\mathscr{R}^d_+$. For the simplex sample space $\mathscr{S}^d$ such classes of distributions have been extremely slow to emerge, with only the Dirichlet class and a few simple generalizations being considered until recently. Unfortunately these Dirichlet classes turn out to be totally inadequate for the description of the variability of compositional data. To avoid technical complications we shall defer consideration of the generalizations until Chapter 13 and confine attention here to the original Dirichlet class. We first define, and establish a notation for, the Dirichlet class. It is then convenient to identify, and comment on, the difficulties of this form of parametric modelling within the framework of a list of properties required for easy reference later in the monograph.

Definition 3.2 The *Dirichlet distribution* with parameter $\boldsymbol{\alpha}\in\mathscr{R}^D_+$, and written as $\mathscr{D}^d(\boldsymbol{\alpha})$, where $d = D - 1$, is the distribution on $\mathscr{S}^d$ with density function

$$\frac{\Gamma(\alpha_1 + \cdots + \alpha_D)}{\Gamma(\alpha_1)\cdots\Gamma(\alpha_D)} x_1^{\alpha_1 - 1} \cdots x_d^{\alpha_d - 1}(1 - x_1 - \cdots - x_d)^{\alpha_D - 1} \quad (\mathbf{x}\in\mathscr{S}^d),$$

and the *Dirichlet class* consists of all the distributions produced by allowing the parameter vector $\boldsymbol{\alpha}$ to vary over the parameter space $\mathscr{R}^D_+$.

In the definition we have deliberately emphasized the dimensionality d of the sample space by writing x_D as $1 - x_1 - \cdots - x_d$.

We now state, without proof, and comment on a number of well-established or easily proved properties of the Dirichlet class.

Property 3.1 The isoprobability contours of $\mathscr{D}^d(\boldsymbol{\alpha})$ are convex for $\alpha_i > 1\,(i = 1, \ldots, D)$.

An isoprobability contour plays a role similar to other iso-lines, such as isobars, isotherms; in relation to a distribution with density function $p(\mathbf{x})$ it is simply the locus of points $\mathbf{x} \in \mathscr{S}^d$ which satisfy $p(\mathbf{x}) = \text{constant}$. Any attempt to fit a Dirichlet distribution $\mathscr{D}^d(\boldsymbol{\alpha})$ to the data of Fig. 1.9 would certainly require all the α_i larger than 1 so that there is clearly no hope of a satisfactory fit to such concave patterns commonly occurring in practice.

Property 3.2 If $\mathbf{x}$ has a $\mathscr{D}^d(\boldsymbol{\alpha})$ distribution and α_+ denotes $\alpha_1 + \cdots + \alpha_D$, then

(a) $\mathrm{E}(x_i) = \alpha_i/\alpha_+$,
(b) $\mathrm{var}(x_i) = \alpha_i(\alpha_+ - \alpha_i)/\{\alpha_+^2(\alpha_+ + 1)\}$,
(c) $\mathrm{cov}(x_i, x_j) = -\alpha_i\alpha_j/\{\alpha_+^2(\alpha_+ + 1)\}\,(i \neq j)$,
(d) $\mathrm{corr}(x_i, x_j) = -(\alpha_i\alpha_j)^{1/2}\{(\alpha_+ - \alpha_i)(\alpha_+ - \alpha_j)\}^{-1/2} \qquad (i \neq j)$.

The correlation structure of a Dirichlet composition is therefore completely negative, with $\mathrm{corr}(x_i, x_j) < 0$ for every $i \neq j$, and so cannot be appropriate to data patterns for which some such correlations are definitely positive.

For the next property we require the following definition.

Definition 3.3 A positive random variable w has a *scaled gamma distribution* with *index* α and *scale parameter* β, written $\mathrm{Ga}(\alpha, \beta)$, if it has density function

$$\frac{\beta^\alpha w^{\alpha-1}\exp(-\beta w)}{\Gamma(\alpha)} \qquad (w > 0),$$

where $\alpha > 0$, $\beta > 0$.

Property 3.3 Every Dirichlet composition may be visualized as the composition formed from a basis of independent, equally scaled gamma-distributed components: more specifically, if $w_i\,(i = 1, \ldots, D)$ are independently distributed as $\mathrm{Ga}(\alpha_i, \beta)$ then $\mathbf{x} = \mathscr{C}(\mathbf{w})$ has a $\mathscr{D}^d(\boldsymbol{\alpha})$ distribution.

It is thus clear that every Dirichlet composition has a very strong implied independence structure and so the Dirichlet class is unlikely to be of any great use for describing compositions whose components have even weak forms of dependence.

Property 3.4 Let $\mathscr{P}(\mathbf{x}) = (\mathbf{t}; \mathbf{s}_1, \ldots, \mathbf{s}_C)$ be the general partition, as specified in Definition 2.12, of order c of a $\mathscr{D}^d(\boldsymbol{\alpha})$ composition. Then $\mathbf{t}$, $\mathbf{s}_1, \ldots, \mathbf{s}_C$ are independent and distributed as $\mathscr{D}^c(\boldsymbol{\gamma})$, $\mathscr{D}^{d_1}(\gamma_1 \boldsymbol{\beta}_1), \ldots, \mathscr{D}^{d_C}(\gamma_C \boldsymbol{\beta}_C)$, respectively, where $\mathscr{P}(\boldsymbol{\alpha}) = (\boldsymbol{\gamma}; \boldsymbol{\beta}_1, \ldots, \boldsymbol{\beta}_C)$.

This independence property, which holds for every partition of every Dirichlet composition, is again extremely strong, and unlikely to be possessed by many compositions in practice. For example, one implication of it is that each ratio x_i/x_j of two components is independent of any other ratio x_k/x_l formed from two other components.

The above criticisms throw substantial doubt on the viability of the Dirichlet class as a tool for the study and analysis of compositions and we shall see this view reinforced when we enter into greater technical detail, particularly of the independence concepts touched upon in Properties 3.3 and 3.4. In this later development we shall require a few more properties of the Dirichlet class and these we now catalogue without further comment.

Property 3.5 If $\mathbf{x}$ is a $\mathscr{D}^d(\boldsymbol{\alpha})$ composition and $\psi(t) = \mathrm{d}\log\Gamma(t)/\mathrm{d}t$ and $\psi'(t)\,(t > 0)$ are the digamma and trigamma functions, respectively, then

(a) $\mathrm{E}\{\log(x_i/x_j)\} = \psi(\alpha_i) - \psi(\alpha_j)$,
(b) $\mathrm{var}\{\log(x_i/x_j)\} = \psi'(\alpha_i) + \psi'(\alpha_j)$,
(c) $\mathrm{cov}\{\log(x_i/x_k), \log(x_j/x_k)\} = \psi'(\alpha_k)$.

We require a further definition to describe the next property.

Definition 3.4 A positive random variable w is *beta-distributed* over $\mathscr{R}^1_+$ with indices α and β, written $\mathrm{Be}^+(\alpha, \beta)$, if it has density function

$$\frac{w^{\alpha-1}}{\mathrm{B}(\alpha, \beta)(1+w)^{\alpha+\beta}} \qquad (w > 0),$$

where $\alpha > 0$, $\beta > 0$ and $B(\cdot,\cdot)$ is the beta function (Abramowitz and Stegun, 1972, Section 6.2.1).

Property 3.6 If $\mathbf{x}$ is a $\mathscr{D}^d(\boldsymbol{\alpha})$ composition then

$$\frac{x_1}{1-x_1}, \frac{x_2}{1-x_1-x_2}, \ldots, \frac{x_d}{1-x_1-\cdots-x_d}$$

are independent and $x_i/(1-x_1-\cdots-x_i)$ has a beta distribution $\mathrm{Be}^+(\alpha_i, \mathbf{j}'\boldsymbol{\alpha}_{(i)})$ over $\mathscr{R}^1_+$ $(i = 1, \ldots, d)$.

3.5 The mixture variation difficulty

In standard factorial experiments we are free in our experimental design to vary the levels of the factors independently of each other; moreover, in the associated analysis of variance and its linear models this freedom of variation allows us to pose sensible questions as to whether the factors are additive in their effect on the response or whether there is some specific form of interaction.

The situation is very different when the factor or design space is a simplex $\mathscr{S}^d$ and we are dealing with response to a mixture of ingredients, as in the fireworks mixture problem of Section 1.12. The factors are then the D parts of a composition and, corresponding to the levels of factors in standard factorial experiments, we have the components $x_1, \ldots, x_D$ of the composition. There is no longer the previous freedom of variation because if we change one component we are forced to change at least one other component because of the unit-sum constraint. A puzzling aspect in the study of the design and analysis of such experiments with mixtures has therefore been the difficulty of devising forms of variation of components of compositions which will allow meaningful hypotheses to be formulated about the nature of the effect of the mixture on the response.

We shall return to this problem in Chapter 12 after we have resolved some of the other difficulties of compositional data analysis.

3.6 Bibliographic notes

This chapter has been concerned with a brief account of the history of the realization of the inherent difficulties of compositional data analysis, and the main references have already been cited. It remains only to direct the reader to the extensive literature which catalogues ways in which a great variety of standard statistical analysis goes wrong when applied to compositional data. Some examples are to be found in Aitchison (1983, 1984a, b), Butler (1975, 1976, 1978, 1979a, b, d, 1981), Chayes (1964, 1972), Chayes and Trochimczyk (1978), Krumbein and Watson (1972), Miesch (1980), Saha, Bhattacharyya and Lakshmipathy (1974), Sarmanov (1961), Skala (1977), Trochimczyk and Chayes (1977, 1978).

In a recent paper on buying behaviour, Goodhardt, Ehrenberg and Chatfield (1984) criticize the view expressed in Section 3.4 that the Dirichlet class is totally inadequate for the description of compositional data, claiming that 'in a strictly unsegmented market where the multinomial choice probabilities for each individual are fixed over time, the Dirichlet model is the only possible model'. Their use of the Dirichlet model of Definition 3.2 is, however, not directed towards the description of patterns of variability of actual, observable, data but only at a model-building stage to describe the unobservable brand-choice probabilities of individuals, as a step towards mixing or compounding with other distributions. Nowhere in the paper is a compositional data set either presented or analysed. There is in fact no conflict here. Failure of a class of distributions, such as the Dirichlet, to provide a reasonable fit to actual data patterns does not prevent it having a substantial role in other statistical problems, for example, as a stepping-stone in the model-building cited above or as the appropriate tool in the construction of distribution-free tolerance intervals (Wilks, 1941, 1942).

Problems

3.1 Review the 4-part compositions of Data 7, and the ternary diagrams you constructed in Problem 2.3. Do you consider that it is possible to infer from these diagrams anything concerning the dependence of the 4-part composition on supervisee or time?

3.2 For the ABCDE kongite compositions of Data 2 estimate the crude covariance matrix, confirm the negative bias property of at least one negative covariance in each row, and confirm also the subcompositional difficulty by comparison with the crude covariance matrices for some subcompositions.

3.3 For the urinary steroid excretions of the 30 children of Data 9 estimate the crude correlations between parts in both the basis and the composition and examine them with respect to the basis difficulty reported in Section 3.3 for the adult excretions.

3.4 If $\mathbf{x}$ is a $\mathscr{D}^d(\boldsymbol{\alpha})$ composition and $\mathbf{S}$ is a $C \times D$ selecting matrix, what are the distributions of

(a) $\mathscr{C}(\mathbf{Sx})$,
(b) $(\mathbf{Sx}, 1 - \mathbf{j}'\mathbf{Sx})$?

3.5 If $\mathbf{x}$ is a $\mathscr{D}^d(\boldsymbol{\alpha})$ composition and $\mathbf{A}$ is a $C \times D$ amalgamation matrix, what is the distribution of $\mathbf{Ax}$?

3.6 A positive random variable w is said to have a lognormal distribution $\Lambda(\mu, \sigma^2)$ when $\log w$ has a normal distribution $\mathscr{N}(\mu, \sigma^2)$. Show that if w_1, w_2, w_3 are independently distributed as $\Lambda(\mu_1, \sigma_1^2)$, $\Lambda(\mu_2, \sigma_2^2)$, $\Lambda(\mu_3, \sigma_3^2)$, then

$$\operatorname{corr}(w_1/w_3, w_2/w_3) = \frac{\exp(\sigma_3^2) - 1}{\{\exp(\sigma_1^2 + \sigma_3^2) - 1\}^{1/2}\{\exp(\sigma_2^2 + \sigma_3^2) - 1\}^{1/2}}.$$

CHAPTER 4

Covariance structure

Things are not 'correlated' in nature. In nature, things are as they are. Period. 'Correlation' is a concept which we use to describe connections which we *perceive. There is no word, 'correlation', apart from people. There is no concept, 'correlation', apart from people. This is because only people use words and concepts.*

Gary Zukav: *The Dancing Wu Li Masters*

4.1 Fundamentals

All the difficulties arising in the traditional approach to covariances and correlations between components of compositions come from a lack of appreciation that to carry over ideas which are highly successful for one particular sample space, such as $\mathscr{R}^d$, into another very different sample space, namely $\mathscr{S}^d$, may be completely inappropriate. The spate of papers, referred to in Chapter 3, setting out the difficulties and the absence of any significant progress surely speak for themselves: the adoption of a crude covariance structure based on $\text{cov}(x_i, x_j)$ causes more confusion about the nature of compositional variability than it removes. Any reader with a lingering nostalgia for an untampered $\text{cov}(x_i, x_j)$ should ponder the quotation (Zukav, 1979, p. 71) at the start of this chapter. It cautions that we should not become hidebound in our approach to new problems to the extent of regarding any modelling concept as some embodiment of nature. Placing the comment in another context, we would not expect that excellent tool of the wide open spaces (or $\mathscr{R}^d$) of North America, namely the barbecue, necessarily to be an appropriate concept for cooking in the confined space (or $\mathscr{S}^d$) of a low-cost housing flatlet in Hong Kong. If our concepts fail to serve us in new situations we must invent new concepts.

A crucial sentence in Zukav's paragraph is surely that correlation is a concept we use to describe *connections we perceive.* The perceived difficulties outlined in Chapter 3 all relate to an absence of a connection where one might have been expected. For example, much has been made of the fact that from subcompositional information, such as the crude variances and covariances, the corresponding compositional quantities cannot be reconstructed. The papers setting out in more and more detail and in increasingly specialized applications what can go wrong with crude covariance interpretation, always highlight some absence of an expected connection.

Yet some do perceive elementary connections, for example that *ratios* of components are the same within a subcomposition and the parent composition, as recorded in Property 2.5, but fail to realize that the embodiment of this single wisp of a connection must surely form the foundation of a sensible study of the relationship between compositions and subcompositions. Recall our comment in Section 2.2 that a composition $\mathbf{x}$ can be completely determined by d ratios such as x_i/x_D $(i = 1, \ldots, d)$. This realization that the study of compositions is essentially concerned with the relative magnitudes of ingredients rather than their absolute values leads naturally to a conclusion that we should think in terms of ratios, and to a concept of correlation structure based on product-moment covariances of ratios such as

$$\operatorname{cov}(x_i/x_k, x_j/x_l). \tag{4.1}$$

Experience with mathematical and statistical modelling has, however, led us to another perception, that in first attempts at modelling new situations there is little harm in opting for the tractable. Now variances and covariances of ratios are awkward to manipulate, and as any lecturer in statistics must grow weary of telling his or her students, when stuck by complicated products and quotients: take logarithms. Thus as far as compositions are concerned, as far as living in the simplex is concerned, we should be able to think more clearly about relationships if we think in terms of logratios, and adopt a new concept of correlation based on covariances of the form

$$\sigma_{ij.kl} = \operatorname{cov}\{\log(x_i/x_k), \log(x_j/x_l)\}. \tag{4.2}$$

Definition 4.1 The *covariance structure* of a D-part composition $\mathbf{x}$ is the set of all

$$\sigma_{ij.kl} = \operatorname{cov}\{\log(x_i/x_k), \log(x_j/x_l)\}$$

as i, j, k, l run through the values $1, \ldots, D$.

When we wish to refer to the set of covariances $\operatorname{cov}(x_i, x_j)$ of the raw untransformed components we shall use the term *crude covariance structure*, as set out in Definition 3.1.

At first sight our new covariance structure may seem forbiddingly complicated since (4.2) appears to produce an enormous number of different covariances. Moreover, we may wonder how we can hope to gain insight into the dependence structure of a composition if we have to think in terms of how one ratio of two components varies relative to another ratio of two components. Fortunately, the simple properties of the logarithmic function, such as $\log(x_i/x_k) = \log x_i - \log x_k$, allow a massive reduction in the number of covariances of form (4.2) and lead to some remarkably simple and natural expressions of covariance structure. We shall see in Section 4.2 that the number of covariances of form (4.2) which can be independently assigned is just $\frac{1}{2}dD$, identical to the number associated with a d-dimensional random vector in $\mathscr{R}^d$, a reassuring feature of any attempt to define a useful covariance structure for d-dimensional variability. Our purpose in the following sections is to identify sets of $\frac{1}{2}dD$ of the covariances (4.2) which are particularly useful in relation to the motivating examples of Chapter 1.

4.2 Specification of the covariance structure

If we let i, j, k, l run freely through the values $1, \ldots, D$ we shall generate D^4 values of $\sigma_{ij.kl}$. Clearly not all of these are independent of each other. For a start, many are trivially zero and therefore uninteresting: for example, the simple logarithmic property that $\log 1 = 0$ implies that

$$\sigma_{ij.il} = \sigma_{ij.kj} = \sigma_{ij.ij} = 0. \tag{4.3}$$

Moreover, the symmetric property of covariances ensures that

$$\sigma_{ij.kl} = \sigma_{ji.lk}, \tag{4.4}$$

and the simple quotient property of logarithms, for example $\log(x_i/x_k) = -\log(x_k/x_i)$, produces further relationships:

$$\sigma_{ij.kl} = -\sigma_{kj.il} = -\sigma_{il.kj} = \sigma_{kl.ij}. \tag{4.5}$$

Even more complicated relationships can be found: for example,

$$\sigma_{ij.kl} = \operatorname{cov}\{\log(x_i/x_k), \log(x_j/x_h) + \log(x_h/x_l)\}$$

$$= \text{cov}\{\log(x_i/x_k), \log(x_j/x_h)\} + \text{cov}\{\log(x_i/x_k), \log(x_h/x_l)\}$$

$$= \sigma_{ij.kh} + \sigma_{ih.kl}. \tag{4.6}$$

A number of obvious questions immediately spring to mind. With all these relationships between the covariances, can we easily discover the number of independently assignable covariances? If so, can we identify simple and practically useful subsets of covariances which completely determine the covariance structure? When such a subset of covariances has been specified, can we readily evaluate all the remaining covariances? Fortunately the answer to each of these three questions is a resounding yes.

The simplest non-trivial covariance $\sigma_{ij.kl}$ is clearly when i, j, k, l take only two values, so that only two-component variation is involved and a particularly simple variance form results:

$$\sigma_{ii.jj} = \text{var}\{\log(x_i/x_j)\}.$$

Before we investigate the properties of such variances we provide a formal definition and a notation for this measure of 2-part variation.

Definition 4.2 For any two parts i and j of a D-part composition $\mathbf{x}$ the *logratio variance* is

$$\tau_{ij} = \text{var}\{\log(x_i/x_j)\}.$$

The following self-evident property determines the effective dimensionality of the set of logratio variances.

Property 4.1 The D^2 logratio variances τ_{ij} of a D-part composition $\mathbf{x}$ satisfy

$$\tau_{ii} = 0 \qquad (i = 1, \ldots, D),$$

$$\tau_{ij} = \tau_{ji} \qquad (i = 1, \ldots, d; j = i + 1, \ldots, D),$$

and so are determined by the $\frac{1}{2}dD$ values τ_{ij} $(i = 1, \ldots, d;$ $j = i + 1, \ldots, D)$.

The next result establishes a simple expression for each element of the covariance structure of Definition 4.1 in terms of logratio variances.

Property 4.2 For any i, j, k, l from $1, \ldots, D$,

$$\sigma_{ij.kl} = \tfrac{1}{2}(\tau_{il} + \tau_{jk} - \tau_{ij} - \tau_{kl}).$$

Proof. Let $\alpha_{ij} = \text{cov}(\log x_i, \log x_j)$. The left-hand side of the relationship is

$$\begin{aligned}\sigma_{ij.kl} &= \text{cov}(\log x_i - \log x_k, \log x_j - \log x_l)\\ &= \alpha_{ij} + \alpha_{kl} - \alpha_{il} - \alpha_{jk}.\end{aligned}$$

Moreover, since $\tau_{ij} = \text{var}(\log x_i - \log x_j) = \alpha_{ii} + \alpha_{jj} - 2\alpha_{ij}$, the right-hand side of the relationship can also be expressed as $\alpha_{ij} + \alpha_{kl} - \alpha_{il} - \alpha_{jk}$.

This property can be re-expressed in the following terms.

Property 4.3 The covariance structure of a D-part composition $\mathbf{x}$ is completely determined by the $\frac{1}{2}dD$ logratio variances

$$\tau_{ij} = \text{var}\{\log(x_i/x_j)\} \qquad (i = 1, \ldots, d; j = i+1, \ldots, D).$$

The covariance structure of a d-dimensional composition $(x_1, \ldots, x_D)$ is similar to the familiar covariance structure of a d-dimensional vector in $\mathscr{R}^d$ in its requirement of the specification of $\frac{1}{2}dD$ quantities. Just as in the more familiar case in $\mathscr{R}^d$ we know that any assignment of values to the $\frac{1}{2}dD$ variances and covariances would have to ensure that the covariance matrix is non-negative definite, so we may anticipate some similar requirement here, but its exact form need not trouble us for the present.

Instead of having to face up to thinking in terms of 4-part variation in our study of compositional covariance structure, we already know that, theoretically at least, we can confine our attention completely to 2-part variation, and moreover to certain variances only, not covariances. We must now consider to what extent we can turn this remarkable theoretical finding to practical advantage.

4.3 The compositional variation array

In Section 4.2 we discovered that specification of all $\frac{1}{2}dD$ logratio variances τ_{ij} completely determines the covariance structure of a composition. Since τ_{ij} simply gives a measure of the variability of one component x_i relative to another x_j, the set of all such τ_{ij} should form a useful tool for investigating the pattern of variability of a composition. Because of the symmetry of the measure we can conveniently set out the τ_{ij} as an upper triangular array and at the same time use the lower triangle to display the logratio means

$$\xi_{ij} = \mathrm{E}\{\log(x_i/x_j)\} \qquad (i = 1, \ldots, d; j = i+1, \ldots, D). \tag{4.7}$$

Definition 4.3 For a D-part composition $\mathbf{x}$ the *compositional variation array* is given by

$$\begin{array}{c|cccccc|l}
 & 1 & 2 & 3 & \cdots & d & D & \\
\hline
1 & \cdot & \tau_{12} & \tau_{13} & & & \tau_{1D} & \text{Variances} \\
2 & \xi_{12} & \cdot & \tau_{23} & & & \tau_{2D} & \\
3 & \xi_{13} & \xi_{23} & \cdot & & & & \\
\vdots & & & & & & & \\
d & \xi_{1d} & \xi_{2d} & \xi_{3d} & & \cdot & \tau_{dD} & \\
D & \xi_{1D} & \xi_{2D} & \xi_{3D} & & \xi_{dD} & \cdot & \\
\hline
 & \text{Means} & & & & & &
\end{array}$$

where $\xi_{ij} = \mathrm{E}\{\log(x_i/x_j)\}$ and $\tau_{ij} = \mathrm{var}\{\log(x_i/x_j)\}$.

Since $\xi_{ij} = \xi_{ik} + \xi_{kj}$, not all the means are necessary to determine the mean structure, but it is nevertheless helpful to record them all. For a picture of how x_i and x_j vary relative to each other within the composition we conveniently select the logratio mean ξ_{ij} and variance τ_{ij} symmetrically placed with respect to the diagonal separating the two triangles of the array.

Estimates $\hat{\xi}_{ij}$ and $\hat{\tau}_{ij}$ of ξ_{ij} and τ_{ij} from a data matrix $\mathbf{X}$ follow the standard formulae for means and variances:

$$N\hat{\xi}_{ij} = \sum_{r=1}^{N} \log(x_{ri}/x_{rj}), \tag{4.8}$$

$$\begin{aligned} n\hat{\tau}_{ij} &= \sum_{r=1}^{N} \{\log(x_{ri}/x_{rj}) - \hat{\xi}_{ij}\}^2 \\ &= \sum_{r=1}^{N} \{\log(x_{ri}/x_{rj})\}^2 - N\hat{\xi}_{ij}^2, \end{aligned} \tag{4.9}$$

where $n = N - 1$.

Table 4.1 provides this array of logratio means and variances for the hongite data matrix of Data 1. We can see at once that the array picks out some of the more obvious features of these compositions. The largest relative variation between two minerals is between B and C with $\hat{\tau}_{BC} = 3.00$. Further, although the positive value 0.80 for $\hat{\xi}_{BC}$ indicates that B tends to be larger than C in a certain average sense,

Table 4.1. *Compositional variation array for hongite*

	A	B	C	D	E	
A	.	0.26	1.53	0.08	0.14	Variances
B	0.80	.	3.00	0.55	0.65	
C	1.60	0.80	.	1.11	0.95	
D	1.55	0.75	−0.05	.	0.19	
E	1.72	0.91	0.12	0.17	.	
	Means					

the fact that $\hat{\xi}_{BC} < \sqrt{\hat{\tau}_{BC}}$ suggests that for a substantial number of replicates $\log(x_B/x_C)$ is negative with the corresponding percentage of mineral C exceeding that of mineral B. Inspection of Data 1 corroborates these features which we have managed to reconstruct from knowledge of the summarizing array of Table 4.1. At the other extreme $\hat{\tau}_{AD} = 0.08$ shows that there is little relative variation between A and D. Moreover, the fact that $\hat{\xi}_{AD} = 1.55$ and is substantially greater than $\sqrt{\hat{\tau}_{AD}} = 0.28$ indicates that the percentage of mineral A is consistently and appreciably greater than that of mineral D.

Table 4.2 gives the compositional variation arrays for two further

Table 4.2. *Compositional variation arrays for kongite and boxite*

Kongite

	A	B	C	D	E	
A	.	0.30	1.57	0.12	0.11	Variances
B	0.88	.	3.15	0.64	0.66	
C	1.45	0.56	.	1.06	1.05	
D	1.53	0.65	0.09	.	0.20	
E	1.64	0.75	0.19	0.10	.	
	Means					

Boxite

	A	B	C	D	E	
A	.	0.145	0.053	0.082	0.032	Variances
B	0.41	.	0.204	0.192	0.137	
C	0.99	0.59	.	0.081	0.048	
D	1.15	0.75	0.16	.	0.054	
E	1.70	1.29	0.71	0.55	.	
	Means					

data sets, the kongite compositions of Data 2 and the boxite compositions of Data 3. Comparison of the arrays for hongite and kongite shows a remarkable similarity in the logratio variances with the rank orders of the entries almost the same for each; there is also a similarity in the logratio means, though possibly less so than for the logratio variances. Such an observation may therefore direct our attention later towards searching for a test of whether there are any essential differences in hongite and kongite compositions. On the other hand, there are obviously substantial differences between either of the hongite or kongite variation arrays and the boxite array, with, for example, differences in the rank orders of the logratio means and variances. The most striking difference is the much smaller values of the logratio variances for boxite, a reflection of the lesser extent of the variability between the compositions of Data 3, a feature already noted in Section 1.2. We have clearly made some progress towards answers to Questions 1.2.1 and 1.2.6.

Another intuitive aspect of the compositional variation array is that the sum of all the $\frac{1}{2}dD$ logratio variance entries must provide some measure of total variability for the composition. We shall be able later, in Chapter 8, to make this concept more precise. For the moment we note that for hongite, kongite and boxite these array sums are 8.46, 8.86 and 1.03, respectively, confirming the similarity between hongite and kongite and their substantially greater variability than boxite. We have thus taken a first step towards answering Question 1.2.8.

We see from these simple examples that the compositional variation array can provide a useful descriptive summary of the pattern of variability of compositions. Shortly we shall begin a detailed study of how the new covariance structure has allowed us to overcome the difficulties of the crude covariance structures. First, however, we pause to investigate whether anything can be salvaged from the past practice of publishing only the useless crude covariance structure: can we recover at least some reasonable approximation to the compositional variation array?

4.4 Recovery of the compositional variation array from the crude mean vector and covariance matrix

As we have emphasized in Chapter 3 the traditional way of attempting to communicate the pattern of variability of compositional data has been through the presentation of estimates of the crude mean vector and covariance matrix with typical entries:

$$\lambda_i = \mathrm{E}(x_i) \qquad (i = 1, \ldots, D), \tag{4.10}$$

$$\kappa_{ij} = \mathrm{cov}(x_i, x_j) \qquad (i, j = 1, \ldots, D). \tag{4.11}$$

From these there is no possible way of reconstructing exactly the logratio means and variances, which we require as a starting point for any satisfactory analysis of a compositional data set. To obtain exact estimates of ξ_{ij} and τ_{ij} would require direct computation from the original compositional data set.

If the data set is not available to us, it is often possible to recover a good approximation by use of the standard approximations for means and covariances of functions of random variables (Cramer, 1946):

$$\xi_{ij} \approx \log \frac{\lambda_i}{\lambda_j} - \frac{1}{2}\left[\frac{\kappa_{ii}}{\lambda_i^2} - \frac{\kappa_{jj}}{\lambda_j^2}\right]. \tag{4.12}$$

$$\tau_{ij} \approx \frac{\kappa_{ii}}{\lambda_i^2} - 2\frac{\kappa_{ij}}{\lambda_i \lambda_j} + \frac{\kappa_{jj}}{\lambda_j^2} + \frac{1}{4}\left[\frac{\kappa_{ii}}{\lambda_i^2} - \frac{\kappa_{jj}}{\lambda_j^2}\right]^2. \tag{4.13}$$

Table 4.3. *Approximate compositional variation arrays for kongite and boxite recovered from the crude mean vectors and covariance matrices*

Kongite

	A	B	C	D	E	
A	.	0.21	1.60	0.10	0.10	Variances
B	0.84	.	2.31	0.47	0.45	
C	1.47	0.63	.	1.35	1.20	
D	1.54	0.70	0.07	.	0.20	
E	1.64	0.81	0.18	0.11	.	
	Means					

Boxite

	A	B	C	D	E	
A	.	0.136	0.056	0.088	0.032	Variances
B	0.41	.	0.196	0.190	0.131	
C	0.99	0.59	.	0.098	0.054	
D	1.16	0.75	0.16	.	0.061	
E	1.70	1.30	0.71	0.54	.	
	Means					

As an illustration of this approximation technique Table 4.3 shows the approximate logratio array recovered through (4.12) and (4.13) from the crude means (4.10) and covariance matrices (4.11) constructed from the kongite and boxite compositions of Data 2 and 3. Comparison of these recovered arrays with the exact arrays already computed in Table 4.2 shows good agreement for boxite, the less variable of the two data sets. Even for the very variable kongite compositions, although there are obvious discrepancies between corresponding entries in the two arrays, particularly in the variance part, the general pattern in the relative magnitudes has clearly been recovered.

If the original data set is available then it is obviously simpler to estimate the compositional variation array directly along the lines of Section 4.3.

4.5 Subcompositional analysis

With the covariance structure of Definition 4.1 and its simple specification through Property 4.3 we can immediately appreciate that the subcompositional difficulty of Section 3.3 for the crude covariance structure does not arise. We observed in Property 2.5 that in any subcomposition the ratio of any pair of components is identical to their ratio in the full composition. It follows that the logratio means ξ_{ij} and variances τ_{ij} are the same for the subcomposition as for the composition. Moreover, we can obtain the variation array for any subcomposition by selection of the appropriate entries from the full array. Thus for hongite the variation array for the ABC and ADE subcompositions are respectively

	A	B	C	
A	.	0.26	1.53	Variances
B	0.80	.	3.00	
C	1.60	0.80	.	
	Means			

and

	A	D	E	
A	.	0.08	0.14	Variances
D	1.55	.	0.19	
E	1.72	0.17	.	
	Means			

With such a covariance structure any relevant information which we have about a subcomposition provides direct and exact information about the full composition. This is in sharp contrast to the use of crude covariance structures, for which analysts have tried to find explanations for the absence of relationships between covariances in subcompositions and the full composition. Indeed we now begin to realize the full power of working in terms of the covariance structure of Definition 4.1 through the following obvious result.

Property 4.4 The covariance structure of a composition is completely determined by knowledge of the covariance structure, that is the logratio variances, of all of its 2-part subcompositions.

In Section 4.3 we pointed out that the sum of the logratio variance entries in a compositional variation array provides a measure of total variability of the composition. Since the sum of the array entries associated with a subcomposition similarly provides a measure of total variability for the subcomposition we have a means of comparing subcompositions *of the same dimension* with respect to the extent of the total compositional variability which is retained by the subcomposition. For example, for the ABC subcomposition of the hongite array of Table 4.1 this measure is $0.26 + 1.53 + 3.00 = 4.79$. This is clearly the largest measure of variability obtainable for a 3-part hongite subcomposition, the next largest being easily identified as the BCD subcomposition with measure $3.00 + 0.55 + 1.11 = 4.66$. Moreover, subcomposition ADE has the smallest measure of variability, $0.08 + 0.14 + 0.19 = 0.41$. These extremes of 3-part subcompositional variability have already been shown in the ABC and ADE ternary diagrams in Figs 1.6–1.7. The reader may wish to compare with these extremes in the ternary diagram the variability of a 3-part subcomposition such as ACD with an intermediate measure of variability. In this form of comparison of subcompositions we have gone some way towards answering Question 1.2.5.

At this stage of our investigations it is not clear how we may compare the variability of subcompositions of different dimensions, whether the totals of logratio variances that we have used above should somehow be modified by the dimension of the subcomposition. We shall study this problem in greater depth in Chapter 8.

4.6 Matrix specifications of covariance structures

We have seen in Sections 4.3 and 4.5 how the compositional variation array can prove a useful descriptive tool for patterns of compositional variability. We now build on that experience and consider how we may adapt such tools for the more analytical aspects of our problems.

It is clear from the interrelationships (4.4)–(4.6) that there are many ways in which we could specify the overall covariance structure of a composition. For example, we have seen in Section 4.2 that one way to determine completely the covariance structure of a 3-part composition is by specification of

$$\tau_{12}, \tau_{13}, \tau_{23}.$$

There are, however, many other ways. For example, if we specify

$$\sigma_{11.33}, \sigma_{12.33}, \sigma_{22.33}$$

the covariance structure is determined since

$$\tau_{12} = \sigma_{11.33} - 2\sigma_{12.33} + \sigma_{22.33}, \quad \tau_{13} = \sigma_{11.33}, \quad \tau_{23} = \sigma_{22.33}.$$

As yet another example we may use a more exotic set

$$\tau_{12}, \sigma_{12.33}, \sigma_{11.23},$$

since

$$\tau_{13} = \sigma_{12.33} + \sigma_{11.23}, \quad \tau_{23} = \sigma_{12.33} - \sigma_{11.23} + \tau_{12}.$$

In forms of analysis involving covariance structures in $\mathscr{R}^d$ it is well known that matrix specifications have proved particularly fruitful. It is therefore natural to ask whether such specifications can play a useful role in compositional data analysis. We shall here introduce and record the interrelationships between three different matrix specifications of compositional covariance structures. While these are equivalent in the sense of each determining the same covariance structure, we shall see later that, for the purposes of a particular form of compositional data analysis, one may be simpler to use and interpret than the others. Each has some role to play for some application and it is worth the effort to become thoroughly familiar with all three. We emphasize that we are now concerned only with the covariance structure, with second-order moments, not with the first-order moments or means: once a suitable matrix specification has been determined it will be obvious what the convenient specification of means should be.

The variation matrix

We have already seen in Section 4.3 that inspection of the $\frac{1}{2}dD$ logratio variances

$$\tau_{ij} = \text{var}\{\log(x_i/x_j)\} \qquad (i = 1, \ldots, d;\, j = i+1, \ldots, D)$$

provides, through the limitation to 2-component variation, a simple view of compositional variability. We can exploit Property 4.1, the symmetry property of τ_{ij}, and build up the logratio variances into a square matrix.

Definition 4.4 For a D-part composition $\mathbf{x}$ the $D \times D$ matrix

$$\boldsymbol{T} = [\tau_{ij}] = [\text{var}\{\log(x_i/x_j)\}\colon i, j = 1, \ldots, D]$$

is termed the *variation matrix* and determines the covariance structure by the relationships

$$\sigma_{ij.kl} = \tfrac{1}{2}(\tau_{il} + \tau_{jk} - \tau_{ij} - \tau_{kl})$$

of Property 4.2.

The variation matrix is symmetric with a diagonal of zeros. An apparent difficulty is that $\boldsymbol{T}$, while it adequately specifies the compositional covariance structure, is not in the form of a covariance matrix in the sense that the (i, j) element can be expressed as $\text{cov}(u_i, u_j)$ for all i, j, where $u_1, \ldots, u_D$ is a set of D random variables. We have emphasized this difference in form by referring to $\boldsymbol{T}$ as the compositional *variation* matrix. Despite this difference in form we shall see that $\boldsymbol{T}$ has a number of simple properties which make it a viable tool for many forms of compositional data analysis. Note that $\boldsymbol{T}$ treats the parts of a composition in a symmetrical manner.

The logratio covariance matrix

So far we have considered only specifications of the covariance structure in terms of 2-component variation through the special forms $\tau_{ij} = \sigma_{ii.jj}$. Let us now consider the introduction of 3-part variation where at most three of the parts i, j, k, l in $\sigma_{ij.kl}$ are different. The simplest starting point is to restrict attention to the set of covariances

$$\sigma_{ij} = \sigma_{ij.DD} = \text{cov}\{\log(x_i/x_D), \log(x_j/x_D)\} \qquad (i, j = 1, \ldots, d), \tag{4.14}$$

where part D is held fixed and thus the final component x_D plays the role of a common divisor in all the logratios. Since $\sigma_{ij}=\sigma_{ji}$ this set (4.14) of covariances consists of $\frac{1}{2}dD$ elements, rather than d^2. The question of whether these $\frac{1}{2}dD$ logratio covariances are sufficient to determine the covariance structure is readily answered.

Property 4.5 For any i, j, k, l from $1,\ldots,D$,

$$\sigma_{ij.kl}=\sigma_{ij}+\sigma_{kl}-\sigma_{il}-\sigma_{jk}$$

and so the covariance structure of a D-part composition $\mathbf{x}$ is completely determined by the $\frac{1}{2}dD$ different values of the set of σ_{ij} $(i,j=1,\ldots,d)$.

Proof. Writing the left-hand side as

$$\text{cov}\{\log(x_i/x_D)-\log(x_k/x_D),\log(x_j/x_D)-\log(x_l/x_D)\}$$

and expanding this covariance of linear forms, we immediately obtain the right-hand side of the identity.

We can now incorporate this result into a second matrix specification of covariance structure.

Definition 4.5 For a D-part composition $\mathbf{x}$ the $d\times d$ matrix

$$\boldsymbol{\Sigma}=[\sigma_{ij}]=[\text{cov}\{\log(x_i/x_D),\log(x_j/x_D)\}: i,j=1,\ldots,d]$$

is termed the *logratio covariance matrix* and determines the covariance structure by the relationships

$$\sigma_{ij.kl}=\sigma_{ij}+\sigma_{kl}-\sigma_{il}-\sigma_{jk}.$$

Our use of the term 'covariance matrix' in Definition 4.5 is in the standard sense that $\boldsymbol{\Sigma}$ is the variance-covariance matrix of a d-dimensional random vector $\mathbf{y}$ defined by

$$y_i=\log(x_i/x_D) \qquad (i=1,\ldots,d) \tag{4.15}$$

or, more compactly, by

$$\mathbf{y}=\log(\mathbf{x}_{-D}/x_D). \tag{4.16}$$

Moreover, $\mathbf{y}$ is free to move throughout $\mathscr{R}^d$ since the logratio transformation

$$\mathbf{x} \in \mathscr{S}^d \to \mathbf{y} = \log(\mathbf{x}_{-D}/x_D) \in \mathscr{R}^d$$

is one-to-one.

With this specification we see at once that the negative bias difficulty does not arise. The logratio transformation has effectively taken us out of the constraints of $\mathscr{S}^d$ into the freedom of $\mathscr{R}^d$. The covariance matrix $\boldsymbol{\Sigma}$ has no restrictions on it other than the standard requirements of non-negative definiteness. In using it we are certainly on familiar ground.

We have, of course, yet to see whether all the questions posed in Chapter 1 are translatable into terms of logratios of components. There is, however, another more immediate technical problem which we shall have to face if we wish to analyse compositions with the use of the logratio covariance matrix. Definition 4.5 does not treat the parts of the composition symmetrically: it picks out the final part D for a special role in providing the common component divisor x_D of all the logratios. An obvious question must therefore be posed for any statistical procedure we may propose involving $\boldsymbol{\Sigma}$. If we choose another component x_k as the common divisor and work with the covariance matrix of $\log(\mathbf{x}_{-k}/x_k)$, will we be led to exactly the same conclusion? We can express this question in a slightly more general and more easily investigated form.

Suppose that a first analyst carries out some statistical procedure involving the logratio covariance matrix $\boldsymbol{\Sigma}$ with the compositional parts in their original order $1, \ldots, D$. Suppose that a second analyst first permutes the parts in some way and then with this permuted order uses Definition 4.5 to obtain the logratio covariance matrix for application of the statistical procedure. Will the first and second analysts arrive at precisely the same conclusions? Technically we are here faced with the question of whether all statistical procedures involving parts of a composition asymmetrically are invariant under the group of permutations. In Chapter 5 we shall take steps to establish a number of invariant forms, that is constructs which are equal whatever permutation of parts is used; when we later introduce specific statistical procedures these will allow us to verify easily that order of parts, and in particular the choice of component divisor, makes no difference to any of the analyses of this monograph.

Centred logratio covariance matrix

A way to retain a covariance matrix form for the specification and to obtain a symmetric treatment of all D parts is to replace the single component divisor x_D by the geometric mean

$$g(\mathbf{x}) = (x_1 \ldots x_D)^{1/D}$$

of all D components. In this case it is more convenient to start with the definition and then explain the nature and properties of the covariance matrix specification.

Definition 4.6 For a D-part composition $\mathbf{x}$ the $D \times D$ covariance matrix

$$\boldsymbol{\Gamma} = [\gamma_{ij}] = [\operatorname{cov}[\log\{x_i/g(\mathbf{x})\}, \log\{x_j/g(\mathbf{x})\}]: i,j = 1, \ldots, D]$$

is termed the *centred logratio covariance matrix* and determines the covariance structure by the relationships

$$\sigma_{ij.kl} = \gamma_{ij} + \gamma_{kl} - \gamma_{il} - \gamma_{jk}.$$

We can easily substantiate the claim of the definition about the determining relationship by expressing $\sigma_{ij.kl}$ as

$$\operatorname{cov}[\log\{x_i/g(\mathbf{x})\} - \log\{x_k/g(\mathbf{x})\}, \log\{x_j/g(\mathbf{x})\} - \log\{x_l/g(\mathbf{x})\}]$$

and expanding in the standard way. The validity of the terminology of 'covariance matrix' is immediate from the fact that $\boldsymbol{\Gamma}$ is the covariance matrix of the random vector

$$\mathbf{z} = \log\{\mathbf{x}/g(\mathbf{x})\}. \tag{4.17}$$

At first sight a puzzling feature of $\boldsymbol{\Gamma}$ may be its apparent over-specification of the compositional covariance structure. As a $D \times D$ covariance matrix $\boldsymbol{\Gamma}$ is, of course, symmetric but still apparently requires $\frac{1}{2}D(D+1)$ entries to determine it, D more than the $\frac{1}{2}dD$ which we know are all that are required to specify the compositional covariance structure. The explanation is that our achievement of both a covariance matrix form of specification and a symmetric treatment of parts has been bought at a price: the centred logratio covariance matrix $\boldsymbol{\Gamma}$ is singular. We can most readily appreciate this feature by noting that the D-dimensional vector $\mathbf{z}$ in (4.17) does not have the freedom of $\mathscr{R}^D$ but is confined to the d-dimensional linear subspace

$$\mathbf{j}'\mathbf{z} = z_1 + \cdots + z_D = 0. \tag{4.18}$$

There is a parallel here with the device of embedding $\mathscr{S}^d$ in $\mathscr{R}^D$ to obtain an expression of the sample space which is symmetric in the parts. In order to obtain a symmetric vector of logratios we have to consider a transformation of $\mathbf{x}$ to $\mathbf{z}$, which takes us into a d-dimensional subspace of $\mathscr{R}^D$.

The consequent singularity of the distribution of $\mathbf{z}$ is reflected in the singularity of its covariance matrix $\boldsymbol{\Gamma}$. This singularity in turn means that there are D restrictions on the elements of $\boldsymbol{\Gamma}$ and these can be most conveniently identified as the requirement that all the row sums of $\boldsymbol{\Gamma}$ are zero. For example, the ith row sum

$$\sum_{j=1}^{D} \gamma_{ij} = \sum_{j=1}^{D} \text{cov}(z_i, z_j) = \text{cov}\left(z_i, \sum_{j=1}^{D} z_j \right) = 0.$$

These D constraints, expressible concisely as $\boldsymbol{\Gamma}\mathbf{j}_D = \mathbf{0}$, are exactly what are required to reduce the $\frac{1}{2}D(D+1)$ different elements of $\boldsymbol{\Gamma}$ to the $\frac{1}{2}dD$ of a compositional covariance structure specification. For ease of reference we gather this matrix condition together with an equivalent one into a single statement.

Property 4.6 The centred logratio covariance matrix $\boldsymbol{\Gamma}$ of a D-part composition $\mathbf{x}$ satisfies the following equivalent conditions:

(a) $\boldsymbol{\Gamma}\mathbf{j}_D = \mathbf{0}$,

(b) $\mathbf{G}_D\boldsymbol{\Gamma} = \boldsymbol{\Gamma}$,

where $\mathbf{j}_D$ is the $D \times 1$ vector of units and $\mathbf{G}_D = \mathbf{I}_D - D^{-1}\mathbf{J}_D$ with $\mathbf{J}_D$ the $D \times D$ matrix of units.

A question that must be asked is whether we have merely exchanged the awkward unit-sum constraint (1.1) on the composition $\mathbf{x}$ for an equally awkward zero-sum constraint (4.18) on $\mathbf{z}$. It should first be emphasized that (4.18) is in no way derived from the unit-sum constraint. It would arise from transformation (4.17) even if $\mathbf{x}$ were free to range over the whole of $\mathscr{R}^D_+$ and is simply a consequence of our insistence on a symmetric formulation in terms of logratios: it is analogous to the kind of zero-sum restrictions on the parameters which are standard practice in symmetric versions of analysis-of-variance models. The difficulties of the unit-sum constraint have already been effectively removed at the step (4.17) of transforming to logratios, in moving from $\mathscr{S}^d$ to $\mathscr{R}^D$. Although we are thereby involved with singular distributions, we are again within the framework of standard theory. Having said this, we must acknowledge that we have to face various problems associated with singular covariance matrices, such as finding convenient pseudo-inverses.

Even at this early stage in the development of our tools of compositional data analysis, we thus see that our three matrix specifications have advantages and disadvantages: $\boldsymbol{T}$ in its confinement to 2-part variation is simple, treats parts symmetrically but is not a covariance matrix; $\boldsymbol{\Sigma}$ is a covariance matrix, generally non-singular but it is asymmetric in its treatment of parts; $\boldsymbol{\Gamma}$ is a covariance matrix, treats parts symmetrically but is singular.

4.7 Some important elementary matrices

Five elementary matrices play a central role in both the theory and practice of compositional data analysis. We provide their definitions and notation in this section. For convenience of reference throughout the monograph we collect their simple properties in a compact form in Appendix A. It will pay the reader to make himself familiar at least with definitions and notation at this stage; the properties, whose obvious proofs are omitted, can be left until they are referred to later as they are required.

Definition 4.7 Table 4.4 provides definitions and terminology of *elementary matrices.*

To simplify notation we shall drop suffices whenever there is no possibility of confusion and when the order can readily be deduced from the context. For example, in theoretical considerations of D-part compositions, $\mathbf{F}$ and $\mathbf{H}$ will always have the order $d \times D$ and $d \times d$ as in Table 4.4. For most of our discussion the $\mathbf{G}$ form will be of order $D \times D$, but occasionally we shall require $\mathbf{G}_N$ of order $N \times N$; on those occasions the suffix notation will be retained for clarity.

Table 4.4

Notation	*Definition*	*Order*	*Rank*
$\mathbf{I}_d$	identity matrix	$d \times d$	d
$\mathbf{J}_d$	matrix of units	$d \times d$	1
$\mathbf{j}_d$	column vector of units	$d \times 1$	1
$\mathbf{F}_{d,D}$	$[\mathbf{I}_d : -\mathbf{j}_d]$	$d \times D$	d
$\mathbf{G}_D$	$\mathbf{I}_D - D^{-1}\mathbf{J}_D$	$D \times D$	d
$\mathbf{H}_d$	$\mathbf{I}_d + \mathbf{J}_d$	$d \times d$	d

The properties in Appendix A are labelled by the main elementary matrix involved and will be referred to in the text simply as F3, G6, and so on.

4.8 Relationships between the matrix specifications

It is useful to have the ability to move readily from one matrix representation to another and for this purpose we require a convenient record of the relationships among $\boldsymbol{T}, \boldsymbol{\Sigma}$ and $\boldsymbol{\Gamma}$. The proofs of all these are simple and are omitted. We first give the relationships in element form and then, in terms of the elementary matrices of Section 4.7, we record the relationships in matrix form.

Element forms

In (4.20) and (4.22) below the replacement of a suffix by $\cdot$ indicates summation over all possible values of that suffix followed by division by D. Thus, noting that $\tau_{ii}=0$ and adopting a convention that the undefined $\sigma_{Dj}=\sigma_{iD}=\sigma_{DD}=0$, we have, for example,

$$\tau_{i.}=D^{-1}\sum_{j=1}^{D}\tau_{ij}, \qquad \tau_{..}=D^{-2}\sum_{i=1}^{D}\sum_{j=1}^{D}\tau_{ij},$$

$$\sigma_{i.}=D^{-1}\sum_{j=1}^{D}\sigma_{ij}, \qquad \sigma_{..}=D^{-2}\sum_{i=1}^{D}\sum_{j=1}^{D}\sigma_{ij}.$$

Each element form of relationship is preceded by an indication of the direction of the transformation.

$$\boldsymbol{T}\rightarrow\boldsymbol{\Sigma}:\quad \sigma_{ij}=\tfrac{1}{2}(\tau_{iD}+\tau_{jD}-\tau_{ij}). \tag{4.19}$$

$$\boldsymbol{T}\rightarrow\boldsymbol{\Gamma}:\quad \gamma_{ij}=\tfrac{1}{2}(\tau_{i.}+\tau_{j.}-\tau_{ij}-\tau_{..}). \tag{4.20}$$

$$\boldsymbol{\Sigma}\rightarrow\boldsymbol{T}:\quad \tau_{ij}=\sigma_{ii}+\sigma_{jj}-2\sigma_{ij}. \tag{4.21}$$

$$\boldsymbol{\Sigma}\rightarrow\boldsymbol{\Gamma}:\quad \gamma_{ij}=\sigma_{ij}-\sigma_{i.}-\sigma_{j.}+\sigma_{..}. \tag{4.22}$$

$$\boldsymbol{\Gamma}\rightarrow\boldsymbol{T}:\quad \tau_{ij}=\gamma_{ii}+\gamma_{jj}-2\gamma_{ij}. \tag{4.23}$$

$$\boldsymbol{\Gamma}\rightarrow\boldsymbol{\Sigma}:\quad \sigma_{ij}=\gamma_{ij}-\gamma_{iD}-\gamma_{jD}+\gamma_{DD}. \tag{4.24}$$

Matrix forms

We first state the relationships using the **F**, **G**, **H** and **J** matrices of Section 4.7 and then comment briefly on them.

$$\boldsymbol{T}\rightarrow\boldsymbol{\Sigma}:\quad \boldsymbol{\Sigma}=-\tfrac{1}{2}\mathbf{F}\boldsymbol{T}\mathbf{F}'. \tag{4.25}$$

$$T \to \Gamma: \quad \Gamma = -\tfrac{1}{2}\mathbf{G}T\mathbf{G}. \tag{4.26}$$

$$\Sigma \to T: \quad \text{use successively (4.28) and (4.29).} \tag{4.27}$$

$$\Sigma \to \Gamma: \quad \Gamma = \mathbf{F}'\mathbf{H}^{-1}\Sigma\mathbf{H}^{-1}\mathbf{F}. \tag{4.28}$$

$$\Gamma \to T: \quad T = \mathbf{J}\,\mathrm{diag}(\Gamma) + \mathrm{diag}(\Gamma)\mathbf{J} - 2\Gamma. \tag{4.29}$$

$$\Gamma \to \Sigma: \quad \Sigma = \mathbf{F}\Gamma\mathbf{F}'. \tag{4.30}$$

The element relationships show that it is relatively simple to move from one matrix specification to another. Although some of the matrix relationships appear more complicated, they do turn out to be extremely useful in some developments of the theory. The difficulty of the matrix form (4.27) of the transformation $\Sigma \to T$ is largely due to the increase in order of matrix required, from $d \times d$ for Σ to $D \times D$ for T. The simplicity of the corresponding element form (4.21) has been achieved by the conventional extension of the definition of σ_{ij} $(i, j = 1, \ldots, d)$ to $\sigma_{Dj} = \sigma_{iD} = \sigma_{DD} = 0$ $(i, j = 1, \ldots, d)$. We could have introduced the matrix counterpart of (4.21) by enlarging Σ from order $d \times d$ to order $D \times D$ by bordering with zeros and then producing a relationship similar to (4.29), but we would find little theoretical advantage in such a complication.

4.9 Estimated matrices for hongite compositions

To illustrate the three different matrix specifications and for ease of future reference we collect in Table 4.5 estimates $\hat{T}, \hat{\Sigma}, \hat{\Gamma}$ of T, Σ, Γ for the hongite compositions of Data 1. The estimate $\hat{T}$ can be directly compiled from the compositional variation array of Table 4.1: then $\hat{\Sigma}$ and $\hat{\Gamma}$ can be obtained by application of either the element forms (4.19) and (4.20) or the matrix forms (4.25) and (4.26) of relationships to these T estimates. Alternatively the variance and covariance entries of any of these specifications can be estimated in the standard way from the data set of appropriate logratio-transformed values.

We shall see the comparative usefulness of different forms as we tackle specific problems in the remainder of the monograph.

4.10 Logratios and logcontrasts

Before we embark on the development of properties of the three matrix specifications of covariance structure and attempt to identify their strengths and weaknesses, it is convenient to introduce one

Table 4.5. *Estimates* $\hat{\boldsymbol{T}}, \hat{\boldsymbol{\Sigma}}, \hat{\boldsymbol{\Gamma}}$ *of the variation matrix, logratio covariance matrix, centred logratio covariance matrix for the hongite compositions of Data 1*

$$\hat{\boldsymbol{T}} = \begin{bmatrix} 0 & 0.259 & 1.533 & 0.083 & 0.139 \\ 0.259 & 0 & 3.001 & 0.547 & 0.649 \\ 1.533 & 3.001 & 0 & 1.111 & 0.948 \\ 0.083 & 0.547 & 1.111 & 0 & 0.187 \\ 0.139 & 0.649 & 0.948 & 0.187 & 0 \end{bmatrix}$$

$$\hat{\boldsymbol{\Sigma}} = \begin{bmatrix} 0.139 & 0.264 & -0.223 & 0.121 \\ 0.264 & 0.649 & -0.702 & 0.144 \\ -0.223 & -0.702 & 0.948 & 0.012 \\ 0.121 & 0.144 & 0.012 & 0.187 \end{bmatrix}$$

$$\hat{\boldsymbol{\Gamma}} = \begin{bmatrix} 0.094 & 0.218 & -0.238 & 0.029 & 0.020 \\ 0.218 & 0.601 & -0.719 & 0.050 & 0.018 \\ -0.238 & -0.719 & 0.962 & -0.051 & 0.050 \\ 0.029 & 0.050 & -0.051 & 0.047 & -0.028 \\ 0.020 & 0.018 & 0.050 & -0.028 & 0.084 \end{bmatrix}$$

further concept, a generalization of the logratios of components now seen to be playing a central role in compositional data analysis. In the study of variability of a random vector $\mathbf{y}$ in $\mathscr{R}^d$, interest often focuses on some linear combination $\mathbf{a}'\mathbf{y} = a_1 y_1 + \cdots + a_d y_d$: for example, in principal component analysis we seek to maximize var($\mathbf{a}'\mathbf{y}$) subject to the condition that $\mathbf{a}'\mathbf{a} = 1$. Since we have already taken the step of constructing a vector of logratios, $\mathbf{y} \in \mathscr{R}^d$, to replace $\mathbf{x} \in \mathscr{S}^d$, it seems likely that linear combinations of $\mathbf{y}$ will have a similar role to play in compositional data analysis. This turns out to be the case. From (4.16) we can write the linear combination $\mathbf{a}'\mathbf{y}$ as

$$\begin{aligned} & a_1 \log(x_1/x_D) + \cdots + a_d \log(x_d/x_D) && (4.31) \\ &= a_1 \log x_1 + \cdots + a_d \log x_d - (a_1 + \cdots + a_d) \log x_D \\ &= a_1 \log x_1 + \cdots + a_d \log x_d + a_D \log x_D, && (4.32) \end{aligned}$$

where $a_1 + \cdots + a_d + a_D = 0$. In standard terminology (4.32) is a loglinear contrast in the components of the composition $\mathbf{x}$ and we shall abbreviate this to logcontrast in our formal definition.

Definition 4.8 A *logcontrast* of a D-part composition $\mathbf{x}$ is any loglinear combination

$$a_1 \log x_1 + \cdots + a_D \log x_D = \mathbf{a}' \log \mathbf{x}$$

with

$$a_1 + \cdots + a_D = \mathbf{a}'\mathbf{j} = 0.$$

Logcontrasts have a number of simple properties.

Property 4.7

(a) The property that $\mathbf{a}'\mathbf{j} = 0$ ensures that a logcontrast can always be expressed as a linear combination of logratios with a common component divisor, for example as

$$a_1 \log(x_1/x_D) + \cdots + a_d \log(x_d/x_D) = \mathbf{a}'_{-D} \log(\mathbf{x}_{-D}/x_D).$$

(b) An alternative form is in terms of logratios with the geometric mean $g(\mathbf{x})$ as divisor, since $\mathbf{a}'\mathbf{j}_D = 0$ ensures that

$$\mathbf{a}' \log \mathbf{x} = \mathbf{a}' \log \{\mathbf{x}/g(\mathbf{x})\}.$$

(c) Logcontrasts are scale-free in the sense that

$$\mathbf{a}' \log (k\mathbf{x}) = \mathbf{a}' \log \mathbf{x}.$$

(d) Every logratio is a logcontrast.

Two further definitions, similar in character to standard linear model definitions, will help simplify some of our results concerning compositional logcontrasts.

Definition 4.9 A logcontrast $\mathbf{a}' \log \mathbf{x}$ is *standard* if $\mathbf{a}'\mathbf{a} = 1$ and $a_1 > 0$.

The important condition is the first, namely $\mathbf{a}'\mathbf{a} = 1$. If $\mathbf{a}'\mathbf{a} = 1$ then $(-\mathbf{a})'(-\mathbf{a}) = 1$ and so the logcontrasts $\mathbf{a}' \log \mathbf{x}$ and $-\mathbf{a}' \log \mathbf{x}$ both satisfy the first condition. In some practical procedures, the ambiguity of sign arising in this way can be a nuisance, for example in the comparison of logcontrasts between compositional data sets. The purpose of the second condition is simply to remove this kind of ambiguity.

Definition 4.10 Two logcontrasts $\mathbf{a}' \log \mathbf{x}$, $\mathbf{b}' \log \mathbf{x}$ are *orthogonal* if $\mathbf{a}'\mathbf{b} = 0$.

The following property records useful equivalent conditions.

Property 4.8 The following conditions are equivalent for the orthogonality of logcontrasts $\mathbf{a}'\log\mathbf{x}$, $\mathbf{b}'\log\mathbf{x}$:

(a) $\mathbf{a}'\mathbf{b}=0,$
(b) $\mathbf{a}'\mathbf{G}\mathbf{b}=0,$
(c) $\mathbf{a}'_{-D}\mathbf{H}\mathbf{b}_{-D}=0.$

Proof. Condition (b) is obvious since, by property G11 of the $\mathbf{G}$ matrix (Appendix A), $\mathbf{G}\mathbf{b}=\mathbf{b}$ when $\mathbf{b}'\log\mathbf{x}$ is a logcontrast.

The equivalence of (c) and (a) follows immediately from the observation that $\mathbf{a}=\mathbf{F}'\mathbf{a}_{-D}$, $\mathbf{b}=\mathbf{F}'\mathbf{b}_{-D}$ and property H2 of the $\mathbf{H}$ matrix.

4.11 Covariance structure of a basis

Where a D-part composition $\mathbf{x}$ has arisen from a D-dimensional basis $\mathbf{w}$ through the constraining operation $\mathbf{x}=\mathscr{C}(\mathbf{w})$, it may be of interest to relate the covariance structure of $\mathbf{x}$ to the covariance structure of $\mathbf{w}$. Our discussion in Section 3.3 of the basis difficulty of the crude covariance structure of a composition highlighted the lack of such a relationship. We now define a covariance structure for a D-part basis which is compatible with the covariance structure we have adopted for a D-part composition in the sense that the two covariance structures are simply related.

Definition 4.11 The *covariance structure of a D-part basis* $\mathbf{w}$ is the $D\times D$ covariance matrix $\boldsymbol{\Omega}$ of the vector $\log\mathbf{w}$, so that

$$\boldsymbol{\Omega}=[\omega_{ij}]=[\operatorname{cov}(\log w_i,\log w_j): i,j=1,\ldots,D].$$

Being a covariance matrix, $\boldsymbol{\Omega}$ is symmetric and is determined by $\frac{1}{2}D(D+1)$ values

$$\omega_{ij}\qquad(i=1,\ldots,D;j=i,\ldots,D).$$

Since a basis determines a unique composition we would expect $\boldsymbol{T}$, $\boldsymbol{\Sigma}$ and $\boldsymbol{\Gamma}$ to be determined by $\boldsymbol{\Omega}$. As in Section 4.8 the relationships can be expressed in either element or matrix forms.

Element forms

$$\boldsymbol{\Omega}\to\boldsymbol{T}:\quad \tau_{ij}=\omega_{ii}+\omega_{jj}-2\omega_{ij},\tag{4.33}$$

$$\boldsymbol{\Omega}\to\boldsymbol{\Sigma}:\quad \sigma_{ij}=\omega_{ij}-\omega_{iD}-\omega_{jD}+\omega_{DD},\tag{4.34}$$

$$\boldsymbol{\Omega}\to\boldsymbol{\Gamma}:\quad \gamma_{ij}=\omega_{ij}-\omega_{i.}-\omega_{j.}+\omega_{..},\tag{4.35}$$

where the subscript denotes the same averaging process as occurred in Section 4.8.

Matrix forms

$$\boldsymbol{\Omega} \rightarrow \boldsymbol{T}: \quad \boldsymbol{T} = \mathbf{J}\,\mathrm{diag}(\boldsymbol{\Omega}) + \mathrm{diag}(\boldsymbol{\Omega})\mathbf{J} - 2\boldsymbol{\Omega}, \tag{4.36}$$

$$\boldsymbol{\Omega} \rightarrow \boldsymbol{\Sigma}: \quad \boldsymbol{\Sigma} = \mathbf{F}\boldsymbol{\Omega}\mathbf{F}', \tag{4.37}$$

$$\boldsymbol{\Omega} \rightarrow \boldsymbol{\Gamma}: \quad \boldsymbol{\Gamma} = \mathbf{G}\boldsymbol{\Omega}\mathbf{G}. \tag{4.38}$$

Thus, in contrast to crude covariance structures, the logratio form of compositional covariance structure developed in this chapter is directly and exactly related to any underlying basis covariance structure. Later, in Chapter 9, we shall obtain the counterpart of Property 2.3, the many–one relationship of the set $\mathscr{B}(\mathbf{x})$ of bases corresponding to a given composition $\mathbf{x}$, by finding the set of basis covariance matrices $\boldsymbol{\Omega}$ corresponding to a given $\boldsymbol{T}$, $\boldsymbol{\Sigma}$ or $\boldsymbol{\Gamma}$. There is another form of basis covariance structure, arising from the one-to-one correspondence between basis $\mathbf{w}$ and size, composition $(t, \mathbf{x})$, and proving useful in certain forms of problem involving bases and compositions. Again we delay the introduction of this concept until Chapter 9 when it is required for a specific problem.

4.12 Commentary

We have now seen that the new definition of covariance structure introduced in this chapter is free of three of the difficulties of the crude covariance structure, cited in Section 3.3, namely the negative bias, subcomposition and basis difficulties. To appreciate that we are also free of the fourth difficulty of Section 3.3, the null correlation difficulty, requires more technically detailed analysis and we delay its consideration until Chapter 10.

We are now equipped with three ways in which a compositional covariance structure can be specified, the matrix specifications $\boldsymbol{T}, \boldsymbol{\Sigma}$, $\boldsymbol{\Gamma}$. This may appear to be an embarrassment of riches, an unnecessary complication of an already complicated situation. We shall see in practice, however, that the analysis of a particular problem is much simpler in terms of one of the specifications than of the others. For example, $\boldsymbol{\Sigma}$ is the simplest version for the analyses of Chapters 6 and 7, $\boldsymbol{\Gamma}$ for Sections 8.1–8.4 and 8.7, $\boldsymbol{T}$ for the simple descriptive purposes of Sections 4.3, 4.5, 8.5–8.6 and 10.2. One major reason for the need for this flexibility is, as we have already indicated, that each specification has some apparent disadvantage relative to the others: $\boldsymbol{T}$ is not a covariance matrix, $\boldsymbol{\Sigma}$ is not symmetric in the compositional parts, $\boldsymbol{\Gamma}$ is singular; there is no specification which is free from all of these three disadvantageous features. The next chapter will be devoted to an investigation of technical properties of $\boldsymbol{T}$, $\boldsymbol{\Sigma}$ and $\boldsymbol{\Gamma}$, which largely demonstrate that each disadvantage is more apparent than real. Nevertheless, it does remain true that over the wide range of compositional data problems each of the specifications turns out for some particular situation to be more attractive than the others.

The reader who is prepared to accept that

1. the variation matrix $\boldsymbol{T}$ has as simple and useful properties as its rival covariance matrices;
2. all the statistical procedures involving the logratio covariance matrix $\boldsymbol{\Sigma}$ that are used in this monograph are invariant under the group of permutations of the parts of the compositions;
3. the singularity of the centred logratio covariance matrix $\boldsymbol{\Gamma}$ can be handled by the determination of a simple pseudo-inverse and pseudo-determinant;

may wish to avoid the technicalities of Chapter 5 and proceed directly to Chapter 6. There will be found an introduction of simple parametric classes of distributions for compositional variability which free us from the difficulties described in Section 3.4. From the viewpoint of the practice of compositional data analysis little will be lost by this move: the reader will be able to refer easily to any of the technical formulae of Chapter 5 as they are required in subsequent chapters.

4.13 Bibliographic notes

The idea of a compositional covariance structure based on logratios of components was introduced by Aitchison (1981a), who used the logratio covariance matrix $\boldsymbol{\Sigma}$ as the means of specification. Also established there was the exact relationship of this covariance structure to the basis covariance matrix $\boldsymbol{\Omega}$ in a resolution of the null correlation difficulty of Section 3.3. The properties and uses of $\boldsymbol{\Sigma}$ were further developed in Aitchison (1982). The centred logratio covariance matrix $\boldsymbol{\Gamma}$ as a means of specifying a compositional covariance structure and the concept of logcontrast of a composition were then used by Aitchison (1983) to provide a practical form of compositional principal component analysis, which will be discussed in Chapter 8.

The possibility of using the variation matrix T as a means of specifying covariance structure was indicated, but not developed, in Aitchison (1984a). Thus all the properties and uses of the variation matrix T and the descriptive use of the variation array appear for the first time in this monograph.

Problems

4.1 Show that the covariance structure of a 3-part composition is completely specified by $\tau_{12}, \sigma_{12}, \gamma_{12}$.

4.2 Estimate the variation array of the AFM compositions of Data 6 and evaluate its merits as a summary description of the variability as observed in the AFM diagram of Fig. 1.9.

4.3 For the 3-part fruit compositions of Data 12 estimate the variation arrays for each of the four groupings (present, T), (past, T), (present, U), (past, U). Consider whether these arrays help in answering Questions 1.11.1 and 1.11.2.

4.4 The crude mean vector and covariance matrix for the (present, T) grouping in Problem 4.3 are

$$\boldsymbol{\lambda} = (0.525 \quad 0.225 \quad 0.250),$$

$$\boldsymbol{K} = 10^{-4} \times \begin{bmatrix} 3.46 & -2.00 & -1.45 \\ -2.00 & 1.48 & 0.52 \\ -1.45 & 0.52 & 0.93 \end{bmatrix}.$$

Apply the recovery technique of Section 4.4 to obtain an approximation to the estimated variation array and compare this with the actual estimated array already computed directly from the compositional data.

4.5 Show that for a 5-part composition, knowledge of the variation arrays of the 123, 145, 234 and 235 subcompositions will allow the construction of the variation array for the full composition.

Show that it is possible to construct the variation array of a 6-part composition from knowledge of the variation arrays of only three 4-part subcompositions, and identify a set of such subcompositions.

4.6 For a D-part composition $\mathbf{x}$ show that $\mathbf{y} = \log(\mathbf{x}_{-D}/x_D)$ and $\mathbf{z} = \log\{\mathbf{x}/g(\mathbf{x})\}$ are related by

$$\mathbf{y} = \mathbf{F}\mathbf{z}, \qquad \mathbf{z} = \mathbf{F}'(\mathbf{F}\mathbf{F}')^{-1}\mathbf{y},$$

where $\mathbf{F}$ is the elementary matrix of Definition 4.7. Hence prove the matrix relationships (4.28) and (4.30) between $\boldsymbol{\Sigma}$ and $\boldsymbol{\Gamma}$.

4.7 For a 4-part composition $\mathbf{x}$ show that

$$u_1 = \frac{1}{\sqrt{2}}(\log x_1 - \log x_2),$$

$$u_2 = \frac{1}{\sqrt{6}}(\log x_1 + \log x_2 - 2\log x_3),$$

$$u_3 = \frac{1}{\sqrt{12}}(\log x_1 + \log x_2 + \log x_3 - 3\log x_4),$$

is a set of standard, orthogonal logcontrasts. By expressing the logratio vector $\mathbf{y} = \log(\mathbf{x}_{-4}/x_4)$ in terms of $\mathbf{u} = (u_1, u_2, u_3)$, or otherwise, show that the covariance matrix $\boldsymbol{\Delta}$ of $\mathbf{u}$ completely specifies the compositional covariance structure.

Generalize this vector $\mathbf{u}$ of logcontrasts to the case of a D-part composition, show that the logratio covariance matrix $\boldsymbol{\Sigma}$ is related to $\boldsymbol{\Delta}$ by $\boldsymbol{\Sigma} = \mathbf{M}\boldsymbol{\Delta}\mathbf{M}'$ and determine the matrix $\mathbf{M}$.

4.8 Estimate directly from the household expenditure data of single men in Data 8 the basis covariance matrix $\boldsymbol{\Omega}$, the variation matrix $\boldsymbol{T}$ and the logratio covariance matrix $\boldsymbol{\Sigma}$ and verify the relationships (4.19), (4.36), (4.37).

4.9 For the 37 adult steroid metabolites excretions of Data 9 obtain estimates $\hat{\boldsymbol{\Omega}}$ and $\hat{\boldsymbol{\Gamma}}$ of the basis and centred logratio covariance matrices, and verify the relationship (4.38).

4.10 Show that each of the basis covariance matrices

$$\boldsymbol{\Omega}_1 = \begin{bmatrix} 1.1 & 0.4 & 0.3 \\ 0.4 & 0.4 & 0 \\ 0.3 & 0 & 0.2 \end{bmatrix}, \boldsymbol{\Omega}_2 = \begin{bmatrix} 0.55 & -0.05 & 0.15 \\ -0.05 & 0.05 & -0.05 \\ 0.15 & -0.05 & 0.45 \end{bmatrix}$$

leads through the relationship $\boldsymbol{\Sigma} = \mathbf{F}\boldsymbol{\Omega}\mathbf{F}'$ to the same logratio covariance matrix. Do you find this result surprising?

4.11 Suppose that the only information published about the variability of a D-part basis $\mathbf{w}$ is its crude mean vector $\boldsymbol{\lambda} = \mathrm{E}(\mathbf{w})$ and its crude covariance matrix $\boldsymbol{K} = \boldsymbol{V}(\mathbf{w})$. Show that the technique analogous to that of Section 4.4 for the recovery of covariance structure $\boldsymbol{\Omega}$ is the following:

$$\mathrm{E}(\log w_i) \simeq \log \lambda_i - \frac{\kappa_{ii}}{2\lambda_i^2} \qquad (i = 1, \ldots, D),$$

$$\omega_{ij} = \mathrm{cov}(\log w_i, \log w_j) \simeq \frac{\kappa_{ij}}{\lambda_i \lambda_j} + \frac{\kappa_{ii}\kappa_{jj}}{4\lambda_i^2 \lambda_j^2} \qquad (i, j = 1, \ldots, D).$$

CHAPTER 5

Properties of matrix covariance specifications

Aren't you the chap who sold me this car last week?
Well, tell me about it again, I get so depressed.

Anon

5.1 Logratio notation

We now embark on a study of the properties of the matrices, $\boldsymbol{T}$, $\boldsymbol{\Sigma}$ and $\boldsymbol{\Gamma}$ which are relevant to compositional data analysis. Although it is clear from Section 4.8 that any property of one of the specifications can be converted into an equivalent property of the others, the expression of the property may be much simpler for one particular specification and so promise more tractability in statistical analysis. Our study will therefore be comparative, searching to discover weaknesses and strengths of the specifications and, in particular, assessing the extent to which their known disadvantages, that $\boldsymbol{T}$ is not a covariance matrix, that $\boldsymbol{\Sigma}$ is asymmetric in the parts, that $\boldsymbol{\Gamma}$ is singular, can be discounted.

We shall use already established notation for the logratio vectors $\mathbf{y} \in \mathscr{R}^d$ and $\mathbf{z} \in \mathscr{R}^D$,

$$\mathbf{y} = \log(\mathbf{x}_{-D}/x_D) = \mathbf{F} \log \mathbf{x}, \tag{5.1}$$

$$\mathbf{z} = \log\{\mathbf{x}/g(\mathbf{x})\} = \mathbf{G} \log \mathbf{x}, \tag{5.2}$$

and shall require the facility of transforming one into the other:

$$\mathbf{y} = \mathbf{F}\mathbf{z}, \tag{5.3}$$

$$\mathbf{z} = \mathbf{F}'(\mathbf{F}\mathbf{F}')^{-1}\mathbf{y} = \mathbf{F}'\mathbf{H}^{-1}\mathbf{y}. \tag{5.4}$$

We note that the matrix relationships (4.28) and (4.30) between $\boldsymbol{\Sigma}$ and $\boldsymbol{\Gamma}$ are immediate consequences of (5.4) and (5.3), respectively, and also record the constrained nature of the centred logratio vector:

$$\mathbf{j}'\mathbf{z} = 0, \qquad \mathbf{G}\mathbf{z} = \mathbf{z}, \qquad \mathbf{J}\mathbf{z} = \mathbf{0}. \tag{5.5}$$

5.2 Logcontrast variances and covariances

One of the central requirements of a specification of compositional covariance structure is that it should yield simple and tractable forms for the variance of a logcontrast and the covariance of two logcontrasts. We can expect success with $\boldsymbol{\Sigma}$ and $\boldsymbol{\Gamma}$ since they are covariance matrices, but what of $\boldsymbol{T}$ which is not? The answer is to be found in the following result.

Property 5.1 For logcontrasts $\mathbf{a}'\log\mathbf{x}$, $\mathbf{b}'\log\mathbf{x}$ of a D-part composition $\mathbf{x}$ the following forms for variances and covariances are equivalent:

(a) $\operatorname{var}(\mathbf{a}'\log\mathbf{x}) = -\frac{1}{2}\mathbf{a}'\boldsymbol{T}\mathbf{a} = \mathbf{a}'_{-D}\boldsymbol{\Sigma}\mathbf{a}_{-D} = \mathbf{a}'\boldsymbol{\Gamma}\mathbf{a}$,
(b) $\operatorname{cov}(\mathbf{a}'\log\mathbf{x}, \mathbf{b}'\log\mathbf{x}) = -\frac{1}{2}\mathbf{a}'\boldsymbol{T}\mathbf{b} = \mathbf{a}'_{-D}\boldsymbol{\Sigma}\mathbf{b}_{-D} = \mathbf{a}'\boldsymbol{\Gamma}\mathbf{b}$.

Proof. The results in terms of $\boldsymbol{\Sigma}$ and $\boldsymbol{\Gamma}$ are immediate from standard theory of covariance matrices. The results in terms of $\boldsymbol{T}$ follow immediately by the use of property G11 of Appendix A, which gives $\mathbf{a} = \mathbf{G}\mathbf{a}$, $\mathbf{b} = \mathbf{G}\mathbf{b}$ for logcontrasts, and by the relationship (4.26).

The fact that we obtain as simple quadratic and bilinear forms using $\boldsymbol{T}$ as we do for $\boldsymbol{\Sigma}$ and $\boldsymbol{\Gamma}$ provides a considerable boost to the status of $\boldsymbol{T}$ as an operational tool: as far as any analyses involving variances and covariances are concerned it appears that $\boldsymbol{T}$ is on an equal footing with $\boldsymbol{\Sigma}$ and $\boldsymbol{\Gamma}$.

5.3 Permutations

We now turn our attention to the main disadvantage of the specification $\boldsymbol{\Sigma}$, the question of which statistical procedures, if any, involving $\boldsymbol{\Sigma}$ or its estimate are invariant under the group of permutations of compositional parts. Let $\mathbf{P}$ be any $D \times D$ permutation matrix: for example, for $D = 5$ the permutation matrix

$$\begin{bmatrix} 0 & 0 & 0 & 1 & 0 \\ 1 & 0 & 0 & 0 & 0 \\ 0 & 0 & 1 & 0 & 0 \\ 0 & 0 & 0 & 0 & 1 \\ 0 & 1 & 0 & 0 & 0 \end{bmatrix}$$

applied to the composition $\mathbf{x} = (x_1, x_2, x_3, x_4, x_5)$ yields $\mathbf{x}_P = \mathbf{Px} = (x_4, x_1, x_3, x_5, x_2)$ and applied to the centred logratio vector $\mathbf{z} = \log\{\mathbf{x}/g(\mathbf{x})\}$ yields $\mathbf{z}_P = \mathbf{Pz} = (z_4, z_1, z_3, z_5, z_2)$. The relationships

$$\mathbf{x}_P = \mathbf{Px}, \tag{5.6}$$

$$\mathbf{z}_P = \mathbf{Pz}, \tag{5.7}$$

are clearly true for general D.

The effect of the permutation $\mathbf{P}$ of the D parts is much more complicated for the d-dimensional logratio vector $\mathbf{y}$. Let $\mathbf{y}$ be the logratio vector (5.1) in terms of the original ordering of the parts, and let $\mathbf{y}_P$ be the logratio vector obtained from the permuted components and therefore with the final component of the permuted composition as the common divisor. Then successive application of (5.3), (5.7), (5.4) gives

$$\mathbf{y}_P = \mathbf{Fz}_P = \mathbf{FPz} = \mathbf{FPF'H}^{-1}\mathbf{y}. \tag{5.8}$$

For our simple example, recalling from property H6 that here $\mathbf{H}^{-1} = \mathbf{I} - (1/5)\mathbf{J}$, we can easily verify that

$$\mathbf{FPF'H}^{-1} = \begin{bmatrix} 0 & -1 & 0 & 1 \\ 1 & -1 & 0 & 0 \\ 0 & -1 & 1 & 0 \\ 0 & -1 & 0 & 0 \end{bmatrix},$$

which, when applied to

$$\mathbf{y} = (\log(x_1/x_5), \log(x_2/x_5), \log(x_3/x_5), \log(x_4/x_5)),$$

correctly yields

$$\mathbf{y}_P = (\log(x_4/x_2), \log(x_1/x_2), \log(x_3/x_2), \log(x_5/x_2)).$$

We can now collect our results here into the following statement, using the notation

$$\mathbf{Q}_P = \mathbf{FPF'H}^{-1}. \tag{5.9}$$

Property 5.2 If $\mathbf{P}$ is any $D \times D$ permutation matrix and a subscript P is used to denote the effect on some compositional characteristic arising from the permutation $\mathbf{P}$ of the parts, or equivalently the permutation $\mathbf{x}_P = \mathbf{Px}$ of the components, then

(a) $\boldsymbol{T}_P$, $\mathbf{z}_P$ and $\boldsymbol{\Gamma}_P$ take forms familiar in standard theory:

(i) $\boldsymbol{T}_P = \mathbf{P}\boldsymbol{T}\mathbf{P}'$,
(ii) $\mathbf{z}_P = \mathbf{P}\mathbf{z}$,
(iii) $\boldsymbol{\Gamma}_P = \mathbf{P}\boldsymbol{\Gamma}\mathbf{P}'$,

(b) $\mathbf{y}_P$ and $\boldsymbol{\Sigma}_P$ are given by

(iv) $\mathbf{y}_P = \mathbf{Q}_P\mathbf{y}$,
(v) $\boldsymbol{\Sigma}_P = \mathbf{Q}_P\boldsymbol{\Sigma}\mathbf{Q}_P'$.

5.4 Properties of P and $\mathbf{Q}_P$ matrices

We digress to record for ease of reference some useful properties of **P** and $\mathbf{Q}_P$ matrices.

$$\mathbf{PP}' = \mathbf{P}'\mathbf{P} = \mathbf{I} \tag{5.10}$$

$$|\mathbf{P}| = \pm 1 \tag{5.11}$$

$$\mathbf{PJP}' = \mathbf{J} \tag{5.12}$$

$$\mathbf{PGP}' = \mathbf{G} \tag{5.13}$$

$$\mathbf{Q}_P\mathbf{H}\mathbf{Q}_P' = \mathbf{H} \tag{5.14}$$

$$\mathbf{Q}_P'\mathbf{H}^{-1}\mathbf{Q}_P = \mathbf{H}^{-1} \tag{5.15}$$

$$|\mathbf{Q}_P| = \pm 1 \tag{5.16}$$

$$\mathbf{Q}_P\mathbf{Q}_{P'} = \mathbf{I}. \tag{5.17}$$

Properties (5.10)–(5.13) are either standard or obvious. Applying successively (5.9), G3, (5.13), F1, H2, we have

$$\mathbf{Q}_P\mathbf{H}\mathbf{Q}_P' = \mathbf{FPF}'\mathbf{H}^{-1}\mathbf{FP}'\mathbf{F}' = \mathbf{FPGP}'\mathbf{F}' = \mathbf{FGF}' = \mathbf{FF}' = \mathbf{H}$$

and

$$\begin{aligned}\mathbf{Q}_P\mathbf{Q}_{P'} &= \mathbf{FPF}'\mathbf{H}^{-1}\mathbf{FP}'\mathbf{F}'\mathbf{H}^{-1} = \mathbf{FPGP}'\mathbf{F}'\mathbf{H}^{-1} = \mathbf{FGF}'\mathbf{H}^{-1} \\ &= \mathbf{FF}'\mathbf{H}^{-1} = \mathbf{I},\end{aligned}$$

thus establishing (5.14) and (5.17). Note that (5.15) is equivalent to (5.14). Taking determinants of both sides of (5.14) immediately gives (5.16).

5.5 Permutation invariants involving $\boldsymbol{\Sigma}$

Because of the symmetric treatment of parts by both the variation matrix $\boldsymbol{T}$ and the centred logratio covariance matrix $\boldsymbol{\Gamma}$ and the standard forms of the permutation effects (i)–(iii) of Property 5.2, we can be sure that statistical procedures based on these entities are invariant under the group of permutations of parts. We therefore concentrate our attention on the logratio covariance matrix $\boldsymbol{\Sigma}$. With a view to the easy verification later that all our statistical procedures involving $\boldsymbol{\Sigma}$ have the invariance property, we here identify three important permutation-invariant forms. Such permutation invariants are constructs or expressions which have the same value whatever the permutation of the parts of the composition. We first collect these invariant results together for ease of reference and then provide informal proofs interspersed with comments.

Property 5.3 For any permutation $\mathbf{P}$ of the parts of D-part composition $\mathbf{x}$ the following relationships hold:

(a) $|\boldsymbol{\Sigma}_P| = |\boldsymbol{\Sigma}|$.
(b) $\mathbf{y}_P'\boldsymbol{\Sigma}_P^{-1}\mathbf{y}_P = \mathbf{y}'\boldsymbol{\Sigma}^{-1}\mathbf{y}$.
(c) The eigensolutions $(\lambda_P, \mathbf{b}_P)$ of $(\boldsymbol{\Sigma}_P - \lambda_P\mathbf{H})\mathbf{b}_P = \mathbf{0}$ and $(\lambda, \mathbf{b})$ of $(\boldsymbol{\Sigma} - \lambda\mathbf{H})\mathbf{b} = \mathbf{0}$ are related by

  (i) $\lambda_P = \lambda$,
  (ii) $\mathbf{b}_P = (\mathbf{Q}_P')^{-1}\mathbf{b}$,
  (iii) $\mathbf{b}_P'\mathbf{y}_P = \mathbf{b}'\mathbf{y}$.

Thus the generalized variance $|\boldsymbol{\Sigma}|$, the quadratic form $\mathbf{y}'\boldsymbol{\Sigma}^{-1}\mathbf{y}$ and the eigenvalues λ and the eigenforms $\mathbf{b}'\mathbf{y}$ associated with $(\boldsymbol{\Sigma} - \lambda\mathbf{H})\mathbf{b} = \mathbf{0}$ are all permutation invariants.

(*a*) *Generalized variance.* We shall find in subsequent chapters that the generalized variance $|\boldsymbol{\Sigma}|$ plays an important role in the distributional aspects of compositional data, so that its permutation invariance is important. We can establish (a) by taking determinants of both sides of (v) of Property 5.2:

$$\begin{aligned} |\boldsymbol{\Sigma}_P| &= |\mathbf{Q}_P||\boldsymbol{\Sigma}||Q_P'| \\ &= |\mathbf{Q}_P|^2|\boldsymbol{\Sigma}| \\ &= |\boldsymbol{\Sigma}| \end{aligned}$$

from (5.16).

(*b*) *Quadratic form.* A quadratic form such as $\mathbf{y}'\boldsymbol{\Sigma}^{-1}\mathbf{y}$ appears commonly in standard theory and we are therefore likely to require its permutation invariance in many statistical procedures. The proof follows directly from the relations (iv) and (v) of Property 5.2:

$$\begin{aligned}\mathbf{y}_P'\boldsymbol{\Sigma}_P^{-1}\mathbf{y}_P &= \mathbf{y}'\mathbf{Q}_P'(\mathbf{Q}_P\boldsymbol{\Sigma}\mathbf{Q}_P')^{-1}\mathbf{Q}_P\mathbf{y}\\ &= \mathbf{y}'\mathbf{Q}_P'(\mathbf{Q}_P')^{-1}\boldsymbol{\Sigma}^{-1}\mathbf{Q}_P^{-1}\mathbf{Q}_P\mathbf{y}\\ &= \mathbf{y}'\boldsymbol{\Sigma}^{-1}\mathbf{y}.\end{aligned}$$

(c) *Eigensolutions.* In Chapter 8 we shall make extensive use of eigensolutions associated with compositional covariance structures. Standard eigenanalysis of the covariance matrix $\boldsymbol{\Sigma}$ for variation in $\mathscr{R}^d$ involves finding the eigensolutions of $(\boldsymbol{\Sigma}-\lambda\mathbf{I})\mathbf{b}=\mathbf{0}$ and in a superficial view of the corresponding compositional problem with logratio covariance matrix $\boldsymbol{\Sigma}$ we might expect the eigenvalues to be permutation invariant. That this is not so can be quickly appreciated by a simple example. For a 3-part composition and

$$\mathbf{P}(x_1, x_2, x_3) = (x_2, x_3, x_1),$$

$$\boldsymbol{\Sigma} = \begin{bmatrix} 1 & 3/4 \\ 3/4 & 3 \end{bmatrix}, \quad \boldsymbol{\Sigma}_P = \begin{bmatrix} 5/2 & 1/4 \\ 1/4 & 1 \end{bmatrix}$$

are corresponding logratio covariance matrices, but their eigenvalues are 13/4, 3/4 and $(7+\sqrt{10})/4$, $(7-\sqrt{10})/4$, respectively. In order to obtain a meaningful compositional counterpart to the $\mathscr{R}^d$ form of analysis appropriate to the logratio covariance matrix $\boldsymbol{\Sigma}$, some modification of the eigenanalysis procedure is clearly required. In this section we confine ourselves to the bare algebraic result. At this stage we are not in a position to provide insights into the real nature of the modification.

The proof of this modified result, statement (c) of Property 5.3, is straightforward. By (v) of Property 5.2, (5.14) and (5.16) we have

$$\begin{aligned}|\boldsymbol{\Sigma}_P-\lambda\mathbf{H}| &= |\mathbf{Q}_P(\boldsymbol{\Sigma}-\lambda\mathbf{H})\mathbf{Q}_P'|\\ &= |\boldsymbol{\Sigma}-\lambda\mathbf{H}|,\end{aligned}$$

and (i) follows immediately. Also,

$$(\boldsymbol{\Sigma}_P-\lambda\mathbf{H})\mathbf{b}_P = \mathbf{Q}_P(\boldsymbol{\Sigma}-\lambda\mathbf{H})\mathbf{Q}_P'\mathbf{b}_P$$

and since $\mathbf{Q}_P$ is non-singular the eigenvectors must clearly be related by $\mathbf{Q}_P'\mathbf{b}_P=\mathbf{b}$, which is equivalent to (ii). Then from (iv) of Property 5.2 and (ii) above, (iii) is immediate.

We may easily verify (c) for our simple example by showing that

$$|\boldsymbol{\Sigma} - \lambda \mathbf{H}| = |\boldsymbol{\Sigma}_P - \lambda \mathbf{H}| = 3\lambda^2 - (13/2)\lambda + 39/16$$

with identical eigenvalues $(13 \pm 2\sqrt{13})/12$.

If we write $\boldsymbol{\mu} = \mathrm{E}(\mathbf{y})$, $\boldsymbol{\mu}_P = \mathrm{E}(\mathbf{y}_P)$ then the relation $\mathbf{y}_P = \mathbf{Q}_P\mathbf{y}$ is carried over to $\boldsymbol{\mu}_P = \mathbf{Q}_P\boldsymbol{\mu}$. In standard $\mathscr{R}^d$ theory we may be concerned with a bilinear form $\boldsymbol{\mu}'\boldsymbol{\Sigma}^{-1}\mathbf{y}$, for example in discriminant analysis, rather than a quadratic form. The following simple extension of (b) of Property 5.3 includes this particular bilinear form as a special case.

Property 5.4 For any d-vectors $\mathbf{a}$ and $\mathbf{b}$, which are transformed by permutation to $\mathbf{a}_P = \mathbf{Q}_P\mathbf{a}$, $\mathbf{b}_P = \mathbf{Q}_P\mathbf{b}$, the relationship

$$\mathbf{a}_P'\boldsymbol{\Sigma}_P^{-1}\mathbf{b}_P = \mathbf{a}'\boldsymbol{\Sigma}^{-1}\mathbf{b}$$

holds, and so $\mathbf{a}'\boldsymbol{\Sigma}^{-1}\mathbf{b}$ is a permutation invariant.

Our final permutation-invariance result is simply deduced from (c) of Property 5.3.

Property 5.5 For any permutation $\mathbf{P}$ of the parts of a D-part composition,

$$\mathrm{tr}(\mathbf{H}^{-1}\boldsymbol{\Sigma}_P) = \mathrm{tr}(\mathbf{H}^{-1}\boldsymbol{\Sigma}).$$

A proof follows directly from c(i) of Property 5.3 since the trace of a square matrix is equal to the sum of its eigenvalues. A more direct proof uses (5.15):

$$\begin{aligned}\mathrm{tr}(\mathbf{H}^{-1}\boldsymbol{\Sigma}_P) &= \mathrm{tr}(\mathbf{H}^{-1}\mathbf{Q}_P\boldsymbol{\Sigma}\mathbf{Q}_P') \\ &= \mathrm{tr}(\mathbf{Q}_P'\mathbf{H}^{-1}\mathbf{Q}_P\boldsymbol{\Sigma}) \\ &= \mathrm{tr}(\mathbf{H}^{-1}\boldsymbol{\Sigma}).\end{aligned}$$

All these permutation-invariance properties of $\boldsymbol{\Sigma}$ indicate that there may be little disadvantage in treating the parts of the composition asymmetrically.

5.6 Covariance matrix inverses

In some areas of application of covariance matrices we require to use the inverse or the determinant: we have already seen examples above involving $\boldsymbol{\Sigma}^{-1}$ and $|\boldsymbol{\Sigma}|$. When a covariance matrix, such as the centred logratio covariance matrix $\boldsymbol{\Gamma}$, is singular, then we have to introduce a generalized or pseudo-inverse and pseudo-determinant. We shall be able to find a unique pseudo-inverse $\boldsymbol{\Gamma}^-$ of $\boldsymbol{\Gamma}$ satisfying the Moore–Penrose conditions:

$$\boldsymbol{\Gamma}\boldsymbol{\Gamma}^-\boldsymbol{\Gamma} = \boldsymbol{\Gamma}, \tag{5.18}$$

$$\boldsymbol{\Gamma}^-\boldsymbol{\Gamma}\boldsymbol{\Gamma}^- = \boldsymbol{\Gamma}^-, \tag{5.19}$$

$$(\boldsymbol{\Gamma}\boldsymbol{\Gamma}^-)' = \boldsymbol{\Gamma}\boldsymbol{\Gamma}^-, \tag{5.20}$$

$$(\boldsymbol{\Gamma}^-\boldsymbol{\Gamma})' = \boldsymbol{\Gamma}^-\boldsymbol{\Gamma}, \tag{5.21}$$

and for our definition of pseudo-determinant $|\boldsymbol{\Gamma}|^+$ we take the product of all the positive eigenvalues of $\boldsymbol{\Gamma}$. Our main result is then easily stated.

Property 5.6 For a centred logratio covariance matrix $\boldsymbol{\Gamma}$,

(a) $\boldsymbol{\Gamma}^- = \mathbf{F}'\boldsymbol{\Sigma}^{-1}\mathbf{F} = \mathbf{F}'(\mathbf{F}\boldsymbol{\Gamma}\mathbf{F}')^{-1}\mathbf{F}$,
(b) $\boldsymbol{\Gamma}^- = (\boldsymbol{\Gamma} + k\mathbf{J})^{-1} - (kD^2)^{-1}\mathbf{J}$, whatever value $k \neq 0$ is assigned,
(c) $|\boldsymbol{\Gamma}|^+ = |\boldsymbol{\Gamma} + D^{-1}\mathbf{J}|$.

Proof. (a) From (4.30) the two stated forms for $\boldsymbol{\Gamma}^-$ are identical, and so we can concentrate on establishing the first form. We need only check the Moore–Penrose conditions (5.18)–(5.21). Successive application of (4.28), H2 and G3 gives

$$\boldsymbol{\Gamma}\boldsymbol{\Gamma}^- = \mathbf{F}'\mathbf{H}^{-1}\boldsymbol{\Sigma}\mathbf{H}^{-1}\mathbf{F}\mathbf{F}'\boldsymbol{\Sigma}^{-1}\mathbf{F} = \mathbf{F}'\mathbf{H}^{-1}\mathbf{F} = \mathbf{G}$$

and, since $\mathbf{G}$ is symmetric, (5.20) holds. Then $\boldsymbol{\Gamma}\boldsymbol{\Gamma}^-\boldsymbol{\Gamma} = \mathbf{G}\boldsymbol{\Gamma} = \boldsymbol{\Gamma}$ by Property 4.6b and so (5.18) holds. Also $\boldsymbol{\Gamma}^-\boldsymbol{\Gamma}\boldsymbol{\Gamma}^- = \boldsymbol{\Gamma}^-\mathbf{G} = \mathbf{F}'\boldsymbol{\Sigma}^{-1}\mathbf{F}\mathbf{G} = \mathbf{F}'\boldsymbol{\Sigma}^{-1}\mathbf{F}$ by F1, and so (5.19) holds. By an argument similar to that for (5.20) we have $\boldsymbol{\Gamma}^-\boldsymbol{\Gamma} = \mathbf{G}$ and so (5.21) holds.

(b) By Property 4.6(a), $\boldsymbol{\Gamma}\mathbf{j} = \mathbf{0}$ so that $\boldsymbol{\Gamma}$ has an eigenvalue 0 with corresponding unit-length eigenvector $\mathbf{a}_D = D^{-1/2}\mathbf{j}$. The other eigenvalues $\lambda_1, \ldots, \lambda_d$ of $\boldsymbol{\Gamma}$ are in general positive: let $\mathbf{a}_1, \ldots, \mathbf{a}_d$ denote the corresponding unit-length eigenvectors. Then the spectral representation of $\boldsymbol{\Gamma}$ is

$$\boldsymbol{\Gamma} = \lambda_1\mathbf{a}_1\mathbf{a}_1' + \cdots + \lambda_d\mathbf{a}_d\mathbf{a}_d' + 0\mathbf{a}_D\mathbf{a}_D'$$

with pseudo-inverse

$$\boldsymbol{\Gamma}^- = \lambda_1^{-1}\mathbf{a}_1\mathbf{a}_1' + \cdots + \lambda_d^{-1}\mathbf{a}_d\mathbf{a}_d' + 0\mathbf{a}_D\mathbf{a}_D'.$$

Since $\mathbf{a}_D\mathbf{a}'_D = D^{-1}\mathbf{jj}' = D^{-1}\mathbf{J}$, we have

$$\boldsymbol{\Gamma} + k\mathbf{J} = \lambda_1\mathbf{a}_1\mathbf{a}'_1 + \cdots + \lambda_d\mathbf{a}_d\mathbf{a}'_d + kD\mathbf{a}_D\mathbf{a}'_D$$

and, since $\boldsymbol{\Gamma} + k\mathbf{J}$ is clearly non-singular for $k \neq 0$, we have the inverse

$$(\boldsymbol{\Gamma} + k\mathbf{J})^{-1} = \lambda_1^{-1}\mathbf{a}_1\mathbf{a}'_1 + \cdots + \lambda_d^{-1}\mathbf{a}_d\mathbf{a}'_d + (kD)^{-1}\mathbf{a}_D\mathbf{a}'_D$$
$$= \boldsymbol{\Gamma}^- + (kD^2)^{-1}\mathbf{J}.$$

The result is therefore established.

(c) Using the notation and argument of (b) above, we have

$$\boldsymbol{\Gamma} + D^{-1}\mathbf{J} = \lambda_1\mathbf{a}_1\mathbf{a}'_1 + \cdots + \lambda_d\mathbf{a}_d\mathbf{a}'_d + 1\mathbf{a}_D\mathbf{a}'_D$$

with eigenvalues $\lambda_1, \ldots, \lambda_d$, 1. Since the determinant of any square matrix is the product of its eigenvalues we have

$$|\boldsymbol{\Gamma} + D^{-1}\mathbf{J}| = \lambda_1 \cdots \lambda_d = |\boldsymbol{\Gamma}|^+$$

by the definition of $|\boldsymbol{\Gamma}|^+$.

5.7 Subcompositions

We have already had an indication in Section 4.5 of how well suited the variation matrix $\boldsymbol{T}$ is when some form of subcompositional analysis is involved. The reason for this is that the specification of $\boldsymbol{T}$ involves variances of 2-component logratios and the variation matrix appropriate to any subcomposition requires only the extraction from $\boldsymbol{T}$ of the elements in the rows and columns corresponding to the parts making up the subcomposition. The simplicity of this operation is reflected in the simplicity of its expression in terms of the appropriate selecting matrix $\mathbf{S}$ of Definition 2.8. Let $\mathbf{S}$ denote the $C \times D$ selection matrix and $\boldsymbol{T}_S$ the variation matrix of the resulting subcomposition. Then

$$\boldsymbol{T}_S = \mathbf{S}\boldsymbol{T}\mathbf{S}'. \tag{5.22}$$

The transfer of covariance structure from full composition to subcomposition is more complicated for the covariance matrix specifications $\boldsymbol{\Sigma}$ and $\boldsymbol{\Gamma}$. In constructing $\boldsymbol{\Gamma}_S$, the centred logratio covariance matrix of the subcomposition, we have not only to select the appropriate rows and columns from $\boldsymbol{\Gamma}$ but also to change the common logratio divisor from the geometric mean of all D components to that of the C subcompositional components. In terms of the

$C \times D$ selecting matrix $\mathbf{S}$ the transfer can be described as

$$\boldsymbol{\Gamma}_S = \mathbf{G}_C \mathbf{S} \boldsymbol{\Gamma} \mathbf{S}' \mathbf{G}_C, \tag{5.23}$$

clearly more complex than (5.22).

The construction of the subcompositional logratio covariance matrix $\boldsymbol{\Sigma}_S$ from $\boldsymbol{\Sigma}$ is even more complicated since the strategy depends crucially on whether D is or is not a part of the subcomposition. If the subcomposition includes the Dth part then $\boldsymbol{\Sigma}_S$ is obtained as the rows and columns of $\boldsymbol{\Sigma}$ corresponding to the other $c = C - 1$ parts. If the subcomposition does not include the Dth part then selection of the C rows and columns of $\boldsymbol{\Sigma}$ corresponding to the subcompositional parts has to be followed by a collapsing of the resulting $C \times C$ matrix to the $c \times c$ matrix $\boldsymbol{\Sigma}_S$ through a $\mathbf{F}_{c,C}(\cdot)\mathbf{F}'_{c,C}$ operation to produce appropriate logratios for the subcompositions. It is not easy to express this complicated operation in matrix terms: probably the simplest version is to use the relationships (4.28) and F1 to build on (5.23):

$$\begin{aligned}\boldsymbol{\Sigma}_S &= \mathbf{F}_{c,C} \boldsymbol{\Gamma}_S \mathbf{F}'_{c,C} = \mathbf{F}_{c,C} \mathbf{G}_C \mathbf{S} \mathbf{F}'_{d,D} \mathbf{H}_d^{-1} \boldsymbol{\Sigma} \mathbf{H}_d^{-1} \mathbf{F}_{d,D} \mathbf{S}' \mathbf{G}_C \mathbf{F}'_{c,C} \\ &= \mathbf{F}_{c,C} \mathbf{S} \mathbf{F}'_{d,D} \mathbf{H}_d^{-1} \boldsymbol{\Sigma} \mathbf{H}_d^{-1} \mathbf{F}_{d,D} \mathbf{S}' \mathbf{F}'_{c,C}.\end{aligned} \tag{5.24}$$

Comparison of (5.22)–(5.24) makes it clear that as far as subcompositional analysis is concerned the variation matrix $\boldsymbol{T}$ is an outright winner.

5.8 Equivalence of characteristics of $\boldsymbol{T}$, $\boldsymbol{\Sigma}$, $\boldsymbol{\Gamma}$

We are now in a position to collect together equivalent forms for a number of characteristics and constructs of $\boldsymbol{T}$, $\boldsymbol{\Sigma}$ and $\boldsymbol{\Gamma}$ such as quadratic forms, eigensolutions and traces. The listing of these technical results will help to free later development from otherwise necessary digressions on largely mathematical issues.

Quadratic forms

In our study of permutation invariants of $\boldsymbol{\Sigma}$ in Section 5.5 we noted the importance of the quadratic form $\mathbf{y}'\boldsymbol{\Sigma}^{-1}\mathbf{y}$ and the fact that the value of this did not depend on which permutation of parts is used. Of some interest, therefore, are the equivalent forms involving $\boldsymbol{T}$ and $\boldsymbol{\Gamma}$.

Property 5.7 For a D-part composition with covariance structure specifications $\boldsymbol{T}, \boldsymbol{\Sigma}, \boldsymbol{\Gamma}$, the following quadratic forms are equal:

(a) $-2\mathbf{y}'(\mathbf{F}\boldsymbol{T}\mathbf{F}')^{-1}\mathbf{y}$, $\quad -2\mathbf{z}'\mathbf{F}'(\mathbf{F}\boldsymbol{T}\mathbf{F}')^{-1}\mathbf{F}\mathbf{z}$,
(b) $\mathbf{y}'\boldsymbol{\Sigma}^{-1}\mathbf{y}$,
(c) $\mathbf{z}'(\boldsymbol{\Gamma}+k\mathbf{J})^{-1}\mathbf{z}$, whatever value $k \neq 0$ is assigned.

Proof. The first form of (a) is equal to (b) by (4.25) and the second form of (a) then follows from (5.3). Form (b) can be shown equal to (c) in the following steps.

$$\begin{aligned}
\mathbf{y}'\boldsymbol{\Sigma}^{-1}\mathbf{y} &= \mathbf{z}'\mathbf{F}'\boldsymbol{\Sigma}^{-1}\mathbf{F}\mathbf{z} && \text{by (5.3)},\\
&= \mathbf{z}'\boldsymbol{\Gamma}^{-}\mathbf{z} && \text{by Property 5.6(a)},\\
&= \mathbf{z}'(\boldsymbol{\Gamma}+k\mathbf{J})^{-1}\mathbf{z} - (kD^2)^{-1}\mathbf{z}'\mathbf{J}\mathbf{z} && \text{by Property 5.6(b)},\\
&= \mathbf{z}'(\boldsymbol{\Gamma}+k\mathbf{J})^{-1}\mathbf{z} && \text{by (5.5)}.
\end{aligned}$$

Here the form in $\boldsymbol{\Sigma}$ appears simplest: we shall see, however, in Section 10.2 when we study a problem of subcompositional independence, that form (a), despite its present complicated general form, has a simple and important role to play.

Eigensolutions

In Property 5.3(c) we saw that the solutions $(\lambda, \mathbf{b})$ of the eigenproblem $(\boldsymbol{\Sigma}-\lambda\mathbf{H})\mathbf{b}=\mathbf{0}$ or $(\mathbf{H}^{-1}\boldsymbol{\Sigma}-\lambda\mathbf{I})\mathbf{b}=\mathbf{0}$ are permutation invariant. It is interesting to ask if there are $\boldsymbol{T}$ and $\boldsymbol{\Gamma}$ equivalents to this problem and if either is simpler than the $\boldsymbol{\Sigma}$ version. The comparison is provided in the following statement.

Property 5.8 The three eigenproblems

(a) $(-\tfrac{1}{2}\mathbf{G}\boldsymbol{T}\mathbf{G}-\lambda\mathbf{I})\mathbf{a}=\mathbf{0}$,
(b) $(\boldsymbol{\Sigma}-\lambda\mathbf{H})\mathbf{b}=\mathbf{0}$ or $(\mathbf{H}^{-1}\boldsymbol{\Sigma}-\lambda\mathbf{I})\mathbf{b}=\mathbf{0}$,
(c) $(\boldsymbol{\Gamma}-\lambda\mathbf{I})\mathbf{a}=\mathbf{0}$,

have identical sets of positive eigenvalues and the corresponding eigenvectors $\mathbf{a}$ and $\mathbf{b}$ are related by

$$\mathbf{b}=\mathbf{a}_{-D}, \quad \mathbf{a}=\mathbf{F}'\mathbf{b}.$$

and $\mathbf{a}'\mathbf{a}=1$ if and only if $\mathbf{b}'\mathbf{H}\mathbf{b}=1$.

Proof. We take problem (c) as our starting point. From the proof of Property 5.6(b) we know that (c) has one zero eigenvalue, with eigenvector $\mathbf{j}$, and d positive eigenvalues. Let λ be a positive eigenvalue of (c) and $\mathbf{a}$ an associated eigenvector. Since $\mathbf{a}$ must be orthogonal to $\mathbf{j}$, we have $\mathbf{j}'\mathbf{a} = 0$ and so we can write $\mathbf{a} = \mathbf{F}'\mathbf{a}_{-D}$. Then, after premultiplication by $\mathbf{F}$ and use of (4.28), (c) gives

$$\begin{aligned} \mathbf{0} &= \mathbf{F}(\mathbf{F}'\mathbf{H}^{-1}\boldsymbol{\Sigma}\mathbf{H}^{-1}\mathbf{F} - \lambda\mathbf{I})\mathbf{F}'\mathbf{a}_{-D} \\ &= (\boldsymbol{\Sigma} - \lambda\mathbf{H})\mathbf{a}_{-D} \qquad \text{by H2.} \end{aligned}$$

Thus every non-zero eigensolution of (c) provides an eigensolution of (b) and both have the same number d of solutions. Thus the equivalence of (b) and (c) is established.

The equivalence of (a) and (c) is immediate by the relationship (4.26).

The remainder of the proposition follows from the relationship $\mathbf{a}'\mathbf{a} = \mathbf{b}'\mathbf{F}\mathbf{F}'\mathbf{b} = \mathbf{b}'\mathbf{H}\mathbf{b}$ by H2.

Traces

Property 5.5 established the permutation invariance of $\mathrm{tr}(\mathbf{H}^{-1}\boldsymbol{\Sigma})$. The following property extends the results to connect this characteristic to $\boldsymbol{T}$ and $\boldsymbol{\Gamma}$ and the eigensolutions of Property 5.8.

Property 5.9 The following three constructs of $\boldsymbol{T}$, $\boldsymbol{\Sigma}$ and $\boldsymbol{\Gamma}$ are equal:

(a) $(2D)^{-1}\mathbf{j}'\boldsymbol{T}\mathbf{j}$,
(b) $\mathrm{tr}(\mathbf{H}^{-1}\boldsymbol{\Sigma}) = \mathrm{tr}(\boldsymbol{\Sigma}) - D^{-1}\mathbf{j}'\boldsymbol{\Sigma}\mathbf{j}$,
(c) $\mathrm{tr}(\boldsymbol{\Gamma})$,

and all are equal to the sum of the eigenvalues of Property 5.8.

Proof. From (4.28) we have

$$\begin{aligned} \mathrm{tr}(\boldsymbol{\Gamma}) &= \mathrm{tr}(\mathbf{F}'\mathbf{H}^{-1}\boldsymbol{\Sigma}\mathbf{H}^{-1}\mathbf{F}) \\ &= \mathrm{tr}(\mathbf{H}^{-1}\boldsymbol{\Sigma}\mathbf{H}^{-1}\mathbf{F}\mathbf{F}') \\ &= \mathrm{tr}(\mathbf{H}^{-1}\boldsymbol{\Sigma}) \qquad \text{by H2,} \\ &= \mathrm{tr}\{(\mathbf{I}_d - D^{-1}\mathbf{J}_d)\boldsymbol{\Sigma}\} \qquad \text{by H6} \\ &= \mathrm{tr}(\boldsymbol{\Sigma}) - D^{-1}\mathrm{tr}(\mathbf{j}\mathbf{j}'\boldsymbol{\Sigma}) \qquad \text{by J1} \\ &= \mathrm{tr}(\boldsymbol{\Sigma}) - D^{-1}\mathrm{tr}(\mathbf{j}'\boldsymbol{\Sigma}\mathbf{j}) \\ &= \mathrm{tr}(\boldsymbol{\Sigma}) - D^{-1}\mathbf{j}'\boldsymbol{\Sigma}\mathbf{j}, \end{aligned}$$

establishing the equality of (b) and (c).

From (4.26) we have

$$\mathrm{tr}(\boldsymbol{\Gamma}) = -\tfrac{1}{2}\mathrm{tr}(\mathbf{G}\boldsymbol{T}\mathbf{G})$$

$$\begin{aligned}
&= -\tfrac{1}{2}\operatorname{tr}(\mathbf{G}\boldsymbol{T}) \qquad \text{by} \quad \text{G2},\\
&= -\tfrac{1}{2}\{\operatorname{tr}(\boldsymbol{T}) - D^{-1}\operatorname{tr}(\mathbf{J}\boldsymbol{T})\}\\
&= (2D)^{-1}\operatorname{tr}(\mathbf{j}'\boldsymbol{T}\mathbf{j})\\
&= (2D)^{-1}\mathbf{j}'\boldsymbol{T}\mathbf{j},
\end{aligned}$$

establishing the equality of (a) and (c). The relation of each of (a)–(c) to the sum of the eigenvalues is immediate, since from standard matrix theory $\operatorname{tr}(\boldsymbol{\Gamma})$ is the sum of the eigenvalues of $\boldsymbol{\Gamma}$.

We note that $\mathbf{j}'\boldsymbol{T}\mathbf{j}$ is the sum of all the elements of the variation matrix, and so is an easily computed quantity. We shall see in Sections 8.5–8.6 that this is a useful aspect in the assessment of subcompositional variability.

Generalized variances

In Section 5.5 we also saw that the generalized variance $|\boldsymbol{\Sigma}|$ is a permutation invariant. The following equivalence result also shows that it is the simplest form for the generalized variance.

Property 5.10 For a D-part composition with covariance structure specifications $\boldsymbol{T}, \boldsymbol{\Sigma}, \boldsymbol{\Gamma}$ the following forms for the generalized variance are equal:

(a) $(-1)^D|\mathbf{F}\boldsymbol{T}\mathbf{F}'|/(2^D D)$,
(b) $D^{-1}|\boldsymbol{\Sigma}|$,
(c) $|\boldsymbol{\Gamma} + D^{-1}\mathbf{J}|$,
(d) the product of the d positive eigenvalues of Property 5.8.

Proof. From Property 5.9(b), $|\mathbf{H}^{-1}\boldsymbol{\Sigma}|$ is the product of the d non-zero eigenvalues, and $|\mathbf{H}^{-1}\boldsymbol{\Sigma}| = |\boldsymbol{\Sigma}|/|\mathbf{H}| = D^{-1}|\boldsymbol{\Sigma}|$ by H5, so that (b) and (d) are equivalent. The equality of (c) and (d) then follows from Property 5.6(c). Finally, by (4.25),

$$|\boldsymbol{\Sigma}| = |-\tfrac{1}{2}\mathbf{F}\boldsymbol{T}\mathbf{F}'| = (-\tfrac{1}{2})^D|\mathbf{F}\boldsymbol{T}\mathbf{F}'|$$

and so (a) and (b) are equal.

We note here that the $\boldsymbol{\Sigma}$ form is by far the simplest though again the $\boldsymbol{T}$ form will prove useful in some subcompositional analyses.

5.9 Logratio-uncorrelated compositions

Later in this monograph we shall have to face up to the problem of how meaningful concepts of compositional independence can be introduced for components subject to the unit-sum constraint. We can make a start, however, in this chapter with the concepts we have already introduced and, by taking care at this stage of much of the technical detail, leave the way free for the easier presentation of the essential statistical ideas. The outcome of this technical process will be the identification of special types of covariance structure through the patterns of the various matrix specifications. An obvious first step in this process is to examine the effect on a compositional covariance structure when we insist that certain of its logratios are uncorrelated.

Definition 5.1 A D-part composition $\mathbf{x}$ is *logratio uncorrelated* if

$$\sigma_{ij.kl} = \text{cov}\{\log(x_i/x_k), \log(x_j/x_l)\} = 0$$

for every selection of four different integers i, j, k, l from $\{1, 2, \ldots, D\}$.

Another way of describing this property is to say that every pair of 2-part subcompositions formed from non-overlapping subvectors of $\mathbf{x}$ is uncorrelated.

The special covariance pattern of a logratio-uncorrelated composition is identified in the following property.

Property 5.11 The covariance structure of a logratio-uncorrelated composition can be specified in terms of a D-vector parameter $\boldsymbol{\alpha}$, with the equivalent forms shown in Table 5.1.

Moreover, the parameter α_k is the common value of

$$\text{cov}\{\log(x_i/x_k), \log(x_j/x_k)\}$$

as i, j vary over values not equal to k.

Table 5.1

Element forms	$i=j$	$i \neq j$	*Matrix forms*
γ_{ij}	$\alpha_i - D^{-1}(2\alpha_i - \alpha_.)$	$-D^{-1}(\alpha_i + \alpha_j - \alpha_.)$	$\mathbf{G}\,\text{diag}(\boldsymbol{\alpha})\mathbf{G}$
σ_{ij}	$\alpha_i + \alpha_D$	α_D	$\text{diag}(\boldsymbol{\alpha}_{-D}) + \alpha_D \mathbf{J}_d$
τ_{ij}	0	$\alpha_i + \alpha_j$	$\boldsymbol{\alpha}\mathbf{j}' + \mathbf{j}\boldsymbol{\alpha}' - 2\,\text{diag}(\boldsymbol{\alpha})$

Proof. We prove the element forms for σ_{ij}. The forms for τ_{ij} and γ_{ij} then follow from (4.21) and (4.22), and the matrix forms are easily verified. We have, for i, j, k, D unequal,

$$\begin{aligned}\sigma_{ij} &= \operatorname{cov}\{\log(x_i/x_D), \log(x_j/x_D)\}\\ &= \operatorname{cov}\{\log(x_i/x_D), \log(x_j/x_k) + \log(x_k/x_D)\}\\ &= 0 + \sigma_{ik}.\end{aligned}$$

Thus all the off-diagonal elements of $\boldsymbol{\Sigma}$ are equal, and so the form for σ_{ij} is established.

Reference to Property 3.5 of the Dirichlet class $\mathscr{D}^d$ shows that the logratio covariance matrix $\boldsymbol{\Sigma}$ of $\mathscr{D}^d(\boldsymbol{\alpha})$ has the form of Property 5.11 with α_i replaced by $\psi'(\alpha_i)$. We can thus add a further property of the Dirichlet class to those of Section 3.4

Property 5.12 Every Dirichlet composition is logratio uncorrelated.

In Chapter 10 we shall require some matrix results concerning these special patterns. Here we express these in terms of the $\boldsymbol{\Sigma}$ version of the logratio uncorrelated covariance structure.

Property 5.13 For the $d \times d$ matrix

$$\boldsymbol{A} = \operatorname{diag}(\alpha_1, \ldots, \alpha_d) + \alpha_D \mathbf{J}$$

we have the following properties.

(a) $$|\boldsymbol{A}| = \sum_{i=1}^{D} \prod_{j \neq i} \alpha_j$$
$$= \prod_{i=1}^{D} \alpha_i \sum_{i=1}^{D} \alpha_i^{-1} \qquad \text{when} \quad \alpha_i \neq 0 \quad (i = 1, \ldots, D).$$

(b) $\boldsymbol{A}^{-1} = \operatorname{diag} \boldsymbol{\beta}_{-D} - B^{-1} \boldsymbol{\beta}_{-D} \boldsymbol{\beta}'_{-D}$
where $\boldsymbol{\beta} = (\alpha_1^{-1}, \ldots, \alpha_D^{-1})$ and $B = \boldsymbol{\beta}'\mathbf{j}$.

(c) $\boldsymbol{A}$ is positive-definite if and only if one of the following conditions holds:

(i) $\alpha_i > 0 \, (i = 1, \ldots, D)$,
(ii) $\alpha_i = 0$ for some i and $\alpha_j > 0 \, (i \neq j)$,
(iii) $\alpha_i < 0$ for some i, $\alpha_j > 0 \, (i \neq j)$ and $\alpha_1^{-1} + \cdots + \alpha_D^{-1} < 0$.

5.10 Isotropic covariance structures

It seems natural to attempt to extend the concept of logratio-uncorrelated compositions from logratios to the more general logcontrasts introduced in Section 4.10. The special feature of the logratios $\log(x_i/x_k)$ and $\log(x_j/x_l)$ of Definition 5.1, that i, j, k, l have different integer values, can be equally expressed in terms of Definition 4.10 as the orthogonality of these logratio forms of logcontrasts. This suggests that it may be of interest to employ orthogonality as the conditional requirement for logcontrasts.

Definition 5.2 A D-part composition $\mathbf{x}$ is *logcontrast uncorrelated* if every two orthogonal logcontrasts are uncorrelated: in other words, if

$$\text{cov}(\mathbf{a}' \log \mathbf{x}, \mathbf{b}' \log \mathbf{x}) = 0$$

whenever $\mathbf{a}'\mathbf{j} = \mathbf{b}'\mathbf{j} = \mathbf{a}'\mathbf{b} = 0$.

The property of being logcontrast uncorrelated is stronger than that of being logratio uncorrelated: every logcontrast-correlated composition is logratio uncorrelated, but not conversely. This is easily seen from the special covariance pattern of a logcontrast-uncorrelated composition, as described in the following result.

Property 5.14 A necessary and sufficient condition for a D-part composition $\mathbf{x}$ to be logcontrast uncorrelated is that the matrix specification of its covariance structure has one of the following special and equivalent patterns:

(a) $\boldsymbol{T} = 2\alpha(\mathbf{J} - \mathbf{I})$,
(b) $\boldsymbol{\Sigma} = \alpha\mathbf{H}$,
(c) $\boldsymbol{\Gamma} = \alpha\mathbf{G}$,

where α is a positive parameter.

Proof. Conditions (a)–(c) are clearly equivalent through relationships (4.25)–(4.26), so that we can concentrate on any one of them.

Consider first the sufficiency of (c). For any orthogonal contrasts $\mathbf{a}' \log \mathbf{x}, \mathbf{b}' \log \mathbf{x}$, we have

$$\begin{aligned} \text{cov}(\mathbf{a}' \log \mathbf{x}, \mathbf{b}' \log \mathbf{x}) &= \mathbf{a}'\boldsymbol{\Gamma}\mathbf{b} && \text{by Property 5.1(b)}, \\ &= \alpha\mathbf{a}'\mathbf{G}\mathbf{b} && \text{by (c)}, \\ &= \alpha\mathbf{a}'\mathbf{b} && \text{since } \mathbf{G}\mathbf{b} = \mathbf{b} \text{ for a logcontrast by G11}, \\ &= 0 && \text{by orthogonality}, \end{aligned}$$

and so the composition is logcontrast uncorrelated.

Consider now the necessity of (b). A logcontrast-uncorrelated composition is logratio uncorrelated and so we know from Property 5.11 that $\boldsymbol{\Sigma}$ must have the form $\text{diag}(\boldsymbol{\alpha}_{-D}) + \alpha_D\mathbf{J}$. Further, from Property 5.11, for any unequal i, j, k,

$$\begin{aligned}\alpha_j - \alpha_k &= \text{cov}\{\log(x_i/x_j), \log(x_k/x_j)\} - \text{cov}\{\log(x_i/x_k), \log(x_j/x_k)\}\\ &= \text{cov}(\log x_j - \log x_k, \log x_j - 2\log x_i + \log x_k)\\ &= 0,\end{aligned}$$

since the two logcontrasts are orthogonal. Thus all the α_k are equal, say to α, and so $\boldsymbol{\Sigma} = \alpha\mathbf{H}$, establishing the necessity of (b).

From other concepts already introduced in this chapter there are two other conditions on a compositional covariance structure which may produce interesting patterns. The first condition is the analogue of the isotropic covariance matrix $\mathbf{I}_d$ for variation in $\mathscr{R}^d$, with the property that it is the only covariance matrix which gives identical values of the variance $\text{var}(\mathbf{a}'\mathbf{y})$ for every linear combination $\mathbf{a}'\mathbf{y}$ satisfying $\mathbf{a}'\mathbf{a} = 1$.

Definition 5.3 A D-part composition is *logcontrast isotropic* if the variances of all standard logcontrasts are equal.

The equivalent forms of the covariance patterns of a logcontrast isotropic composition can easily be identified.

Property 5.15 A necessary and sufficient condition for a D-part composition $\mathbf{x}$ to be logcontrast isotropic is that the matrix specification of its covariance structure has one of the following special and equivalent patterns:

(a) $\boldsymbol{T} = 2\alpha(\mathbf{J} - \mathbf{I})$,
(b) $\boldsymbol{\Sigma} = \alpha\mathbf{H}$,
(c) $\boldsymbol{\Gamma} = \alpha\mathbf{G}$,

where α is a positive parameter.

Proof. The sufficiency of the conditions is trivial to establish. To establish the necessity, suppose that the composition is logcontrast isotropic and consider the spectral representation of $\boldsymbol{\Gamma}$. From the proof of Property 5.6(b) we have

$$\boldsymbol{\Gamma} = \lambda_1 \mathbf{a}_1 \mathbf{a}_1' + \cdots + \lambda_d \mathbf{a}_d \mathbf{a}_d'.$$

Moreover, $\mathbf{a}_i' \log \mathbf{x}$ are standard logcontrasts and $\mathrm{var}(\mathbf{a}_i' \log \mathbf{x}) = \mathbf{a}_i' \boldsymbol{\Gamma} \mathbf{a}_i = \lambda_i$. Hence all the positive eigenvalues of $\boldsymbol{\Gamma}$ are equal, say to α, and so $\boldsymbol{\Gamma} = \alpha \mathbf{G}$.

The second condition is that of permutation invariance applied to the entire covariance structure.

Definition 5.4 A D-part composition has a *permutation-invariant* covariance structure if any of the following matrix specification relationships holds for any permutation $\mathbf{P}$ of the D parts:

(a) $\boldsymbol{T}_P = \boldsymbol{T}$,
(b) $\boldsymbol{\Sigma}_P = \boldsymbol{\Sigma}$,
(c) $\boldsymbol{\Gamma}_P = \boldsymbol{\Gamma}$.

Property 5.16 A necessary and sufficient condition for a D-part composition $\mathbf{x}$ to have a permutation-invariant covariance structure is that the matrix specification has one of the special and equivalent patterns:

(a) $\boldsymbol{T} = 2\alpha(\mathbf{J} - \mathbf{I})$,
(b) $\boldsymbol{\Sigma} = \alpha \mathbf{H}$,
(c) $\boldsymbol{\Gamma} = \alpha \mathbf{G}$,

where α is a positive parameter.

Proof. As in the proof of Property 5.14 the conditions are obviously equivalent, and the proof of sufficiency is trivial. For the necessity of the condition we concentrate on (c). The operation $\mathbf{P}\boldsymbol{\Gamma}\mathbf{P}'$ with $\mathbf{P}$ a permutation matrix simply permutes the rows and columns of $\boldsymbol{\Gamma}$ in exactly the same way, so that for $\mathbf{P}\boldsymbol{\Gamma}\mathbf{P}' = \boldsymbol{\Gamma}$ to hold for every permutation $\mathbf{P}$ it is clear that all the diagonal elements of $\boldsymbol{\Gamma}$ must be equal, say to β, and all the off-diagonal must be equal, say to γ. Then $\boldsymbol{\Gamma} = (\beta - \gamma)\mathbf{I} + \gamma \mathbf{J}$. Since, however, $\boldsymbol{\Gamma}\mathbf{j} = \mathbf{0}$ we must have $\beta - \gamma + \gamma D = 0$. Using this condition we can express $\boldsymbol{\Gamma}$ as $-\gamma D \mathbf{I} + \gamma \mathbf{J}$, leading to the form $\alpha \mathbf{G}$, where $\alpha = -\gamma D$.

The necessary and sufficient conditions of Properties 5.14–5.16 are identical and we can record the consequent equivalence as a formal statement.

Property 5.17 For a D-part composition the following properties of the covariance structure are identical:

(a) logcontrast uncorrelated,
(b) logcontrast isotropic,
(c) permutation invariant.

To simplify our terminology we shall adopt the term 'isotropic' for any of these three properties.

Definition 5.5 When a D-part composition has a covariance structure specified by any of the equivalents, $\boldsymbol{T} = 2\alpha(\mathbf{J} - \mathbf{I})$, $\boldsymbol{\Sigma} = \alpha\mathbf{H}$, $\boldsymbol{\Gamma} = \alpha\mathbf{G}$, then it is termed *isotropic*.

5.11 Bibliographic notes

Apart from the results involving eigensolutions which are the substance of the study by Aitchison (1983) of compositional principal component analysis, the properties of the matrix covariance specifications reported in this chapter appear for the first time.

Problems

Problems 5.1–5.4 relate to the AFM compositions of Data 6 and require you first to obtain estimates of the variation matrix $\boldsymbol{T}$, the logratio covariance matrix $\boldsymbol{\Sigma}$ and the centred logratio covariance matrix $\boldsymbol{\Gamma}$.

5.1 Verify the equivalence of the forms for variance and covariance in Property 5.1 by considering the logcontrasts

$$u_1 = \log A + 2\log F - 3\log M,$$
$$u_2 = 3\log A - 2\log F - \log M.$$

Find a logcontrast u_3 which is orthogonal to u_1, and also a logcontrast u_4 which is uncorrelated with u_1. Are there any circumstances under which u_3 and u_4 could be identical?

5.2 Construct the $\mathbf{P}$ and $\mathbf{Q}_P$ matrices corresponding to the permutation AFM → MFA. Verify the relationships of Property 5.2 and the permutation-invariant properties of Property 5.3.

5.3 Verify the results of Property 5.6 by computing $\boldsymbol{\Gamma}^-$ and $|\boldsymbol{\Gamma}|^+$.

5.4 Verify the equivalence of the results of Properties 5.8–5.10.

5.5 Let $\mathbf{y}_k = \log(\mathbf{x}_{-k}/x_k)$, the logratio vector with the kth component as the common divisor, and let $\boldsymbol{\Sigma}_k$ denote the covariance matrix of $\mathbf{y}_k$. Show that, for $k < l$, $\mathbf{y}_k = \mathbf{M}_{kl}\mathbf{y}_l$, where

$$\mathbf{M}_{kl} = \begin{bmatrix} \mathbf{I}_{k-1} & \mathbf{0} & -\mathbf{j}_{k-1} & \mathbf{0} \\ \mathbf{0} & \mathbf{0} & -1 & \mathbf{0} \\ \mathbf{0} & \mathbf{I}_{l-k-1} & -\mathbf{j}_{l-k-1} & \mathbf{0} \\ \mathbf{0} & \mathbf{0} & -\mathbf{j}_{d-l+1} & \mathbf{I}_{d-l+1} \end{bmatrix}.$$

Establish the following properties of $\mathbf{M}_{kl}$ matrices:

(i) $\mathbf{M}_{kl}\mathbf{H}\mathbf{M}'_{kl} = \mathbf{H}$,
(ii) $|\mathbf{M}_{kl}| = \pm 1$.

Using these properties, or otherwise, show that

(a) $|\boldsymbol{\Sigma}_k| = |\boldsymbol{\Sigma}_l|$
(b) $\mathbf{y}'_k\boldsymbol{\Sigma}_k^{-1}\mathbf{y}_k = \mathbf{y}'_l\boldsymbol{\Sigma}_l^{-1}\mathbf{y}_l$
(c) the eigensolutions of $(\boldsymbol{\Sigma}_k - \lambda_k\mathbf{H})\mathbf{a}_k = \mathbf{0}$ and $(\boldsymbol{\Sigma}_l - \lambda_l\mathbf{H})\mathbf{a}_l = \mathbf{0}$ are related by $\lambda_k = \lambda_l$, $\mathbf{a}'_k\mathbf{y}_k = \mathbf{a}'_l\mathbf{y}_l$, and $\mathbf{a}_l = \mathbf{M}'_{kl}\mathbf{a}_k$.

CHAPTER 6

Logistic normal distributions on the simplex

A mathematician, like a painter or poet, is a maker of patterns. If his patterns are more permanent than theirs, it is because they are made with ideas.

G. H. Hardy: *A Mathematician's Apology*

6.1 Introduction

The range of statistical analyses of compositional data is likely to be considerably extended if in addition to a sensible covariance structure we can find a parametric class of distributions on $\mathscr{S}^d$ rich enough to capture the patterns of variability we observe in the simplex sample space. As indicated in Section 3.4 the familiar Dirichlet class and its generalizations to date are not sufficiently rich for the purposes of compositional data analysis, particularly in their inability to support a sufficient degree of compositional dependence. Rather than elaborate these negative points of the Dirichlet class at this stage, we propose to take the more positive step of introducing immediately a practical tool of analysis.

Alternatives to the Dirichlet class have been slow to emerge but there are now a number of practically useful classes based on an old and delightfully simple idea. When trying to find suitable means of describing patterns of variability over the positive real line $\mathscr{R}^1_+$, for which at that time none existed, McAlister (1879) saw that it was sensible to transfer the highly successful normal pattern $\mathscr{N}(\mu, \sigma^2)$ on the whole of the real line $\mathscr{R}^1$ to $\mathscr{R}^1_+$ through the obvious transformation

$$w = \exp(y) \qquad (w \in \mathscr{R}^1_+, y \in \mathscr{R}^1),$$

with inverse transformation

$$y = \log w \qquad (y \in \mathscr{R}^1, w \in \mathscr{R}^1_+),$$

and so the lognormal class of distributions $\Lambda(\mu, \sigma^2)$ was invented. In the same way we can induce a class of distributions on the simplex from the class of multivariate normal distributions on $\mathscr{R}^d$ by use of any one-to-one transformation from $\mathscr{R}^d$ to $\mathscr{S}^d$. There are many such transformations and in this chapter we shall concentrate on only three practically useful forms leading to the additive logistic normal class, which we study in detail in Sections 6.2–6.12, and the multiplicative and partitioned logistic normal classes, briefly described in Sections 6.13 and 6.14, respectively.

6.2 The additive logistic normal class

In attempting to apply the transformation technique just suggested we cannot do better than start with the simplest transformation, already in use in other areas of statistical activity, namely the generalized logistic transformation, or as we shall prefer to term it, the additive logistic transformation between $\mathbf{x} \in \mathscr{S}^d$ and $\mathbf{y} \in \mathscr{R}^d$.

Definition 6.1 The *additive logistic transformation* is the one-to-one transformation from $\mathbf{y} \in \mathscr{R}^d$ to $\mathbf{x} \in \mathscr{S}^d$ defined by

$$x_i = \exp(y_i)/\{\exp(y_1) + \cdots + \exp(y_d) + 1\} \qquad (i = 1, \ldots, d),$$
$$x_D = 1 - x_1 - \cdots - x_d = 1/\{\exp(y_1) + \cdots + \exp(y_d) + 1\},$$

with inverse the *additive logratio transformation*

$$y_i = \log(x_i/x_D) \qquad (i = 1, \ldots, d),$$

and jacobian

$$\text{jac}(\mathbf{y} | \mathbf{x}^{(d)}) = (x_1 \cdots x_D)^{-1}.$$

We now let $\mathscr{N}^d(\boldsymbol{\mu}, \boldsymbol{\Sigma})$ denote the d-dimensional normal distribution with mean vector $\boldsymbol{\mu}$ and covariance matrix $\boldsymbol{\Sigma}$ and define an important transformed normal class of distributions on the simplex $\mathscr{S}^d$.

Definition 6.2 A D-part composition $\mathbf{x}$ is said to have an *additive logistic normal* distribution $\mathscr{L}^d(\boldsymbol{\mu}, \boldsymbol{\Sigma})$ when $\mathbf{y} = \log(\mathbf{x}_{-D}/x_D)$ has a $\mathscr{N}^d(\boldsymbol{\mu}, \boldsymbol{\Sigma})$ distribution.

We are effectively starting with a tractable, rich and well-established class of distributions on $\mathscr{R}^d$, namely the $\mathscr{N}^d(\boldsymbol{\mu}, \boldsymbol{\Sigma})$ class, and, in the manner of McAlister (1879), inducing a corresponding class of distributions on $\mathscr{S}^d$ through the additive logistic transformation from $\mathscr{R}^d$ to $\mathscr{S}^d$ with the logratio transformation as inverse. This explains the terminology of 'additive logistic normal': in Sections 6.13–6.14 we shall introduce two other forms of transformation from $\mathscr{R}^d$ to $\mathscr{S}^d$, with the resulting transformed normal classes distinguished by the names of 'multiplicative' and 'partitioned' logistic normal classes. Until then it is convenient to drop the qualification of 'additive' in relation to the $\mathscr{L}^d$ class.

A particular member of the $\mathscr{L}^d$ class is clearly determined by the $\frac{1}{2}d(d+3)$ parameters making up $(\boldsymbol{\mu}, \boldsymbol{\Sigma})$ but could equally be determined in a centred logratio mode by $\boldsymbol{\lambda} = \mathrm{E}(\mathbf{z})$, $\boldsymbol{\Gamma} = \mathrm{V}(\mathbf{z})$, where $\mathbf{z} = \log\{\mathbf{x}/g(\mathbf{x})\}$, $\mathrm{V}(\cdot)$ denotes covariance matrix, and $\boldsymbol{\lambda}'\mathbf{j} = 0$, $\boldsymbol{\Gamma}\mathbf{j} = \mathbf{0}$; or in a variation mode by the variation array of Definition 4.3, consisting of the means $\xi_{ij} = \mathrm{E}\{\log(x_i/x_j)\}$ and variances $\tau_{ij} = \mathrm{var}\{\log(x_i/x_j)\}$ $(i = 1, \ldots, d; j = i+1, \ldots, D)$. It would thus be possible to introduce different notations for the $\mathscr{L}^d$ class to suit the parametrization which is selected for the analysis of a particular problem. This seems an unnecessary complication with the facility for switching between the different specifications which was developed in Section 4.8. We therefore confine our specifications to the $\mathscr{L}^d(\boldsymbol{\mu}, \boldsymbol{\Sigma})$ form, leaving it to the reader to carry out any necessary conversion to the other two forms.

An important point to note is that the covariance parameter $\boldsymbol{\Sigma}$ of the $\mathscr{L}^d$ distribution is precisely the logratio covariance matrix $\boldsymbol{\Sigma}$ which we arrived at in Definition 4.5. Thus if we can adopt a logistic normal model for the description of our pattern of variability we shall be in a position to test parametrically hypotheses concerning the covariance structure.

The great advantage of this transformation approach to compositional data analysis is that it makes available the whole range of statistical procedures based on multivariate normality. All we have to do is to transform each composition $\mathbf{x}$ to its logratio composition $\mathbf{y}$ and then work within $\mathscr{R}^d$ and on multivariate normal assumptions. In particular, we can test the reasonableness of the logistic normality assumptions through the battery of tests for multivariate normality. It would be preposterous to believe that all compositional data will turn out to be logistic normal and no doubt modifications to statistical procedures will have to be made in the light of further experience. There do, however, appear to be a sufficient number of data sets that are reasonably logistic normal in pattern and it is necessary to make a start with some parametric form. In addition to investigation of covariance structure, particularly attractive procedures from the viewpoint of the statistical analysis of compositions are discriminant analysis for classification purposes, and linear modelling of the mean for investigating the dependence of composition on other explanatory variables or covariates. Moreover, Bayesian methods, including statistical prediction analysis, are readily available through the substantial multivariate normal counterpart. These procedures will all be developed and applied in Chapter 7.

In Sections 6.3–6.12 we set out the main properties of the logistic normal class. Most of the properties derive from corresponding properties of multinormal distributions but usually require adjustment to provide useful practical results. For example, normality of the marginal distribution of $\mathbf{y}^{(c)}$ over $\mathscr{R}^c$ does not provide a simple result about the distribution of $\mathbf{x}^{(c)}$ over $\mathscr{S}^c$, that is about $x_1, \ldots, x_c$ and the fill-up $1 - x_1 - \cdots - x_c$, but rather about the relative structure within the subvector $(x_1, \ldots, x_c, x_D)$, namely the subcomposition $\mathscr{C}(\mathbf{x}^{(c)}, x_D)$. The proofs of the properties are all relatively straightforward and are therefore omitted.

6.3 Density function

Starting with the density function

$$(2\pi)^{-d/2}|\boldsymbol{\Sigma}|^{-1/2}\exp\{-\tfrac{1}{2}(\mathbf{y}-\boldsymbol{\mu})'\boldsymbol{\Sigma}^{-1}(\mathbf{y}-\boldsymbol{\mu})\} \qquad (\mathbf{y}\in\mathscr{R}^d) \quad (6.1)$$

of the $\mathscr{N}^d(\boldsymbol{\mu}, \boldsymbol{\Sigma})$ distribution, we can immediately apply the logistic transformation $\mathbf{y} \to \mathbf{x}$ of Definition 6.1 with its jacobian $\text{jac}(\mathbf{y}|\mathbf{x}^{(d)}) = (x_1 \cdots x_D)^{-1}$, to obtain the following form for the density function of $\mathscr{L}^d(\boldsymbol{\mu}, \boldsymbol{\Sigma})$:

$$(2\pi)^{-d/2}|\boldsymbol{\Sigma}|^{-1/2}(x_1 \cdots x_D)^{-1}\exp[-\tfrac{1}{2}\{\log(\mathbf{x}_{-D}/x_D)-\boldsymbol{\mu}\}'\boldsymbol{\Sigma}^{-1} \times \{\log(\mathbf{x}_{-D}/x_D)-\boldsymbol{\mu}\}] \qquad (\mathbf{x}\in\mathscr{S}^d), \quad (6.2)$$

or

$$(2\pi)^{-d/2}|\boldsymbol{\Sigma}|^{-1/2}(x_1 \cdots x_D)^{-1}\exp\{-\tfrac{1}{2}(\mathbf{y}-\boldsymbol{\mu})'\boldsymbol{\Sigma}^{-1}(\mathbf{y}-\boldsymbol{\mu})\}, \quad (6.3)$$

where $\mathbf{y} = \log(\mathbf{x}_{-D}/x_D)$.

Notes

1. We re-emphasize that the density function is defined on the strictly positive simplex $\mathscr{S}^d$. This is necessary because of the logarithmic transformation involved.

2. To illustrate the ease with which we can move to other specifications, we give here the corresponding density functions in $\boldsymbol{T}$ and $\boldsymbol{\Gamma}$ forms. The form in $\boldsymbol{T}$ follows from Properties 5.7(a) and 5.10(a):

$$(2\pi)^{-d/2}(x_1 \cdots x_D)^{-1}|-\tfrac{1}{2}\mathbf{F}\boldsymbol{T}\mathbf{F}'|^{-1/2} \\ \times \exp[-\tfrac{1}{2}\mathrm{tr}\{\mathbf{F}'(\mathbf{F}\boldsymbol{T}\mathbf{F}')^{-1}\mathbf{F}\boldsymbol{\Delta}(\mathbf{x},\boldsymbol{\Xi})\}], \quad (6.4)$$

where

$$\boldsymbol{\Delta}(\mathbf{x},\boldsymbol{\Xi}) = [\{\log(x_i/x_j) - \xi_{ij}\}^2].$$

The version in $(\boldsymbol{\lambda},\boldsymbol{\Gamma})$ is easily derived from Properties 5.6(c) and 5.7(c):

$$(2\pi)^{-d/2}D^{-1/2}(x_1 \cdots x_D)^{-1}|\boldsymbol{\Gamma} + D^{-1}\mathbf{J}|^{-1/2} \\ \times \exp[-\tfrac{1}{2}(\mathbf{z}-\boldsymbol{\lambda})'\{\boldsymbol{\Gamma} + D^{-1}\mathbf{J}\}^{-1}(\mathbf{z}-\boldsymbol{\lambda})], \quad (6.5)$$

where $\mathbf{z} = \log\{\mathbf{x}/g(\mathbf{x})\}$.

6.4 Moment properties

Although moments and logarithmic moments of all positive orders,

$$\mathrm{E}\left(\prod_{i=1}^{D} x_i^{a_i}\right), \quad \mathrm{E}\left\{\prod_{i=1}^{D} (\log x_i)^{a_i}\right\} \qquad (a_i > 0 : i = 1, \ldots, D), \quad (6.6)$$

exist, the integral expressions are not reducible to any simple form. This is no great loss since interest in practice is more naturally in the ratios x_i/x_j or their logarithms. From normal-lognormal theory, with σ_{ij} denoting the (i,j) element of $\boldsymbol{\Sigma}$ and with the convention that $\mu_D = 0$ and $\sigma_{i,D} = 0$ $(i = 1, \ldots, D)$, we have that

$$\mathrm{E}\{\log(x_i/x_k)\} = \mu_i - \mu_k, \quad (6.7)$$

$$\mathrm{cov}\{\log(x_i/x_k), \log(x_j/x_l)\} = \sigma_{ij} + \sigma_{kl} - \sigma_{il} - \sigma_{jk}, \quad (6.8)$$

$$\mathrm{E}(x_i/x_k) = \exp\{\mu_i - \mu_k + \tfrac{1}{2}(\sigma_{ii} - 2\sigma_{ik} + \sigma_{kk})\}, \quad (6.9)$$

$$\mathrm{cov}(x_i/x_k, x_j/x_l) \\ = \mathrm{E}(x_i/x_k)\mathrm{E}(x_j/x_l)\{\exp(\sigma_{ij} + \sigma_{kl} - \sigma_{il} - \sigma_{jk}) - 1\}. \quad (6.10)$$

Although simple closed forms for moments (6.6) are not available for general $\mathscr{L}^d(\boldsymbol{\mu},\boldsymbol{\Sigma})$, numerical values for a specific distribution are readily obtainable by Hermitian integration, at least for moderate values of d. We do not digress here to discuss this since the Hermitian technique required forms a natural part of the discussion of Chapter

13, where the interested reader will find methods appropriate to the numerical evaluation of (6.6).

6.5 Composition of a lognormal basis

One way in which a logistic normal distribution in $\mathscr{S}^d$ can arise in practice is as the composition of a basis which has a multivariate lognormal distribution in $\mathscr{R}_+^D$.

Definition 6.3 A D-part basis $\mathbf{w} \in \mathscr{R}_+^D$ is distributed as $\boldsymbol{\Lambda}^D(\boldsymbol{\zeta}, \boldsymbol{\Omega})$, *multivariately lognormal* with mean parameter $\boldsymbol{\zeta}$ and covariance parameter $\boldsymbol{\Omega}$, if $\log \mathbf{w}$ is distributed as $\mathscr{N}^D(\boldsymbol{\zeta}, \boldsymbol{\Omega})$.

The compositional property can then be expressed as follows.

Property 6.1 If a basis $\mathbf{w}$ is $\Lambda^D(\boldsymbol{\zeta}, \boldsymbol{\Omega})$ distributed then its composition $\mathbf{x} = \mathscr{C}(\mathbf{w})$ is $\mathscr{L}^d(\boldsymbol{\mu}, \boldsymbol{\Sigma})$ distributed, where

$$\boldsymbol{\mu} = \mathbf{F}\boldsymbol{\zeta}, \qquad \boldsymbol{\Sigma} = \mathbf{F}\boldsymbol{\Omega}\mathbf{F}'.$$

Proof. The result is immediate from multivariate normal theory since $\mathbf{y} = \log(\mathbf{x}_{-D}/x_D) = \mathbf{F} \log \mathbf{w}$ and $\mathbf{F} \log \mathbf{w}$ is $\mathscr{N}^d(\mathbf{F}\boldsymbol{\zeta}, \mathbf{F}\boldsymbol{\Omega}\mathbf{F}')$.

Notes

1. The result states that any D-dimensional lognormal basis gives rise to a logistic normal composition. This is the logistic normal result which corresponds to the basis–composition relationship (Property 3.3) of gamma and Dirichlet distributions, with one substantial difference. For the Dirichlet composition the gamma basis has to have components which are independent and equally scaled: here the multivariate lognormal is quite general with any covariance structure.
2. This connection of the additive logistic normal class of compositions in $\mathscr{S}^d$ with the class of lognormal bases in $\mathscr{R}_+^D$ is so obvious that there could be a tendency to assume that this is the only, or the natural, way for logistic normal compositions to arise from bases. We shall see later in Chapter 9 that such an attitude can lead to the exclusion of what proves to be a very useful tool of analysis of the relationships of compositions to the actual bases from which they are derived.

6.6 Class-preserving properties

The well-known linear transformation property of multinormal distributions, that if $\mathbf{y}$ is $\mathcal{N}^d(\boldsymbol{\mu}, \boldsymbol{\Sigma})$ and $\mathbf{K} = [k_{ij}]$ is a $c \times d$ matrix of constants then $\mathbf{Ky}$ is $\mathcal{N}^c(\mathbf{K}\boldsymbol{\mu}, \mathbf{K\Sigma K}')$, has the following counterpart in logistic normal theory.

If $\mathbf{x}$ is $\mathcal{L}^d(\boldsymbol{\mu}, \boldsymbol{\Sigma})$ then

$$\mathscr{C}\left\{\prod_{j=1}^{d}(x_j/x_D)^{k_{1j}}, \ldots, \prod_{j=1}^{d}(x_j/x_D)^{k_{cj}}, 1\right\} \quad \text{is} \quad \mathcal{L}^c(\mathbf{K}\boldsymbol{\mu}, \mathbf{K\Sigma K}'). \tag{6.11}$$

Carrying over properties of the multinormal class in an unmotivated way is a pointless exercise, and such general results as (6.11) are useless until directed towards the particular types of problem which arise in compositional data analysis. We shall focus on three particular class-preserving properties involving permutations, subcompositions and power transformations.

Permutation property

Suppose that, of two compositional data analysts, the first works with the parts of the composition $\mathbf{x}$ in their original order $1, \ldots, D$, whereas for the second the parts have been subjected to a new ordering, for example by a permutation matrix $\mathbf{P}$ to obtain a composition $\mathbf{x}_P = \mathbf{Px}$. Suppose that the first analyst assumes that $\mathbf{x}$ is distributed as $\mathcal{L}^d(\boldsymbol{\mu}, \boldsymbol{\Sigma})$, and the second analyst assumes that $\mathbf{x}_P$ is distributed as $\mathcal{L}^d(\boldsymbol{\mu}, \boldsymbol{\Sigma}_P)$. Are the assumptions consistent and, if so, how are $(\boldsymbol{\mu}_P, \boldsymbol{\Sigma}_P)$ related to $(\boldsymbol{\mu}, \boldsymbol{\Sigma})$? With our groundwork on permutations in Sections 5.3–5.4 the answer is easily provided.

Property 6.2 If a D-part composition $\mathbf{x}$ is distributed as $\mathcal{L}^d(\boldsymbol{\mu}, \boldsymbol{\Sigma})$ and if $\mathbf{x}_P = \mathbf{Px}$ is the composition with the parts reordered by the permutation matrix $\mathbf{P}$, then $\mathbf{x}_P$ is distributed as $\mathcal{L}^d(\boldsymbol{\mu}_P, \boldsymbol{\Sigma}_P)$ with

$$\boldsymbol{\mu}_P = \mathbf{Q}_P\boldsymbol{\mu}, \qquad \boldsymbol{\Sigma}_P = \mathbf{Q}_P\boldsymbol{\Sigma}\mathbf{Q}_P',$$

where

$$\mathbf{Q}_P = \mathbf{FPF}'\mathbf{H}^{-1}.$$

Proof. Let $\mathbf{y} = \mathbf{F}\log\mathbf{x}$ and $\mathbf{y}_P = \mathbf{F}\log\mathbf{x}_P$ be the logratio vectors obtained from compositions $\mathbf{x}$ and $\mathbf{x}_P$, respectively. Then $\mathbf{y}$ is $\mathcal{N}^d(\boldsymbol{\mu}, \boldsymbol{\Sigma})$. The result follows immediately from multinormal theory, since we have already established in Property 5.2(iv) that $\mathbf{y}_P = \mathbf{Q}_P\mathbf{y}$.

Notes

1. From the distributional point of view the particular order which the analyst uses is of no consequence. Note also that since, by Property 5.3(a), the determinant $|\boldsymbol{\Sigma}|$, and, by Property 5.4, the quadratic form $(\mathbf{y}-\boldsymbol{\mu})'\boldsymbol{\Sigma}^{-1}(\mathbf{y}-\boldsymbol{\mu})$ are permutation invariant, the density function (6.1) itself is permutation invariant.
2. The other parametrizations are simpler. In terms of (6.4),

$$\boldsymbol{\Delta}_P(\mathbf{x},\boldsymbol{\Xi}) = \mathbf{P}\boldsymbol{\Delta}(\mathbf{x},\boldsymbol{\Xi})\mathbf{P}', \qquad \boldsymbol{T}_P = \mathbf{P}\boldsymbol{T}\mathbf{P}'. \tag{6.12}$$

In terms of (6.5),

$$\boldsymbol{\lambda}_P = \mathbf{P}\boldsymbol{\lambda}, \qquad \boldsymbol{\Gamma}_P = \mathbf{P}\boldsymbol{\Gamma}\mathbf{P}'. \tag{6.13}$$

Subcompositional property

From our preliminary examination in Chapter 1 of compositional data problems, we have seen that subcompositional analysis is an important aspect. A question of particular interest therefore is the nature of the distribution of a subcomposition of a logistic normal composition. Do logistic normal compositions have logistic normal subcompositions? A happily positive answer to this question is provided in the following result.

Property 6.3 If a D-part composition $\mathbf{x}$ is distributed as $\mathscr{L}^d(\boldsymbol{\mu},\boldsymbol{\Sigma})$ and if $\mathbf{x}_S = \mathscr{C}(\mathbf{S}\mathbf{x})$ is the subcomposition with parts selected by the $C \times D$ selecting matrix $\mathbf{S}$, then $\mathbf{x}_S$ is distributed as $\mathscr{L}^c(\boldsymbol{\mu}_S,\boldsymbol{\Sigma}_S)$, with $c = C-1$ and

$$\boldsymbol{\mu}_S = \mathbf{Q}_S\boldsymbol{\mu}, \qquad \boldsymbol{\Sigma}_S = \mathbf{Q}_S\boldsymbol{\Sigma}\mathbf{Q}_S',$$

where

$$\mathbf{Q}_S = \mathbf{F}_{c,C}\mathbf{S}\mathbf{F}_{d,D}'\mathbf{H}^{-1}.$$

Proof. Let $\mathbf{y} = \mathbf{F}_{d,D}\log\mathbf{x}$ and $\mathbf{y}_S = \mathbf{F}_{c,C}\log\mathbf{x}_S$ be the logratio vectors obtained from composition $\mathbf{x}$ and subcomposition $\mathbf{x}_S$, respectively. Then $\mathbf{y}$ is $\mathscr{N}^d(\boldsymbol{\mu},\boldsymbol{\Sigma})$. The result follows from multinormal theory, since we have established that $\mathbf{y}_S = \mathbf{Q}_S\mathbf{y}$ in (5.24).

Note

1. Although the $(\boldsymbol{\mu},\boldsymbol{\Sigma})$ specification is convenient for establishing the fact that every subcomposition of a logistic normal composition has

also a logistic normal distribution, the parameters of the subcompositional distribution are more conveniently obtained in terms of the variation array $\boldsymbol{\Delta}$ and variation matrix $\boldsymbol{T}$:

$$\boldsymbol{\Delta}_S = \mathbf{S}\boldsymbol{\Delta}\mathbf{S}', \qquad \boldsymbol{T}_S = \mathbf{S}\boldsymbol{T}\mathbf{S}'. \tag{6.14}$$

Power transformation property

In some applications, as for example in modal analysis (Chayes, 1956) in geology, an initial mineral composition $\mathbf{x}$ of a rock may be expressed in proportions by volume, and then transformed into a new composition $\mathbf{X}$ expressed in proportions by weight. In such a transformation $X_i \propto a_i x_i$, where a_i is the density of the ith part $(i = 1, \dots, D)$. Expressed in terms of the familiar constraining operator, this relation becomes

$$\mathbf{X} = \mathscr{C}(a_i x_i : i = 1, \dots, D).$$

This is no more than a perturbation

$$\mathbf{X} = \mathbf{a} \circ \mathbf{x},$$

where the perturbing vector $\mathbf{a}$ is a known vector of positive constants. The transformation is, however, a special case of a more general transformation, the power transformation, and we discuss it within this wider context.

Definition 6.4 From a D-part composition $\mathbf{x}$ any new D-part composition $\mathbf{X}$ formed by the operation

$$\mathbf{X} = \mathscr{C}(a_i x_i^b : i = 1, \dots, D)$$

is termed a *power-transformed composition* of $\mathbf{x}$.

Notes

1. The modal transformation is the special case $b = 1$.
2. The original composition $\mathbf{x}$ is itself a power-transformed composition of the power-transformed $\mathbf{X}$, since

$$\mathbf{x} = \mathscr{C}(a_i^{-1/b} X_i^{1/b} : i = 1, \dots, D).$$

Within the class of logistic normal distributions there is a simple relationship between $\mathbf{x}$ and $\mathbf{X}$.

Property 6.4 If the D-part composition $\mathbf{x}$ is distributed as $\mathscr{L}^d(\boldsymbol{\mu}, \boldsymbol{\Sigma})$

then its power-transformed composition $\mathbf{X} = \mathscr{C}(a_i x_i^b : i = 1, \ldots, D)$ is distributed as

$$\mathscr{L}^d\{\log(\mathbf{a}_{-D}/a_D) + b\boldsymbol{\mu}, b^2\boldsymbol{\Sigma}\}.$$

Proof. From the definition of $\mathbf{X}$ in terms of $\mathbf{x}$ we obtain

$$\log(X_i/X_D) = \log(a_i/a_D) + b\log(x_i/x_D)$$

so that the logratio vectors $\mathbf{y} = \log(\mathbf{x}_{-D}/x_D)$ and $\mathbf{Y} = \log(\mathbf{X}_{-D}/X_D)$ formed from $\mathbf{x}$ and $\mathbf{X}$, respectively, are related by

$$\mathbf{Y} = \log(\mathbf{a}_{-D}/a_D) + b\mathbf{y}.$$

The logistic normal distribution for $\mathbf{Y}$ follows from the linear transformation property of multinormal theory.

The fact that the logratio covariance matrix of $\mathbf{X}$ is a constant multiple of that of $\mathbf{x}$ allows us to make the following statements.

Property 6.5

(a) Any power-transformed composition of a logratio uncorrelated composition is logratio uncorrelated.

(b) Any power-transformed composition of an isotropic composition is isotropic.

Amalgamation and partition difficulty

Suppose that a D-part composition $\mathbf{x}$ is distributed as $\mathscr{L}^d(\boldsymbol{\mu}, \boldsymbol{\Sigma})$ and that we are interested in the distribution of the general partition $(\mathbf{t}; \mathbf{s}_1, \ldots, \mathbf{s}_C)$ of Definition 2.12. As far as the subcompositions of the partition are concerned, there is no difficulty in arriving at the joint distribution of $(\mathbf{s}_1, \ldots, \mathbf{s}_C)$. For example, the logratio vector $\mathbf{y}_j$ formed from $\mathbf{s}_j$ can be related to the logratio vector $\mathbf{y}$ of $\mathbf{x}$ by

$$\mathbf{y}_j = \mathbf{F}_j\mathbf{S}_j\mathbf{F}'\mathbf{H}^{-1}\mathbf{y} \qquad (j = 1, \ldots, C),$$

where $\mathbf{F}_j = \mathbf{F}_{d_j, d_j+1}$, $\mathbf{F} = \mathbf{F}_{d,D}$ and $\mathbf{S}_j$ is the selecting matrix for $\mathbf{s}_j$. Then

$$\begin{bmatrix} \mathbf{y}_1 \\ \vdots \\ \mathbf{y}_C \end{bmatrix} = \begin{bmatrix} \mathbf{F}_1\mathbf{S}_1 \\ \vdots \\ \mathbf{F}_C\mathbf{S}_C \end{bmatrix} \mathbf{F}'\mathbf{H}^{-1}\mathbf{y} = \mathbf{Q}\mathbf{y}, \tag{6.15}$$

say. The joint distribution of $(\mathbf{s}_1, \ldots, \mathbf{s}_C)$ is then effectively determined by the resulting $\mathscr{N}^{d-c}(\mathbf{Q}\boldsymbol{\mu}, \mathbf{Q}\boldsymbol{\Sigma}\mathbf{Q}')$ distribution for $(\mathbf{y}_1, \ldots, \mathbf{y}_C)$.

Any attempt, however, to find a tractable form for the amalgamation $\mathbf{t}$ of the partition is fraught with difficulty. The reason for this difficulty is simply that there is no way of expressing the logarithm of a sum of components in terms of the logarithms of the components. In contrast, we may recall the elegant distribution property (Property 3.4) for the complete partition of a Dirichlet composition, but also remind ourselves that the associated strong independence makes the Dirichlet class unattractive as a practical tool. Thus neither the $\mathscr{D}^d$ nor the $\mathscr{L}^d$ class is suited to partition analysis. We shall see in Sections 6.14 and 10.3 that there is a way of exploiting the fundamental ideas underlying logistic normal classes to provide a modelling tool for the analysis of partitions.

6.7 Conditional subcompositional properties

We may sometimes wish to consider the variability of the relative proportions of some parts of a composition for fixed relative proportions of other parts: in other words, we may be interested in the conditional distribution of one subcomposition $\mathbf{s}_1$ given another subcomposition $\mathbf{s}_2$. In particular, we may ask what is the form of this conditional distribution $\mathbf{s}_1 | \mathbf{s}_2$ if the full composition is distributed as $\mathscr{L}^d(\boldsymbol{\mu}, \boldsymbol{\Sigma})$.

We need consider only the case where $\mathbf{s}_1$ and $\mathbf{s}_2$ contain at most one part in common. Otherwise, if for example $\mathbf{s}_1$ and $\mathbf{s}_2$ have two parts i and j in common, then the ratio x_i/x_j is fixed by the given $\mathbf{s}_2$ and so is also known and fixed in $\mathbf{s}_1$ and there is no need to study its variability. One general approach to the conditional problem would be through (6.15). We prefer a simpler approach in terms of two cases, first where the two subcompositions have exactly one part in common, second where there is no part in common.

When there is exactly one part in common we can suppose the parts to be already ordered, without loss of generality, so that the final part D is this common one. We can then consider

$$\mathbf{s}_1 = \mathscr{C}(\mathbf{x}^{(c)}, x_D) = \mathscr{C}(x_1, \ldots, x_c, x_D),$$

$$\mathbf{s}_2 = \mathscr{C}(\mathbf{x}_{(c)}) = \mathscr{C}(x_C, \ldots, x_D).$$

Suppose that $\mathbf{x}$ is distributed as $\mathscr{L}^d(\boldsymbol{\mu}, \boldsymbol{\Sigma})$, and that $\mathbf{y}_1$ and $\mathbf{y}_2$ are the c- and $(D-C)$-dimensional logratio vectors of $\mathbf{s}_1$ and $\mathbf{s}_2$. Then $(\mathbf{y}_1, \mathbf{y}_2) = \mathbf{y}$, the d-dimensional logratio vector of the full composition $\mathbf{x}$. The following conditional result then follows by a standard multinormal move from joint to conditional distribution.

Property 6.6 Suppose that the D-part composition $\mathbf{x}$ be distributed as $\mathscr{L}^d(\boldsymbol{\mu}, \boldsymbol{\Sigma})$ and let $\mathbf{s}_1 = \mathscr{C}(\mathbf{x}^{(c)}, x_D)$ and $\mathbf{s}_2 = \mathscr{C}(\mathbf{x}_{(c)})$ be subcompositions with common part D. Then the conditional distribution of $\mathbf{s}_1$ for given $\mathbf{s}_2$ is $\mathscr{L}^c(\boldsymbol{\mu}_{1.2}, \boldsymbol{\Sigma}_{1.2})$, with $c = C - 1$ and

$$\boldsymbol{\mu}_{1.2} = \boldsymbol{\mu}_1 + \boldsymbol{\Sigma}_{12}\boldsymbol{\Sigma}_{22}^{-1}(\mathbf{y}_2 - \boldsymbol{\mu}_2), \qquad \boldsymbol{\Sigma}_{1.2} = \boldsymbol{\Sigma}_{11} - \boldsymbol{\Sigma}_{12}\boldsymbol{\Sigma}_{22}^{-1}\boldsymbol{\Sigma}_{21},$$

where

$$\begin{bmatrix} \boldsymbol{\mu}_1 \\ \boldsymbol{\mu}_2 \end{bmatrix}, \qquad \begin{bmatrix} \boldsymbol{\Sigma}_{11} & \boldsymbol{\Sigma}_{12} \\ \boldsymbol{\Sigma}_{21} & \boldsymbol{\Sigma}_{22} \end{bmatrix}$$

are the $(c, D - C)$ partitions of $\boldsymbol{\mu}, \boldsymbol{\Sigma}$, and $\mathbf{y}_2$ is the logratio vector of $\mathbf{s}_2$.

Note

1. If $\mathbf{s}_1$ contains part D and only some of parts $1, \ldots, C$, we can obtain the appropriate conditional distribution by an application of Property 6.3.

A corresponding result for the case when $\mathbf{s}_1$ and $\mathbf{s}_2$ contain no common parts can be obtained simply from Property 6.3.

Property 6.7 For a D-part composition $\mathbf{x}$ distributed as $\mathscr{L}^d(\boldsymbol{\mu}, \boldsymbol{\Sigma})$ and for subcompositions $\mathbf{s}_1 = \mathscr{C}(\mathbf{x}^{(c)})$, $\mathbf{s}_2 = \mathscr{C}(\mathbf{x}_{(c)})$ without common parts, the conditional distribution of $\mathbf{s}_1$ for given $\mathbf{s}_2$ is, in the notation of Property 6.6, $\mathscr{L}^{c-1}(\mathbf{F}\boldsymbol{\mu}_{1.2}, \mathbf{F}\boldsymbol{\Sigma}_{1.2}\mathbf{F}')$, where $\mathbf{F} = \mathbf{F}_{c-1,c}$.

6.8 Perturbation properties

Now that we have established a method of overcoming the awkwardness of the unit-sum constraint on compositions, we can reformulate the perturbation operation of Definition 2.13. In terms of the logratio vectors $\mathbf{y} = \log(\mathbf{x}_{-D}/x_D)$, $\mathbf{v} = \log(\mathbf{u}_{-D}/u_D)$ and $\mathbf{Y} = \log(\mathbf{X}_{-D}/X_D)$ of the original composition $\mathbf{x}$, the perturbing vector $\mathbf{u}$ and the perturbed composition $\mathbf{X}$, the perturbation

$$\mathbf{X} = \mathbf{u} \circ \mathbf{x} \tag{6.16}$$

can be re-expressed as

$$\mathbf{Y} = \mathbf{v} + \mathbf{y}, \tag{6.17}$$

an additive relation of the logratio vectors. Since addition of multinormal logratio vectors $\mathbf{v}$ and $\mathbf{y}$ in (6.17) will yield a multinormal logratio vector $\mathbf{Y}$, we can translate these multinormal relationships into compositional distributional properties of perturbations.

Property 6.8 A D-part composition $\mathbf{x}$, which is $\mathscr{L}^d(\boldsymbol{\mu}, \boldsymbol{\Sigma})$ distributed, is perturbed by a vector $\mathbf{u}$ of D positive components, distributed independently of $\mathbf{x}$. The distribution of the perturbed vector $\mathbf{X} = \mathbf{u} \circ \mathbf{x}$ is as given below for three different distributional assumptions about $\mathbf{u}$.

	Distribution of $\mathbf{u}$	*Distribution of* $\mathbf{X}$
(a)	$\Lambda^D(\boldsymbol{\zeta}, \boldsymbol{\Omega})$	$\mathscr{L}^d(\boldsymbol{\mu} + \mathbf{F}\boldsymbol{\zeta}, \boldsymbol{\Sigma} + \mathbf{F}\boldsymbol{\Omega}\mathbf{F}')$
(b)	$\mathscr{L}^d(\boldsymbol{\theta}, \boldsymbol{\Theta})$	$\mathscr{L}^d(\boldsymbol{\mu} + \boldsymbol{\theta}, \boldsymbol{\Sigma} + \boldsymbol{\Theta})$
(c)	Constant vector	$\mathscr{L}^d\{\boldsymbol{\mu} + \log(\mathbf{u}_{-D}/u_D), \boldsymbol{\Sigma}\}$

Notes

1. Properties (b) and (c) can be regarded as class-preserving properties and could accordingly have been added to the list in Section 6.6. We have preferred to treat them separately because of their close relationship to central limit theorems and to the modelling of measurement error and imprecision in Chapter 11.
2. The independence of $\mathbf{u}$ and $\mathbf{x}$ is not an essential condition for the logistic normality of $\mathbf{X}$. From (6.17), if the dependence of $\mathbf{u}$ and $\mathbf{x}$ is expressed in terms of the joint normality of $\mathbf{v}$ and $\mathbf{y}$ with mean vector and covariance matrix

$$\begin{bmatrix} \boldsymbol{\theta} \\ \boldsymbol{\mu} \end{bmatrix}, \quad \begin{bmatrix} \boldsymbol{\Theta} & \boldsymbol{\Psi} \\ \boldsymbol{\Psi}' & \boldsymbol{\Sigma} \end{bmatrix},$$

then $\mathbf{X}$ is $\mathscr{L}^d(\boldsymbol{\mu} + \boldsymbol{\theta}, \boldsymbol{\Sigma} + \boldsymbol{\Theta} + \boldsymbol{\Psi} + \boldsymbol{\Psi}')$.

6.9 A central limit theorem

Just as the central limit theorem in its additive form leads to normal distributions and in its multiplicative form leads to lognormal distributions, so it is possible to establish a central limit theorem leading to logistic normal distributions through the operation of perturbation. We shall not attempt to state the theorem in any rigorous form since our only interest in it is simply to demonstrate that there could be processes in nature which would lead to logistic normal distributions.

Property 6.9 If $\mathbf{x}_n\,(n = 1, 2, \ldots)$ is a sequence of compositions, generated by successive independent, but not necessarily logistic normal nor lognormal, perturbations $\mathbf{u}_n$ $(n = 1, 2, \ldots)$:

$$\mathbf{x}_n = \mathbf{u}_n \circ \mathbf{x}_{n-1} \qquad (n = 1, 2, \ldots),$$

then, under certain regularity conditions and for large n, $\mathbf{x}_n$ will follow an additive logistic normal distribution.

Sketch proof. Let $\mathbf{y}_n$ denote the logratio vector of $\mathbf{x}_n\,(n = 0, 1, \ldots)$ and $\mathbf{v}_n$ the logratio vector of $\mathbf{u}_n$ $(n = 1, 2, \ldots)$. Then

$$\begin{aligned}\mathbf{y}_n &= \mathbf{v}_n + \mathbf{y}_{n-1}\\ &= \mathbf{y}_0 + \mathbf{v}_1 + \cdots + \mathbf{v}_n\end{aligned}$$

and the asymptotic multinormality of $\mathbf{y}_n$, and hence the asymptotic logistic normality of $\mathbf{x}_n$, follows from the familiar additive version of the central limit theorem.

Note

1. As we pointed out in Section 2.8, such successive perturbations may be a natural process of differential wastage of parts of a composition. One implicit occurrence of this concept of modelling is in genetic selection, where the 3-part composition has as its components the proportions x_{n1}, x_{n2}, x_{n3} of the three genotypes AA, Aa, aa at the nth generation and there is genetic selection, or differential survival rates u_{n1}, u_{n2}, u_{n3}, during the period to the next $(n+1)$th generation. Another example is in the relative growth of parts of an animal or plant. For example, for some purposes the composition of a plant may be regarded as the proportions by weight or volume of root, shoot, leaf, fruit and the differential growth of these may be regarded as a series of perturbations to the attained composition.

6.10 A characterization by logcontrasts

The following logcontrast property of the $\mathscr{L}^d$ class is obvious, since

$$\mathbf{a}' \log \mathbf{x} = \mathbf{a}'_{-D} \log(\mathbf{x}_{-D}/x_D) \tag{6.18}$$

when $\mathbf{a}'\mathbf{j} = 0$.

Property 6.10 If a D-part composition $\mathbf{x}$ is distributed as $\mathscr{L}^d(\boldsymbol{\mu}, \boldsymbol{\Sigma})$

then every logcontrast of $\mathbf{x}$ has a univariate normal distribution: $\mathbf{a}'\log\mathbf{x}$ is $\mathcal{N}^1(\mathbf{a}'_{-D}\boldsymbol{\mu}, \mathbf{a}'_{-D}\boldsymbol{\Sigma}\mathbf{a}_{-D})$.

The converse of this provides a form of characterization of logistic normality. The multinormal characterization (Rao, 1965, p. 435), which states that $\mathbf{y}\in\mathcal{R}^d$ is multivariate normal if every linear combination $\mathbf{b}'\mathbf{y}$ is univariate normal, can be adapted into the following characterization of the $\mathcal{L}^d$ class.

Property 6.11 If a D-part composition $\mathbf{x}$ is such that every logcontrast of $\mathbf{x}$ is univariate normal then $\mathbf{x}$ has a logistic normal distribution.

Proof. The proof follows from the multivariate normal characterization by (6.18) with $\mathbf{y} = \log(\mathbf{x}_{-D}/x_D)$.

6.11 Relationships with the Dirichlet class

The logistic normal class $\mathcal{L}^d$ and the Dirichlet class $\mathcal{D}^d$ of Definition 3.2 are separate in the sense that it is not possible to find two distributions, one from each class, which are arbitrarily close, in any meaningful sense, to each other. We recall our comments in Section 3.4 indicating that the $\mathcal{D}^d$ class has the ultimate independence structure for compositions. Although the $\mathcal{L}^d$ class in its separateness cannot compete with this independence structure, and although the investigation of such a strong independence structure may seldom be of any practical importance, it is of some theoretical interest to ask to what extent any given Dirichlet distribution $\mathcal{D}^d(\boldsymbol{\alpha})$ may be reasonably approximated by a $\mathcal{L}^d(\boldsymbol{\mu}, \boldsymbol{\Sigma})$ distribution.

Kullback–Leibler logistic normal approximation to $\mathcal{D}^d(\boldsymbol{\alpha})$

Let $p(\mathbf{x}|\boldsymbol{\alpha})$ and $q(\mathbf{x}|\boldsymbol{\mu}, \boldsymbol{\Sigma})$ denote the density functions of the $\mathcal{D}^d(\boldsymbol{\alpha})$ and $\mathcal{L}^d(\boldsymbol{\mu}, \boldsymbol{\Sigma})$ distributions, respectively. The Kullback–Leibler (1951) measure $K(p,q)$ or $K(\boldsymbol{\alpha}; \boldsymbol{\mu}, \boldsymbol{\Sigma})$ of directed divergence, a measure of how much the approximating q misses the target p, is

$$K(p,q) = \int_{\mathcal{S}^d} p(\mathbf{x}|\boldsymbol{\alpha}) \log\{p(\mathbf{x}|\boldsymbol{\alpha})/q(\mathbf{x}|\boldsymbol{\mu}, \boldsymbol{\Sigma})\}\, \mathrm{d}\mathbf{x}. \tag{6.19}$$

Then the Kullback–Leibler logistic normal approximation to $\mathcal{D}^d(\boldsymbol{\alpha})$ is the $\mathcal{L}^d(\boldsymbol{\mu}, \boldsymbol{\Sigma})$ distribution which minimizes $K(p,q)$ with respect to $\boldsymbol{\mu}$

and $\boldsymbol{\Sigma}$. In so far as $K(p, q)$ depends on $\boldsymbol{\mu}$ and $\boldsymbol{\Sigma}$, we can write (6.19) as

$$\tfrac{1}{2}\log|\boldsymbol{\Sigma}| + \tfrac{1}{2}\int_{\mathscr{S}^d} \{\log(\mathbf{x}_{-D}/x_D) - \boldsymbol{\mu}\}'\boldsymbol{\Sigma}^{-1} \times \{\log(\mathbf{x}_{-D}/x_D) - \boldsymbol{\mu}\} p(\mathbf{x}|\boldsymbol{\alpha})\mathrm{d}\mathbf{x},$$

which, by standard multivariate theory, is minimized by

$$\boldsymbol{\mu} = \mathrm{E}_{\alpha}\{\log(\mathbf{x}_{-D}/x_D)\}, \qquad \boldsymbol{\Sigma} = \mathrm{V}_{\alpha}\{\log(\mathbf{x}_{-D}/x_D)\},$$

where $\mathrm{E}_{\boldsymbol{\alpha}}$ and $\mathrm{V}_{\boldsymbol{\alpha}}$ denote mean vector and covariance matrix with respect to the $p(\mathbf{x}|\boldsymbol{\alpha})$ distribution. From Property 3.5 the minimizing $\boldsymbol{\mu}$ and $\boldsymbol{\Sigma}$ are given by

$$\mu_i = \psi(\alpha_i) - \psi(\alpha_D) \qquad (i = 1, \ldots, d), \tag{6.20}$$

$$\sigma_{ii} = \psi'(\alpha_i) + \psi'(\alpha_D) \qquad (i = 1, \ldots, d), \tag{6.21}$$

$$\sigma_{ij} = \psi'(\alpha_D) \qquad (i \neq j = 1, \ldots, d). \tag{6.22}$$

The minimized value of K can then be expressed as

$$\tfrac{1}{2}d\{1 + \log(2\pi)\} - \log\Delta(\boldsymbol{\alpha}) + \sum_{i=1}^{D} \alpha_i\psi(\alpha_i) - A\psi(A) + \tfrac{1}{2}\sum_{i=1}^{D} \log\psi'(\alpha_i) + \tfrac{1}{2}\log\left[\sum_{i=1}^{D} 1/\psi'(\alpha_i)\right], \tag{6.23}$$

where

$$A = \boldsymbol{\alpha}'\mathbf{j}, \tag{6.24}$$

$$\Delta(\boldsymbol{\alpha}) = \Gamma(\alpha_1)\cdots\Gamma(\alpha_D)/\Gamma(A). \tag{6.25}$$

Using asymptotic expansions (Abramowitz and Stegun, 1972, Chapter 6) for ψ and ψ', we can obtain an approximation to this for large α_i $(i = 1, \ldots, D)$. After some tedious algebra K can be expressed as

$$(1/12)\left(\sum_{i=1}^{D} \alpha_i^{-1} + 2/A\right). \tag{6.26}$$

For $\mathscr{D}^d(\boldsymbol{\alpha})$ with $d = 1, 2, 3$ and with components of $\boldsymbol{\alpha}$ in the range 5 to 100, the minimized divergences range from 2×10^{-6} for $\boldsymbol{\alpha} = (5, 5)$ to 5×10^{-2} for $\boldsymbol{\alpha} = (5, 5, 5, 100)$.

Some indication of the degree of closeness of these approximations can be provided by directed divergences for more familiar situations. For example, the directed divergence of a $\mathscr{N}(\lambda, 1)$ from the $\mathscr{N}(0, 1)$

distribution also ranges from 2×10^{-6} to 5×10^{-2} as λ ranges from 0.002 to 0.316.

We can also try to judge success for given $\mathscr{D}^d(\boldsymbol{\alpha})$ by finding a neighbouring $\mathscr{D}^d(\boldsymbol{\beta})$ distribution with the same Kullback–Leibler directed divergence from $\mathscr{D}^d(\boldsymbol{\alpha})$ as the minimized logistic normal divergence (6.23). Confining attention to neighbouring distributions with $\boldsymbol{\alpha}$ differing from $\boldsymbol{\beta}$ in a single component, we have the following results. When the components of $\boldsymbol{\alpha}$ are equal, the increase in a single α_i never exceeds 0.6. When the components are unequal, the more asymmetrical the component values are, the greater, roughly speaking, is the increase in these single component values, the greatest increases being 0.7 in 5, 1.4 in 20 and 8.6 in 100, the last occurring for $d = 3$ and $\boldsymbol{\alpha} = (5, 5, 5, 100)$.

Logistic normal distribution as a limiting form of Dirichlet distributions

From our discussion above we have seen that for any given Dirichlet distribution $\mathscr{D}^d(\boldsymbol{\alpha})$ with large $\alpha_i\,(i = 1, \ldots, D)$ there is a very close logistic normal distribution. We can look at this relationship in another way and ask if the Dirichlet distribution has any convergence property as $\alpha_i \to \infty$ $(i = 1, \ldots, D)$.

Let $\mathbf{x}$ be $\mathscr{D}^d(\boldsymbol{\alpha})$ and consider the logcontrast $u = \mathbf{a}' \log \mathbf{x}$ with $\mathbf{a}'\mathbf{j} = 0$. The moment generating function of u is given by

$$M_u(t) = \mathrm{E}(e^{tu}) = \mathrm{E}\left(\prod_{i=1}^{D} x_i^{ta_i} \right)$$
$$= \Delta(\boldsymbol{\alpha} + t\mathbf{a})/\Delta(\boldsymbol{\alpha}),$$

where $\Delta(\cdot)$ is defined by (6.25). It follows that

$$\mathrm{E}(u) = \sum_{i=1}^{D} a_i \psi(\alpha_i), \qquad \mathrm{var}(u) = \sum_{i=1}^{D} a_i^2 \psi'(\alpha_i),$$

so that the moment generating function of $\{u - \mathrm{E}(\mathbf{u})\}/\sqrt{\mathrm{var}(u)}$, the standardized logcontrast, is given by

$$\exp\left[-\sum_{i=1}^{D} a_i \psi(\alpha_i) \bigg/ \left\{ \sum_{i=1}^{D} a_i^2 \psi'(\alpha_i) \right\}^{1/2} \right]$$
$$\times \Delta\left[\boldsymbol{\alpha} + t\mathbf{a} \bigg/ \left\{ \sum_{i=1}^{D} a_i^2 \psi'(\alpha_i) \right\}^{1/2} \right] \bigg/ \Delta(\boldsymbol{\alpha}),$$

which can be shown to tend to $\exp(\frac{1}{2}t^2)$ as $\alpha_i \to \infty$ $(i = 1, \ldots, D)$. Thus every logcontrast of $\mathbf{x}$ tends to univariate normality and so, by the characterization of Property 6.11, $\mathbf{x}$ tends to logistic normality, with parameters $\boldsymbol{\mu}$ and $\boldsymbol{\Sigma}$ given by (6.20)–(6.22).

6.12 Potential for statistical analysis

Some comparisons of the Dirichlet and the logistic normal classes with respect to their properties and their appropriateness for statistical analysis are now possible.

First it must be clear that the Dirichlet class has some very elegant mathematical properties. The processes of permutation of components, of focusing on a subcomposition, of amalgamation and of partition, all lead to Dirichlet distributions in an appropriately dimensioned simplex, with parameters simply related to the parent composition parameter. These features all stem from the single fact that a Dirichlet composition can be regarded as the composition derived from a basis $(w_1, \ldots, w_D)$ of gamma-distributed components. The fact that the gamma components are independent and have equal scale parameters indicates that the components of a Dirichlet composition have a special and near-independence structure, with correlations between components arising solely from the division by the common $t = w_1 + \cdots + w_D$ at the constraining operation $\mathscr{C}(\mathbf{w})$. Indeed, the weakness of Dirichlet modelling lies in this strength of structure which leads also to the variety of automatic independence properties. As we have already seen in Section 3.4, Dirichlet modelling may be too simple to be realistic in the analysis of compositional data. If there is an underlying basis to the composition under study then clearly the Dirichlet model is unlikely to be appropriate if the basis has dependent components.

Although the logistic normal class also has some nice class-preserving properties, it has a richness which does not lead automatically to the kinds of independence displayed by the Dirichlet class. To take a specific example, recall that with $\mathbf{x}$ distributed as $\mathscr{D}^d(\boldsymbol{\alpha})$ the subcompositions $\mathbf{s}_1 = \mathscr{C}(\mathbf{x}^{(c)})$ and $\mathbf{s}_2 = \mathscr{C}(\mathbf{x}_{(c)})$ are independent, by Property 3.4, whereas with $\mathbf{x}$ distributed as $\mathscr{L}^d(\boldsymbol{\mu}, \boldsymbol{\Sigma})$ the conditional distribution property of $\mathbf{s}_1 \mid \mathbf{s}_2$ in Property 6.7 shows that $\mathbf{s}_1$ and $\mathbf{s}_2$ are independent if and only if $\mathbf{F}\boldsymbol{\Sigma}_{12}\mathbf{F}' = \mathbf{0}$. Thus if this form of independence is under scrutiny the logistic normal model provides a framework within which to test for independence through the parametric hypothesis $\mathbf{F}\boldsymbol{\Sigma}_{12}\mathbf{F}' = \mathbf{0}$, whereas this is an impossible procedure through Dirichlet modelling which subsumes the independence.

Thus the logistic normal class appears to have much in its favour compared with the Dirichlet class. These advantages are as follows.

1. The logistic normal class through its richness is better able to describe actual data patterns.
2. The logistic normal class can accommodate both dependent and independent structures so that, not only can such independence be tested as a parametric hypothesis but in the event of rejection of the hypothesis there is the possibility of fitting a model to describe the dependence established.
3. The logistic normal class provides, through its close relationship to the multinormal class, a ready means of tractable statistical analysis. In addition to simple estimation and hypothesis testing of the $\boldsymbol{\mu}$ and $\boldsymbol{\Sigma}$ parameters, tests of logistic normality, linear modelling of expectations to take account of experimental design and possible concomitant factors, and all the special multivariate techniques such as discriminant analysis, Bayesian statistical analysis is also directly available through the normal-Wishart class of conjugate prior distributions for $\boldsymbol{\mu}$ and $\boldsymbol{\tau} = \boldsymbol{\Sigma}^{-1}$. In particular, predictive density functions are easily derived in terms of new and easily defined classes of distributions such as logistic Student distributions over $\mathscr{S}^d$.

All these points will become much clearer as we begin to apply logistic normal modelling to real compositional data in Chapter 7.

6.13 The multiplicative logistic normal class

The additive logistic transformation of Definition 6.1 is by no means the only one-to-one transformation from $\mathscr{S}^d$ to $\mathscr{R}^d$, though its simplicity has certainly been responsible for the many useful properties we have found for the additive logistic normal class. We have seen, however, that the additive logistic normal class is not well suited to compositional problems involving partitions when more than the subcompositions of the partition is involved. In this and the next section we shall discuss briefly two other transformations – multiplicative logistic transformations and partitioned logistic transformations – which are more suited to this purpose.

For the D-part composition $\mathbf{x}$ and the single division

$$(\mathbf{x}^{(c)}|\mathbf{x}_{(c)}) = (x_1, \ldots, x_c | x_C, \ldots, x_D), \tag{6.27}$$

the additive logistic normal class allows us easily to discuss $\mathbf{s}_1 = \mathscr{C}(\mathbf{x}^{(c)})$ and $\mathbf{s}_2 = \mathscr{C}(\mathbf{x}_{(c)})$ but has little to offer concerning, for example, the relationship of $\mathbf{s}_2$ to $\mathbf{x}^{(c)}$, the actual components of the first c parts. We now introduce a new transformation which allows this facility.

Definition 6.5 The *multiplicative logistic transformation* is the one-to-one transformation from $\mathbf{y} \in \mathscr{R}^d$ to $\mathbf{x} \in \mathscr{S}^d$ defined by

$$x_i = \exp(y_i)/[\{1 + \exp(y_1)\} \ldots \{1 + \exp(y_i)\}] \qquad (i = 1, \ldots, d)$$
$$x_D = 1/[\{1 + \exp(y_1)\} \ldots \{1 + \exp(y_d)\}]$$

with inverse the *multiplicative logratio transformation*,

$$y_i = \log\{x_i/(1 - x_1 - \cdots - x_i)\} \qquad (i = 1, \ldots, d)$$

and jacobian

$$\mathrm{jac}(\mathbf{y}|\mathbf{x}^{(d)}) = (x_1 \cdots x_D)^{-1}.$$

Note

1. Note that $\mathbf{x}^{(c)}$ determines $\mathbf{y}^{(c)} = (y_1, \ldots, y_c)$ uniquely and $\mathbf{y}^{(c)}$ determines $\mathbf{x}^{(c)}$ uniquely. Also the $D - c$ components of the subcomposition $\mathscr{C}(\mathbf{x}_{(c)})$ are given by

$$\exp(y_{c+j})/[\{1 + \exp(y_{c+1})\} \cdots \{1 + \exp(y_{c+j})\}]$$
$$(j = 1, \ldots, D - c - 1)$$

and

$$1/[\{1 + \exp(y_{c+1})\} \cdots \{1 + \exp(y_d)\}],$$

so that $\mathscr{C}(\mathbf{x}_{(c)})$ is determined by $\mathbf{y}_{(c)} = (y_{c+1}, \ldots, y_d)$. We shall find this property useful in Chapters 10 and 12.

The multiplicative logistic transformation can be used in exactly the same way as the additive logistic transformation to induce a class of distributions on the simplex $\mathscr{S}^d$ from the multinormal class on $\mathscr{R}^d$.

Definition 6.6 A D-part composition $\mathbf{x}$ is said to have a *multiplicative* *logistic normal* distribution $\mathscr{M}^d(\boldsymbol{\mu}, \boldsymbol{\Sigma})$ when $\mathbf{y}^{(d)}$, defined by

$$y_i = \log\{x_i/(1 - x_1 - \cdots - x_i)\} \qquad (i = 1, \ldots, d),$$

has a $\mathscr{N}^d(\boldsymbol{\mu}, \boldsymbol{\Sigma})$ distribution.

Notes

1. The definition of $\mathscr{M}^d(\boldsymbol{\mu}, \boldsymbol{\Sigma})$ is clearly dependent on a particular ordering of the parts, and so the $\mathscr{M}^d$ class will have none of the permutation properties of the $\mathscr{L}^d$ class. Indeed the strength of the $\mathscr{M}^d$ class is in its direction to problems where there is a specific ordering in mind.
2. The $\mathscr{M}^d$ class has very few nice properties compared with the $\mathscr{L}^d$ class, but the following one is important for forms of analysis with ordered parts.

Property 6.12 Suppose that a D-part composition $\mathbf{x}$ is distributed as $\mathscr{M}^d(\boldsymbol{\mu}, \boldsymbol{\Sigma})$ and let the $(c, D - C)$ partition of $\boldsymbol{\mu}$ and $\boldsymbol{\Sigma}$ be

$$\begin{bmatrix} \boldsymbol{\mu}_1 \\ \boldsymbol{\mu}_2 \end{bmatrix}, \qquad \begin{bmatrix} \boldsymbol{\Sigma}_{11} & \boldsymbol{\Sigma}_{12} \\ \boldsymbol{\Sigma}_{21} & \boldsymbol{\Sigma}_{22} \end{bmatrix}$$

The following distribution properties then hold.

(a) The amalgamation $(\mathbf{x}^{(c)}, \mathbf{j}'\mathbf{x}_{(c)})$ is distributed as $\mathscr{M}^c(\boldsymbol{\mu}_1, \boldsymbol{\Sigma}_{11})$.
(b) The subcomposition $\mathscr{C}(\mathbf{x}_{(c)})$ is distributed as $\mathscr{M}^{D-C}(\boldsymbol{\mu}_2, \boldsymbol{\Sigma}_{22})$.
(c) The conditional distribution of $\mathscr{C}(\mathbf{x}_{(c)})$ given $\mathbf{x}^{(c)}$ is given by $\mathscr{M}^{D-C}(\boldsymbol{\mu}_{2.1}, \boldsymbol{\Sigma}_{2.1})$, where

$$\boldsymbol{\mu}_{2.1} = \boldsymbol{\mu}_2 + \boldsymbol{\Sigma}_{21}\boldsymbol{\Sigma}_{11}^{-1}(\mathbf{y}^{(c)} - \boldsymbol{\mu}_1),$$
$$\boldsymbol{\Sigma}_{2.1} = \boldsymbol{\Sigma}_{22} - \boldsymbol{\Sigma}_{21}\boldsymbol{\Sigma}_{11}^{-1}\boldsymbol{\Sigma}_{12},$$

and $\mathbf{y}^{(c)} = (y_1, \ldots, y_c)$, the first c of the logratios of Definition 6.5.

6.14 Partitioned logistic normal classes

If our interest is in the subcompositions and amalgamation associated with the single division (6.27), then we require to look for a transformation which is appropriate to handling simultaneously $\mathbf{t} = (\mathbf{j}'\mathbf{x}^{(c)}, \mathbf{j}'\mathbf{x}_{(c)}), \mathbf{s}_1 = \mathscr{C}(\mathbf{x}^{(c)}), \mathbf{s}_2 = \mathscr{C}(\mathbf{x}_{(c)})$. If we view the transformation problem here along the lines of Property 2.10 then we see the partition transformation as a one-to-one transformation taking us out from the original simplex sample space to a product space

$$\mathscr{S}^d \leftrightarrow \mathscr{S}^1 \times \mathscr{S}^{c-1} \times \mathscr{S}^{d-c}, \tag{6.28}$$

where the three simplexes on the right refer in sequence to $\mathbf{t}, \mathbf{s}_1, \mathbf{s}_2$. If we are to exploit the technique of transforming multinormal distributions from $\mathscr{R}^d$ to suit this partition problem it is clear that we have to involve this product space. Fortunately we can achieve this very simply. For example, if we apply separate logistic transformations between $\mathscr{R}^1$ and $\mathscr{S}^1$, between $\mathscr{R}^{c-1}$ and $\mathscr{S}^{c-1}$, between $\mathscr{R}^{d-c}$ and $\mathscr{S}^{d-c}$, the composite transformation

$$\begin{array}{ccccc} \mathscr{S}^1 & \times \mathscr{S}^{c-1} & \times \mathscr{S}^{d-c} & & \\ \updownarrow & \updownarrow & \updownarrow & & \\ \mathscr{R}^1 & \times \mathscr{R}^{c-1} & \times \mathscr{R}^{d-c} & = \mathscr{R}^d & \end{array} \tag{6.29}$$

is one-to-one between the product of simplex spaces and the product of three real spaces which, unlike its simplex counterpart, is simply d-dimensional real space $\mathscr{R}^d$.

For example, if we take these three transformations to be of the additive logistic and logratio form of Definition 6.1 then the three logratio transformations in (6.29) take the form

$$\begin{aligned} y_0 &= \log\{t/(1-t)\}, \\ y_{1i} &= \log(s_{1i}/s_{1c}) \quad (i=1,\ldots,c-1), \\ y_{2i} &= \log(s_{2i}/s_{2,D-c}) \quad (i=1,\ldots,d-c). \end{aligned} \tag{6.30}$$

The corresponding transformation from $\mathscr{R}^d$ to the product simplex space is easily obtained as

$$\begin{aligned} t &= \exp(y_0)/\{\exp(y_0)+1\} \\ s_{1i} &= \exp(y_{1i})/\{\exp(y_{11})+\cdots+\exp(y_{1,c-1})+1\} \\ &\qquad\qquad (i=1,\ldots,c-1), \\ s_{1c} &= 1/\{\exp(y_{11})+\cdots+\exp(y_{1,c-1})+1\}, \\ s_{2i} &= \exp(y_{2i})/\{\exp(y_{21})+\cdots+\exp(y_{2,d-c})+1\} \\ &\qquad\qquad (i=1,\ldots,d-c), \\ s_{2,D-c} &= 1/\{\exp(y_{21})+\cdots+\exp(y_{2,d-c})+1\}. \end{aligned} \tag{6.31}$$

Using now the inverse of the partition transformation of Definition 2.12 in the special form at (6.28), we obtain the one-to-one transformation from $\mathscr{R}^d$ to $\mathscr{S}^d$:

$$x_i = \frac{\exp(y_0)}{\exp(y_0)+1} \cdot \frac{\exp(y_{1i})}{\exp(y_{11})+\cdots+\exp(y_{1,c-1})+1} \qquad (i=1,\ldots,c-1),$$

$$x_c = \frac{\exp(y_0)}{\exp(y_0)+1} \cdot \frac{1}{\exp(y_{11}) + \cdots + \exp(y_{1,c-1}) + 1}$$

$$x_{c+i} = \frac{1}{\exp(y_0)+1} \cdot \frac{\exp(y_{2i})}{\exp(y_{21}) + \cdots + \exp(y_{2,d-c}) + 1} \qquad (i = 1, \ldots, d-c),$$

$$x_D = \frac{1}{\exp(y_0)+1} \cdot \frac{1}{\exp(y_{21}) + \cdots + \exp(y_{2,d-c}) + 1}. \tag{6.32}$$

If we now place an $\mathcal{N}^d(\boldsymbol{\mu}, \boldsymbol{\Sigma})$ distribution on the d-dimensional vector

$$\mathbf{y} = (y_0, \mathbf{y}_1, \mathbf{y}_2) = (y_0, y_{11}, \ldots, y_{1,c-1}, y_{21}, \ldots, y_{2,d-c}) \in \mathscr{R}^d$$

we induce a distribution for the corresponding composition $\mathbf{x}$ in $\mathscr{S}^d$.

This may appear a complex operation but in practice it is extremely simple to apply. Transformation (6.30) is the key to the whole process: all that we are really doing is to introduce separate logratio transformations for the amalgamation and each of the subcompositions of the partition, and work on the assumption that the d-dimensional vector made out of these logratios is distributed as $\mathcal{N}^d(\boldsymbol{\mu}, \boldsymbol{\Sigma})$. Within this multinormal variability these logratios may be dependent, and so allowance is made for the description and testing of dependence between the subcompositions and also between the amalgamation and the subcompositions. This is in complete contrast to a Dirichlet-distributed composition where, by Property 3.4, only independence is possible.

In our example above we have used additive logistic transformations on each of the subcompositions $\mathbf{s}_1$ and $\mathbf{s}_2$, but clearly either or both of these can be replaced by the multiplicative logistic transformation if such a transformation is suggested by the nature of the problem. We finally therefore extend the concepts in an obvious way to the general partition and introduce a terminology and notation for describing partitioned logistic normal distributions.

Definition 6.7 Suppose that the general partition of Definition 2.12 is applied to the D-part composition $\mathbf{x}$ to give $(\mathbf{t}; \mathbf{s}_1, \ldots, \mathbf{s}_C)$. Suppose also that logratio transformations $l_0, l_1, \ldots, l_C$, which may be of either of the types associated with additive or multiplicative logistic transformations, are applied to $\mathbf{t}$, $\mathbf{s}_1, \ldots, \mathbf{s}_C$ to produce logratios

$$\mathbf{y}_0 = l_0(\mathbf{t}); \qquad \mathbf{y}_i = l_i(\mathbf{s}_i) \qquad (i = 1, \ldots, C).$$

If $\mathbf{y} = (\mathbf{y}_0, \mathbf{y}_1, \ldots, \mathbf{y}_C)$ is distributed as $\mathcal{N}^d(\boldsymbol{\mu}, \boldsymbol{\Sigma})$ then we say that the composition $\mathbf{x}$ has a *partitioned logistic normal* distribution

$$\mathscr{P}^d(l_0; l_1, \ldots, l_C; \boldsymbol{\mu}, \boldsymbol{\Sigma}).$$

Note

1. In our example above we obtained

$$\mathscr{P}^d(l_0, l_1, l_2; \boldsymbol{\mu}, \boldsymbol{\Sigma})$$

where l_0, l_1, l_2 were of additive form taking $\mathscr{S}^1 \to \mathscr{R}^1$, $\mathscr{S}^{c-1} \to \mathscr{R}^{c-1}$ and $\mathscr{S}^{d-c} \to \mathscr{R}^{d-c}$, respectively.

6.15 Some notation

It is useful to have a simple notation for logratio and logistic transformations. For logratio transformations we define two vector functions alr(·) and mlr(·) corresponding to the additive and multiplicative versions of Definitions 6.1 and 6.5, respectively.

Definition 6.8 The *additive logratio function* alr(·) from $\mathscr{R}_+^D$ to $\mathscr{R}^D$ is defined by

$$\operatorname{alr}(\mathbf{w}) = (\log(w_i/w_D)\colon i = 1, \ldots, d) \qquad (\mathbf{w} \in \mathscr{R}_+^D).$$

The *multiplicative logratio function* mlr(·) from $\mathscr{R}_+^D$ to $\mathscr{R}^d$ is defined by

$$\operatorname{mlr}(\mathbf{w}) = (\log\{w_i/(w_{i+1} + \cdots + w_D)\}\colon i = 1, \ldots, d) \quad (\mathbf{w} \in \mathscr{R}_+^D).$$

Although our applications of logratio transformations have so far been confined to compositions and subcompositions, it is convenient to have this extension of domain from $\mathscr{S}^d$ to $\mathscr{R}_+^D$. For example, instead of writing $\operatorname{alr}\{\mathscr{C}(\mathbf{x}^{(c)})\}$ we can write $\operatorname{alr}(\mathbf{x}^{(c)})$ since

$$\operatorname{alr}(\mathbf{x}^{(c)}) = (\log(x_i/x_c)\colon i = 1, \ldots, c-1).$$

Similarly,

$$\begin{aligned}\operatorname{alr}(\mathbf{x}_{(c)}) &= (\log(x_{c+i}/x_D)\colon i = 1, \ldots, d-c) \\ &= \operatorname{alr}\{\mathscr{C}(\mathbf{x}_{(c)})\}.\end{aligned}$$

Note that the common divisor is always the final component of the positive vector.

Similarly,

$$\begin{aligned}\text{mlr}(\mathbf{x}) &= (\log\{x_i/(x_{i+1}+\cdots+x_D)\}:i=1,\ldots,d)\\ &= (\log\{x_i/(1-x_1-\cdots-x_i)\}:i=1,\ldots,d),\end{aligned}$$

in conformity with Definition 6.5. Also

$$\begin{aligned}\text{mlr}(\mathbf{x}^{(c)}) &= (\log\{x_i/(x_{i+1}+\cdots+x_c)\}:i=1,\ldots,c-1),\\ &= \text{mlr}\{\mathscr{C}(\mathbf{x}^{(c)})\},\end{aligned}$$

and similarly,

$$\text{mlr}(\mathbf{x}_{(c)}) = \text{mlr}\{\mathscr{C}(\mathbf{x}_{(c)})\}.$$

For logistic transformations we similarly have additive and multiplicative forms, the vector functions agl(·) and mgl(·), corresponding to Definitions 6.1 and 6.5, respectively.

Definition 6.9 The *additive generalized logistic function* agl(·) from $\mathscr{R}^d$ to $\mathscr{S}^d$ is defined by

$$\text{agl}(\mathbf{y}) = \mathbf{x} \qquad (\mathbf{y}\in\mathscr{R}^d),$$

where

$$\begin{aligned}x_i &= \exp(y_i)/\{\exp(y_1)+\cdots+\exp(y_d)+1\} \qquad (i=1,\ldots,d),\\ x_D &= 1/\{\exp(y_1)+\cdots+\exp(y_d)+1\}.\end{aligned}$$

The *multiplicative generalized logistic function* mgl(·) from $\mathscr{R}^d$ to $\mathscr{S}^d$ is defined by

$$\text{mgl}(\mathbf{y}) = \mathbf{x} \qquad (\mathbf{y}\in\mathscr{R}^d),$$

where

$$\begin{aligned}x_i &= \exp(y_i)/[\{1+\exp(y_1)\}\cdots\{1+\exp(y_i)\}] \qquad (i=1,\ldots,d),\\ x_D &= 1/[\{1+\exp(y_1)\}\cdots\{1+\exp(y_d)\}].\end{aligned}$$

6.16 Bibliographic notes

The idea of the additive logistic normal distribution seems to have made its first appearance in an unpublished technical report by Obenchain (1970), who was concerned with a practical problem involving compositions in the form of activity patterns of technicians. The idea was communicated to, and mentioned by, Johnson and Kotz

(1972, p. 20), but Obenchain did not develop his ideas further and did not seek publication because of his concern with the problem of zero components (Aitchison, 1982, Discussion p. 170). The $\mathscr{L}^d$ class of distributions and its properties and uses and comparison with the $\mathscr{D}^d$ class were then studied by Aitchison and Shen (1980), who trace a number of earlier implicit uses.

1. For the case $d = 1$ as a Johnson (1949) four-parameter lognormal distribution with the two range parameters determining the interval (0, 1).
2. In the Bayesian analysis of multinomial and contingency table data in the use of normal approximations to logcontrasts by Lindley (1964), Swe (1964), Bloch and Watson (1967); and as the first stage in the construction of exchangeable prior distributions by Leonard (1973).
3. In studies of size and shape in biological allometry, for example in Mosimann (1975a), as the distribution of ratios of lognormally distributed measurements.
4. In statistical diagnosis where classification of the basic cases is subject to uncertainty, as in Aitchison and Begg (1976), who provide an explicit definition of the class of logistic normal distributions.
5. In the reconciliation of subjective probability assessments, where Lindley, Tversky and Brown (1979) use normal log-odds models to describe assessments.

The operation of perturbation and its relevance to a central limit theorem, together with the multiplicative and partitioned logistic normal classes, were introduced by Aitchison (1982).

Problems

6.1 Show, for $d > 1$, that the transformation

$$y_i = \log \frac{x_i}{1 - x_i} \qquad (i = 1, \ldots, d)$$

maps $\mathscr{S}^d$ onto a proper subset of $\mathscr{R}^d$ and not onto the whole of $\mathscr{R}^d$.

Sketch this proper subset for the case $d = 2$ and show that this transformation is not suitable for defining a transformed-normal class of distributions on the simplex.

6.2 Show that for a 2-part composition $\mathbf{x}$ distributed as $\mathscr{L}^1(\mu, \sigma^2)$ the first moment $E(x_1)$ can be expressed as the integral

$$\int_{-\infty}^{\infty} \frac{\exp(\mu + \sigma v)}{\exp(\mu + \sigma v) + 1} \phi(v) dv,$$

where $\phi(\cdot)$ is the density function of a $\mathcal{N}(0, 1)$ distribution.

6.3 The 4-part composition $\mathbf{x}$ has a $\mathscr{L}^3(\boldsymbol{\mu}, \boldsymbol{\Sigma})$ distribution with

$$\boldsymbol{\mu} = \begin{bmatrix} 0.7 \\ 0.8 \\ 0.9 \end{bmatrix}, \quad \boldsymbol{\Sigma} = \begin{bmatrix} 0.4 & -0.2 & -0.3 \\ -0.2 & 0.5 & 0.1 \\ -0.3 & 0.1 & 0.6 \end{bmatrix}.$$

What are the distributions of

(i) (x_2, x_4, x_3, x_1),
(ii) $\mathscr{C}(x_1, x_2, x_3)$,
(iii) $\mathscr{C}(x_1, x_2, x_4)$ conditioned on $\mathscr{C}(x_1, x_3)$,
(iv) $\mathscr{C}(x_2, x_4)$ conditioned on $\mathscr{C}(x_1, x_3)$?

6.4 A 4-part basis has a $\Lambda^4(\boldsymbol{\zeta}, \boldsymbol{\Omega})$ distribution with

$$\boldsymbol{\zeta} = \begin{bmatrix} 0.6 \\ 0.1 \\ -0.2 \\ 0.5 \end{bmatrix}, \quad \boldsymbol{\Omega} = \begin{bmatrix} 0.1 & 0.1 & 0.3 & -0.1 \\ 0.1 & 0.8 & 0.7 & 0.3 \\ 0.3 & 0.7 & 1.1 & 0.5 \\ -0.1 & 0.3 & 0.5 & 0.2 \end{bmatrix}.$$

Find the distribution of the composition of this basis.

Find a lognormal basis with independent components which could also have given rise to this compositional distribution.

6.5 A 2-part basis (w_1, w_2) is distributed with density function

$$(2\pi w_1 w_2)^{-1} \exp[-\tfrac{1}{2}\{\log(w_1/w_2)\}^2 - \tfrac{1}{2}\{\log(w_1 + w_2)\}^2] \qquad (w_1 > 0, w_2 > 0).$$

Show that its composition is distributed as $\mathscr{L}^1(0, 1)$. Show further that such a compositional distribution could also have arisen from a basis distributed as $\Lambda^2(\mathbf{0}, \mathbf{I})$, or from a basis with density function

$$(2\pi w_1 w_2)^{-1} \exp[-\{\log(w_1/w_2)\}^2 - \log(w_1/w_2)\log(w_1 + w_2) - \tfrac{1}{2}\{\log(w_1 + w_2)\}^2].$$

Do such coincidences surprise you?

6.6 A D-part composition $\mathbf{x}$ has a $\mathscr{D}^d(\boldsymbol{\alpha})$ distribution and its

Kullback–Leibler logistic normal approximation is $\mathscr{L}^d(\boldsymbol{\mu}, \boldsymbol{\Sigma})$. The subcomposition $\mathscr{C}(\mathbf{x}^{(c)})$ has a $\mathscr{D}^{c-1}(\boldsymbol{\alpha}^{(c)})$ distribution and its Kullback–Leibler logistic normal approximation is $\mathscr{L}^{c-1}(\boldsymbol{\mu}_s, \boldsymbol{\Sigma}_s)$. Show that this subcompositional approximation is identical to the subcompositional distribution for $\mathscr{C}(\mathbf{x}^{(c)})$ obtained directly from $\mathscr{L}^d(\boldsymbol{\mu}, \boldsymbol{\Sigma})$.

6.7 The 5-part composition $\mathbf{x}$ has a $\mathscr{M}^4(\boldsymbol{\mu}, \boldsymbol{\Sigma})$ distribution with

$$\boldsymbol{\mu} = \begin{bmatrix} 0.1 \\ 0.2 \\ 0.3 \\ 0.4 \end{bmatrix}, \qquad \boldsymbol{\Sigma} = \begin{bmatrix} 0.4 & 0.1 & 0.2 & -0.3 \\ 0.1 & 0.5 & -0.1 & -0.2 \\ 0.2 & -0.1 & 0.6 & 0.3 \\ -0.3 & -0.2 & 0.3 & 0.7 \end{bmatrix}.$$

Find the distributions of

(i) the amalgamation $(x_1, x_2, x_3 + x_4 + x_5)$,
(ii) the subcomposition $\mathscr{C}(x_3, x_4, x_5)$,
(iii) the subcomposition $\mathscr{C}(x_3, x_4, x_5)$ conditioned on the subvector (x_1, x_2),
(iv) the subcomposition $\mathscr{C}(x_4, x_5)$ conditioned on the subvector (x_1, x_2).

6.8 For a D-part composition distributed as $\mathscr{D}^d(\boldsymbol{\alpha})$ show that the Kullback–Leibler multiplicative logistic-normal approximation $\mathscr{M}^d(\boldsymbol{\mu}, \boldsymbol{\Sigma})$ is given by

$$\mu_i = \psi(\alpha_i) - \psi(\alpha_{i+1} + \cdots + \alpha_D) \qquad (i = 1, \ldots, d),$$
$$\sigma_{ii} = \psi'(\alpha_i) + \psi'(\alpha_{i+1} + \cdots + \alpha_D) \qquad (i = 1, \ldots, d),$$
$$\sigma_{ij} = 0 \qquad (i \neq j).$$

6.9 For the partition $(x_1, x_2, x_3 | x_4, x_5, x_6)$ and the partitioned logistic normal distribution $\mathscr{P}^5(\text{alr}; \text{mlr}, \text{alr}; \boldsymbol{\mu}, \boldsymbol{\Sigma})$ construct the composite logratio transformation from $\mathbf{x} \in \mathscr{S}^5$ to $\mathbf{y} \in \mathscr{R}^5$, and determine its inverse.

6.10 Show that the transformation from $\mathbf{x} \in \mathscr{R}^d$ to $\mathbf{y} \in \mathscr{S}^d$ defined by

$$x_i = \frac{\exp\left(\sum_{j=i}^{d} y_j\right)}{\prod_{j=i-1}^{d} \{1 + \exp(y_j)\}} \qquad (i = 2, \ldots, d),$$

$$x_D = \frac{1}{1 + \exp(y_d)}, \quad x_1 = 1 - x_2 - \cdots - x_D,$$

is one-to-one, and determine its inverse.

Show further that $\mathscr{C}(\mathbf{x}^{(c)})$ and $\mathbf{x}_{(c)}$ are in one-to-one correspondence with $\mathbf{y}^{(c-1)}$ and $y_{(c-1)}$, and consider how you might use this to investigate the independence of $\mathscr{C}(\mathbf{x}^{(c)})$ and $\mathbf{x}_{(c)}$.

6.11 Find the inverse and the jacobian of the transformation from $\mathscr{S}^d$ to $\mathscr{R}^d$ defined by

$$y_1 = \log\frac{x_1}{1-x_1}$$

$$y_i = \log\frac{x_i}{(1-x_1-\cdots-x_{i-1})(1-x_1-\cdots-x_i)} \qquad (i=2,\ldots,d).$$

Derive the density function in the simplex induced by the selection of $\mathscr{N}^d(\boldsymbol{\mu},\boldsymbol{\Sigma})$ for $\mathbf{y}$.

Show that

$$x_{c+1}+\cdots+x_D = 1/\{1+\exp(y_1)+\cdots+\exp(y_c)\}.$$

Under what circumstances will $x_d/\{x_D(x_d+x_D)\}$ be independent of $x_{c+1}+\cdots+x_D$?

6.12 Consider the following transformation between $\mathscr{S}^d$ and $\mathscr{R}^d$:

$$x_1 = \frac{f(y_1)}{1+f(y_1)},$$

$$x_2 = \frac{f(y_2)}{\{1+f(y_1)\}\{1+f(y_1)+f(y_2)\}},$$

$$\vdots$$

$$x_d = \frac{f(y_d)}{\{1+f(y_1)+\cdots+f(y_{d-1})\}\{1+f(y_1)+\cdots+f(y_d)\}},$$

where f is a one-to-one transformation from $\mathscr{R}^1$ to $\mathscr{R}^1_+$. Show that the jacobian $\operatorname{jac}(\mathbf{x}\,|\,\mathbf{y})$ can be expressed as

$$\frac{g_1 g_2 \cdots g_d}{(1+f_1)^2(1+f_1+f_2)^2\cdots(1+f_1+\cdots+f_d)^2},$$

where $f_i = f(y_i)$, $g_i = f'(y_i)$.

Express the inverse transformation from $\mathscr{R}^d$ to $\mathscr{S}^d$ in terms of h, the inverse function of f.

CHAPTER 7

Logratio analysis of compositions

A judge is not supposed to know anything about the facts of life until they have been presented in evidence and explained to him at least three times.

Lord Chief Justice Parker

7.1 Introduction

We now turn our attention towards finding answers to some of the questions raised in Chapter 1, in particular in relation to the hongite and kongite compositions of Data 1 and 2, the Arctic lake sediment data of Data 5 and the behavioural compositions of Data 7. We have already seen in Section 4.3 the use of the compositional variation array, consisting of logratio means and variances, as a descriptive tool for the study of compositional variability. Now we concentrate on building up a repertoire of more analytical tools involving estimation, hypothesis testing, and other statistical procedures, based on the additive logistic normal class of distributions. Our strategy for such analyses is simple and straightforward: convert the compositions to additive logratio compositions, reformulate the underlying problem in terms of these logratio compositions and make use, wherever possible, of standard multivariate statistical procedures. The examples of this chapter will illustrate this strategy.

A few comments on some expository decisions may be helpful. In making use of the logistic normal class we have to face the choice of which of the three specifications, $\boldsymbol{T}$, $\boldsymbol{\Sigma}$ and $\boldsymbol{\Gamma}$, of covariance structure we adopt. To avoid the complications of singular distributions and the computation of pseudo-inverses of singular matrices and to retain the familiar idea of a covariance matrix specification of covariance structure, we shall concentrate on the asymmetric form of logratio transformation,

$$\mathbf{y} = \text{alr}(\mathbf{x}) = \log(\mathbf{x}_{-D}/x_D), \tag{7.1}$$

and the associated form $\mathscr{L}^d(\boldsymbol{\mu}, \boldsymbol{\Sigma})$ of logistic normal distribution. The $N \times D$ compositional data matrix $\mathbf{X}$ is thereby transformed to an $N \times d$ matrix

$$\mathbf{Y} = [y_{ri}: r = 1, \ldots, N; i = 1, \ldots, d] = [\log(x_{ri}/x_{rD})]. \tag{7.2}$$

We remind the reader that the use of the logratio composition (7.1) with its choice of the special divisor x_D raises the question of the invariance of any statistical procedure used. Since all the statistical quantities involved in this chapter take only forms known from the discussion of Section 5.5 to be invariant under the group of permutations of the parts, we are assured that the results of the analyses do not depend on the particular order of the parts or the choice of common component divisor.

For the reader who has avoided the technicalities, we briefly record that, under a permutation $\mathbf{P}$ of the parts $1, \ldots, D$, the new logratio $\mathbf{y}_P$, formed by using the final component of the permuted composition as divisor, is related to the original $\mathbf{y}$ by

$$\mathbf{y}_P = \mathbf{Q}_P \mathbf{y}, \tag{7.3}$$

where

$$\mathbf{Q}_P = \mathbf{FPF'H}^{-1}, \tag{7.4}$$

and $\mathbf{F}$ and $\mathbf{H}$ are elementary matrices defined in Appendix A. The corresponding mean vector $\boldsymbol{\mu}_P = \text{E}(\mathbf{y}_P)$ and covariance matrix $\boldsymbol{\Sigma}_P = \text{V}(\mathbf{y}_P)$ are then related to $\boldsymbol{\mu} = \text{E}(\mathbf{y})$ and $\boldsymbol{\Sigma} = \text{V}(\mathbf{y})$ by

$$\boldsymbol{\mu}_P = \mathbf{Q}_P \boldsymbol{\mu}, \tag{7.5}$$

$$\boldsymbol{\Sigma}_P = \mathbf{Q}_P \boldsymbol{\Sigma} \mathbf{Q}'_P. \tag{7.6}$$

The constructs which occur in the statistical procedures of this chapter are $|\boldsymbol{\Sigma}|$ and quadratic or bilinear forms such as $\mathbf{a}'\boldsymbol{\Sigma}^{-1}\mathbf{b}$, where permuted forms $\mathbf{a}_P$ and $\mathbf{b}_P$ of $\mathbf{a}$ and $\mathbf{b}$ are related in the same way as $\mathbf{y}_P$ to $\mathbf{y}$, namely $\mathbf{a}_P = \mathbf{Q}_P\mathbf{a}$, $\mathbf{b}_P = \mathbf{Q}_P\mathbf{b}$. Then

$$|\boldsymbol{\Sigma}_P| = |\boldsymbol{\Sigma}|,$$

by Property 5.3(a), and

$$\mathbf{a}'_P \boldsymbol{\Sigma}_P^{-1} \mathbf{b}_P = \mathbf{a}'\boldsymbol{\Sigma}^{-1}\mathbf{b},$$

by Property 5.4, ensure the invariance of the statistical procedures.

Table 7.1. *Estimates of $\boldsymbol{\mu}$ and $\boldsymbol{\Sigma}$ for hongite and kongite*

Data set	$\hat{\boldsymbol{\mu}}$	$\hat{\boldsymbol{\Sigma}}$
Hongite	$\begin{bmatrix} 1.715 \\ 0.914 \\ 0.115 \\ 0.167 \end{bmatrix}$	$\begin{bmatrix} 0.139 & 0.264 & -0.223 & 0.121 \\ 0.264 & 0.649 & -0.702 & 0.144 \\ -0.233 & -0.702 & 0.948 & 0.012 \\ 0.121 & 0.144 & 0.012 & 0.187 \end{bmatrix}$
Kongite	$\begin{bmatrix} 1.639 \\ 0.754 \\ 0.191 \\ 0.104 \end{bmatrix}$	$\begin{bmatrix} 0.113 & 0.235 & -0.201 & 0.096 \\ 0.235 & 0.655 & -0.723 & 0.106 \\ -0.201 & -0.723 & 1.050 & 0.091 \\ 0.096 & 0.106 & 0.091 & 0.195 \end{bmatrix}$

7.2 Estimation of $\boldsymbol{\mu}$ and $\boldsymbol{\Sigma}$

With the assumption that the pattern of compositional variability is of $\mathscr{L}^d(\boldsymbol{\mu}, \boldsymbol{\Sigma})$ form, the problem of estimation of $\boldsymbol{\mu}$ and $\boldsymbol{\Sigma}$ follows the standard multivariate normal procedure in terms of the logratio data matrix $\mathbf{Y}$, with estimates $\hat{\boldsymbol{\mu}} = [\hat{\mu}_i]$, $\hat{\boldsymbol{\Sigma}} = [\hat{\sigma}_{ij}]$ given by

$$\hat{\mu}_i = N^{-1} \sum_{r=1}^{N} y_{ri} \qquad (i = 1, \ldots, d), \tag{7.7}$$

$$\begin{aligned} \hat{\sigma}_{ij} &= n^{-1} \sum_{r=1}^{N} (y_{ri} - \hat{\mu}_i)(y_{rj} - \hat{\mu}_j) \\ &= n^{-1} \left[\sum_{r=1}^{N} y_{ri} y_{rj} - N \hat{\mu}_i \hat{\mu}_j \right], \end{aligned} \tag{7.8}$$

where $n = N - 1$, or equivalently in matrix form by

$$\hat{\boldsymbol{\mu}} = N^{-1} \mathbf{Y}'\mathbf{j}, \qquad \hat{\boldsymbol{\Sigma}} = n^{-1} \mathbf{Y}' \mathbf{G}_N \mathbf{Y}. \tag{7.9}$$

Table 7.1 records the results of these computations for the hongite and kongite compositions of Data 1 and 2.

7.3 Validation: tests of logistic normality

For any compositional data set we can readily test the validity of the distributional assumption of logistic normality by subjecting the logratio compositions to tests of multivariate normality. There is indeed an embarrassingly large battery of such tests with varying claims to appropriateness, and it is no easy matter to make a sensible selection. In our selection here we have depended heavily on Andrews, Gnadadesikan and Warner (1973) and the recommendations of Stephens (1974) for the use of empirical distribution function tests, in their Anderson–Darling, Cramer–von Mises, and Watson forms. For D-part compositions these are applied as tests associated with different dimensions for the d-dimensional logratio compositions, as follows:

1. all d marginal, univariate distributions,
2. all $\frac{1}{2}d(d-1)$ bivariate angle distributions,
3. the d-dimensional radius distribution.

Taking account of all three forms and of all the different dimensions we thus have a battery of $\frac{3}{2}dD+3$ tests for D-part compositions $(D \geqslant 3)$.

The computational forms of the empirical distribution function test statistics are the same for all the different dimensions and are presented for easy reference in Table 7.2. What differs from dimension to dimension is the meaning of $z_r\,(r=1,\ldots,N)$, and we now proceed to describe the various computations of these.

Marginal univariate distributions

For the ith marginal distribution of the logratio composition the observations are

$$y_{ri} = \log(x_{ri}/x_{rD}) \qquad (r = 1,\ldots,N). \tag{7.10}$$

With $\hat{\mu}_i$ and $\hat{\sigma}_{ii}$ the estimates of marginal mean and variance as computed by (7.7) and (7.8), evaluate

$$\Phi\{(y_{ri}-\hat{\mu}_i)/\sqrt{\hat{\sigma}_{ii}}\} \qquad (r = 1,\ldots,N), \tag{7.11}$$

Table 7.2. *Computational forms of empirical distribution function test statistics*

Anderson–Darling	$Q_A = -N^{-1}\sum_{r=1}^{N}(2r-1)\{\log z_r + \log(1-z_{N+1-r})\} - N$
Cramer–von Mises	$Q_C = \sum_{r=1}^{N}\{z_r-(2r-1)/(2N)\}^2 + 1/(12N)$
Watson	$Q_W = Q_C - N(\bar{z}-\frac{1}{2})^2$, where $N\bar{z} = \sum_{r=1}^{N} z_r$

where $\Phi(\cdot)$ is the distribution function of the $\mathcal{N}(0, 1)$ distribution. Then z_r $(r = 1, \ldots, N)$ denote the values of (7.11) arranged in ascending order of magnitude. The idea underlying the test here is that if the observations are indeed normally distributed then the z_r should be approximately the order statistics of a uniform distribution over the interval $(0, 1)$. The tests make such comparisons, making due allowance for the fact that in (7.11) the mean and variance are estimated.

Bivariate angle distributions

Corresponding to each (i, j) with $i < j$ we can form a set of bivariate observations (y_{ri}, y_{rj}) $(r = 1, \ldots, N)$. The test approach here is based on the following idea: if (u_1, u_2) is circular normal, that is distributed as $\mathcal{N}^2(\mathbf{0}, \mathbf{I}_2)$ then the radian angle between the vector from $(0, 0)$ to (u_1, u_2) and the u_1-axis is distributed uniformly over the interval $(0, 2\pi)$. Since any bivariate normal distribution can be reduced to a circular normal distribution by a suitable transformation, we can apply such a transformation to the bivariate observations and ask if the resulting angles have the uniform property. The transformation has, of course, to be built on the estimates $\hat{\boldsymbol{\mu}}$ and $\hat{\boldsymbol{\Sigma}}$ of the mean vector and covariance matrix.

We arrive at the z_r by the following sequence.

1. First compute $\hat{\delta}_{ij} = \sqrt{(\hat{\sigma}_{ii}\hat{\sigma}_{jj} - \hat{\sigma}_{ij}^2)}$ and then u_r, v_r $(r = 1, \ldots, N)$ from

$$\hat{\delta}_{ij}u_r = (y_{ri} - \hat{\mu}_i)\sqrt{\hat{\sigma}_{jj}} - (y_{rj} - \hat{\mu}_j)\hat{\sigma}_{ij}/\sqrt{\hat{\sigma}_{jj}}, \tag{7.12}$$

$$\hat{\delta}_{ij}v_r = (y_{rj} - \hat{\mu}_j)\sqrt{(\hat{\sigma}_{ii} - \hat{\sigma}_{ij}^2/\hat{\sigma}_{jj})}. \tag{7.13}$$

2. Compute the radian angles $\hat{\theta}_r$ required to rotate the u_r-axis anticlockwise about the origin to reach the points (u_r, v_r). If $\arctan(t)$ denotes the angle between $-\frac{1}{2}\pi$ and $\frac{1}{2}\pi$ whose tangent is t, then

$$\hat{\theta}_r = \arctan(v_r/u_r) + \tfrac{1}{2}\{1 - \operatorname{sgn}(u_r)\}\pi + \tfrac{1}{4}\{1 + \operatorname{sgn}(u_r)\}\{1 - \operatorname{sgn}(v_r)\}\pi. \tag{7.14}$$

3. The z_r $(r = 1, \ldots, N)$ are then simply the values of $\hat{\theta}_r/(2\pi)$ in ascending order of magnitude.

The test statistics are then aimed at measuring departures of the z_r from the order statistics of a uniform distribution on $(0, 1)$ in a manner similar to that for the marginal distribution tests.

Table 7.3. *Modified empirical distribution function test statistics and their critical values*

Significance level (per cent)	*Anderson–Darling*	*Cramer–von Mises*	*Watson*
	Marginal tests		
	$Q_A\left[1+\frac{4}{N}-\frac{25}{N^2}\right]$	$Q_C\left[1+\frac{1}{2N}\right]$	$Q_W\left[1+\frac{1}{2N}\right]$
10	0.656	0.104	0.096
5	0.787	0.126	0.116
2.5	0.918	0.148	0.136
1	1.092	0.178	0.163
	Bivariate angle and radius tests		
	Q_A	$\left[Q_C-\frac{0.4}{N}+\frac{0.6}{N^2}\right]\times\left[1+\frac{1}{N}\right]$	$\left[Q_W-\frac{0.1}{N}+\frac{0.1}{N^2}\right]\times\left[1+\frac{0.8}{N}\right]$
10	1.933	0.347	0.152
5	2.492	0.461	0.187
2.5	3.070	0.581	0.221
1	3.857	0.743	0.267

Radius tests

We first compute the 'radii'

$$u_r = (\mathbf{y}_r - \hat{\boldsymbol{\mu}})'\hat{\boldsymbol{\Sigma}}^{-1}(\mathbf{y}_r - \hat{\boldsymbol{\mu}}) \qquad (r = 1, \ldots, N).$$

Under the assumption of multivariate normality of the $\mathbf{y}_r$ these radii are approximately distributed as $\chi^2(d)$. Then $z_r\,(r = 1, \ldots, N)$ are the values in ascending order of $F(u_r)$ $(r = 1, \ldots, N)$, where F is the distribution function of the $\chi^2(d)$ distribution, and the z_r are again to be compared against the order statistics of a uniform distribution on $(0, 1)$.

To test significant departures from additive logistic normality we then compare the modified test statistics as defined in Table 7.3 against the specified critical values (Stephens, 1974).

The testing procedures are illustrated by application to the 5-part hongite compositions of Data 1 and 3-part Arctic lake sediment compositions of Data 5. For the hongite data the computed values of the 33 modified empirical distribution function test statistics are shown in Table 7.4. Comparison of these with the corresponding critical values of Table 7.3 shows no significant departure from additive logistic normality at the 5 per cent level by any of these tests. We have thus found a satisfactory answer to Question 1.2.1. For the Arctic sediment data the results for the 12 tests are presented in Table 7.5. Here the picture is very different, with significant departures from additive logistic normality for both the marginal tests and for the Watson form of the bivariate angle test. We return to this data

Table 7.4. *Values of the modified test statistics for the hongite compositions*

		Anderson–Darling	Cramer–von Mises	Watson
Marginal	i			
	1	0.330	0.057	0.055
	2	0.256	0.033	0.032
	3	0.232	0.033	0.033
	4	0.325	0.038	0.037
Bivariate angle	i j			
	1 2	0.220	0.011	0.016
	1 3	0.750	0.090	0.027
	1 4	0.282	0.020	0.020
	2 3	0.701	0.090	0.024
	2 4	0.247	0.016	0.022
	3 4	0.236	0.012	0.023
Radius		0.600	0.092	0.099

Table 7.5. *Values of the modified test statistics for the Arctic lake sediment compositions*

		Anderson–Darling	Cramer–von Mises	Watson
Marginal	i			
	1	1.744	0.289	0.256
	2	3.294	0.549	0.471
Bivariate angle	i j			
	1 2	1.781	0.395	0.268
Radius		1.088	0.183	0.078

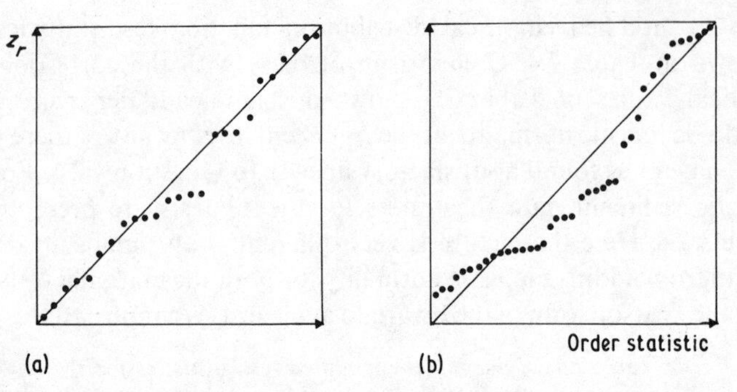

Fig. 7.1. *Plots of z_r against uniform order statistics for marginal y_1 tests of (a) hongite compositions, (b) Arctic lake sediments.*

set in Sections 7.6–7.8, where we shall investigate whether, by taking the depth of the sediment into consideration, we may yet describe the pattern of variability in logistic normal terms.

A visual representation of each test can be given in the form of a plot in the unit square of the z_r against the associated order statistic $(2r-1)/(2N)$ $(r = 1, \ldots, N)$ of the uniform distribution. Conformity with logistic normality corresponds to a pattern of points along the diagonal of the square. Figure 7.1(a) illustrates this conformity in the plot for the marginal y_1 test for the hongite compositions; Fig. 7.1(b) shows departure from logistic normality for the y_1 marginal test for the Arctic lake sediments.

7.4 Hypothesis testing strategy and techniques

Once the assumption of logistic normality is accepted or validated, the way is open to convert compositional data analysis with its sample space $\mathscr{S}^d$ into multinormal analysis with its sample space $\mathscr{R}^d$. Conversion of compositions into logratio vectors, with some care in reformulating the original compositional problem, will allow the reader to use any favourite multivariate statistical methodology. This is a monograph on compositional data analysis, not on the fundamentals of statistical inference, and we shall not attempt to dissuade anyone from any particular statistical philosophy. We have, however, the duty of illustrating how the logistic and logratio transformations allow us to resolve the compositional problems of Chapter 1 and it is only possible to achieve this within some specific statistical framework. In order that the reader may more easily follow the methodology in this and subsequent chapters we give some pointers to the particular hypothesis testing strategy and techniques underlying our analysis. The reader should then have no difficulty in converting the methodology to any preferred system.

Lattice of hypotheses

In most of our applications we shall be assuming that there is a sufficiently general parametric model which is the most complex we would consider as capable of explaining, or useful in explaining, the experienced pattern of variability. We are hesitant, however, to believe that the complexity of the model with its many parameters is really necessary and so postulate a number of hypotheses which provide a simpler explanation of the variability than the model. These hypotheses place constraints on the parameters of the model or equivalently allow a reparametrization of the situation in terms of fewer parameters than in the model. We can then usually show the hypotheses of interest and their relations of implication with respect to each other and the model in diagrammatic form in a lattice. The idea is most simply conveyed by a simple example.

Example

Suppose that our data set consists of the measurements of some characteristic of a sediment, such as specific gravity or, in a compositional problem, logratio of sand to clay components, at different depths in a lake bed. Suppose that our aim is to explore the nature of the dependence, if any, of characteristic y on depth u, and that we are prepared to assume that the most complex possible dependence is with expected characteristic of the form $\alpha + \beta u + \gamma u^2 + \delta \log u$. The lattice of Fig. 7.2 provides a number of possible hypotheses for investigation. Note the following features of such a lattice. The hypotheses and model have been arranged in a series of levels. At the highest level is the model with its four parameters; at the lowest level is the hypothesis of no dependence on depth, of essentially random unexplained variation of the characteristic, with only one parameter α representing the mean of the random variation. At intermediate levels are hypotheses of the same intermediate complexity, requiring the same number of parameters for their description: for example, the two hypotheses at level 2 correspond to a logarithmic dependence $\alpha + \delta \log u$ and linear dependence $\alpha + \beta u$ on depth. When

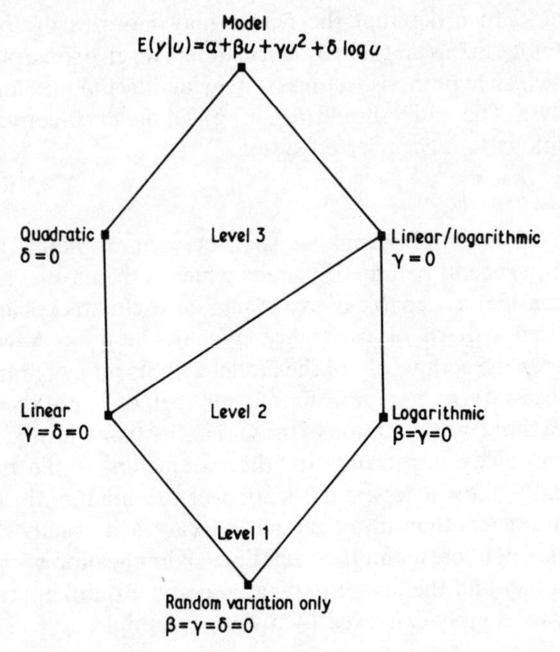

Fig. 7.2. *Lattice of hypotheses within the model with expected characteristic of the form* $\alpha + \beta u + \gamma u^2 + \delta \log u$.

a hypothesis at a lower level implies one at a higher level, the lattice shows a line joining the two hypotheses: for example, the hypothesis $\gamma = \delta = 0$ at level 2 implies $\gamma = 0$ and implies $\delta = 0$ at level 3 and so the two associated joins are made, whereas $\beta = \gamma = 0$ at level 2 does not imply $\delta = 0$ at level 3 and so no join is made. In short, the lattice displays clearly the relative simplicities and the hierarchy of implication of the hypotheses and their relation to the model.

Testing within a lattice

Once the model and relevant hypotheses have been set out in a lattice, how should we proceed to test the various hypotheses? The problem is clearly one of multiple hypotheses with no optimum solution unless we can frame it as a decision problem with a complete loss structure, a situation seldom realized for such problems. Some more *ad hoc* procedure is usually adopted. In our approach here we adopt the simplicity postulate of Jeffreys (1961), which within our context may be expressed as follows: we prefer a simple explanation, with few parameters, to a more complicated explanation, with many parameters. In terms of the lattice of hypotheses, therefore, we will want to see positive evidence before we are prepared to move from a hypothesis at a lower level to one at a higher level. In terms of standard Neyman–Pearson testing, the setting of the significance level ε at some low value may be viewed as placing some kind of protection on the hypothesis under investigation: if the hypothesis is true our test has only a small probability, at most ε, of rejecting it. With this protection, rejection of a hypothesis is a fairly positive act: we believe that we really have evidence against it. This is ideal for our view of hypothesis testing within a lattice of hypotheses under the simplicity postulate. In moving from a lower level to a higher level we are seeking a mandate to complicate the explanation, to introduce further parameters. The rejection of a hypothesis gives us a positive reassurance that we have reasonable grounds for moving to this more complicated explanation.

Our lattice testing procedure can then be expressed in terms of the following rules.

1. In every test of a hypothesis within the lattice, regard the model as the alternative hypothesis.
2. Start the testing procedure at the lowest level, by testing each hypothesis at that level within the model.
3. Move from one level to the next higher level only if all hypotheses at the lower level have been rejected.
4. Stop testing at the level at which the first non-rejection of a hypothesis occurs. All non-rejected hypotheses at that level are acceptable as 'working models' on which further analysis such as estimation and prediction may be based.

Construction of tests

For the construction of tests of a hypothesis h within a model m in an unfamiliar situation, we shall adopt the generalized likelihood ratio principle. In simple terms, let $L(\boldsymbol{\theta}|\mathbf{X})$ denote the likelihood of the parameter $\boldsymbol{\theta}$ for data $\mathbf{X}$ and $\hat{\boldsymbol{\theta}}_{\mathrm{h}}(\mathbf{X})$ and $\hat{\boldsymbol{\theta}}_{\mathrm{m}}(\mathbf{X})$ denote the maximum likelihood estimates and $L_{\mathrm{h}}(\mathbf{X}) = L\{\hat{\boldsymbol{\theta}}_{\mathrm{h}}(\mathbf{X})|\mathbf{X}\}$ and $L_{\mathrm{m}}(\mathbf{X}) = L\{\hat{\boldsymbol{\theta}}_{\mathrm{m}}(\mathbf{X})|\mathbf{X}\}$ denote the maximized likelihood under the hypothesis h and the model m, respectively. The generalized likelihood ratio test statistic is then

$$R(\mathbf{X}) = L_{\mathrm{m}}(\mathbf{X})/L_{\mathrm{h}}(\mathbf{X}), \tag{7.15}$$

and the larger this is, the more critical of the hypothesis h we shall be. When the exact distribution of this test statistic under the hypothesis h is not known, we shall make use of the Wilks (1938) asymptotic approximation: under the hypothesis h, which places c constraints on the parameters, the test statistic

$$Q(\mathbf{X}) = 2\log\{R(\mathbf{X})\} \tag{7.16}$$

is distributed approximately as $\chi^2(c)$.

Significance probabilities

All the tests which we shall apply, whether constructed especially as above or whether existing standard multivariate normal textbook tests, are of one of two types, with the test statistic $Q \equiv Q(\mathbf{X})$ distributed either as $\chi^2(\nu)$ or as $F(\nu_1, \nu_2)$. For each application of such a test we shall quote the significance probability $\Pr\{Q \geqslant q\}$ where q is the computed value of the test statistic. For the computation of these significance probabilities we have used the simple relationships of the χ^2 and F distributions to the gamma and beta distributions and then made use of simple algorithms for the incomplete gamma and incomplete beta distributions. We prefer this procedure for two reasons: first, we may require significant probabilities for fractional degrees of freedom of χ^2 or F and tabular values may not be so widely available; second, we shall find that the incomplete beta and gamma functions are convenient for other purposes. Before proceeding we require two definitions.

Definition 7.1 A positive random variable w is *gamma-distributed with index* k, written $\mathrm{Ga}(k)$, if it has density function

$$\frac{w^{k-1}e^{-w}}{\Gamma(k)} \qquad (w > 0).$$

Definition 7.2 A random variable w is *beta-distributed over the interval* (0, 1) *with indices* k *and* l, written $\mathrm{Be}(k, l)$, if it has density function

$$\frac{w^{k-1}(1-w)^{l-1}}{B(k,l)} \qquad (0 < w < 1).$$

The distributional relationships are then as follows.

Property 7.1
(a) If u is $\chi^2(v)$ then $\frac{1}{2}u$ is $\mathrm{Ga}(\frac{1}{2}v)$.
(b) If u is $\mathrm{F}(v_1, v_2)$ then $v_1 u/(v_2 + v_1 u)$ is $\mathrm{Be}(\frac{1}{2}v_1, \frac{1}{2}v_2)$.

Definition 7.3 The *incomplete gamma* and *beta functions* are defined as follows.

(a) $$\mathscr{I}_t(b) = \{\Gamma(b)\}^{-1} \int_0^t w^{b-1} \mathrm{e}^{-w} \mathrm{d}w \qquad (t > 0),$$

(b) $$\mathscr{I}_t(b, c) = \{B(b, c)\}^{-1} \int_0^t w^{b-1}(1-w)^{c-1} \mathrm{d}w \qquad (0 < t \leqslant 1).$$

We can then express the significance probabilities in the following way.

Property 7.2 If the computed value of a test statistic Q is q, then the significance probability $\Pr(Q \geqslant q)$ is given by

(a) $1 - \mathscr{I}_{q/2}(\frac{1}{2}v)$, when Q is distributed as $\chi^2(v)$,
(b) $1 - \mathscr{I}_{v_1 q/(v_2 + v_1 q)}(\frac{1}{2}v_1, \frac{1}{2}v_2) = \mathscr{I}_{v_2/(v_2 + v_1 q)}(\frac{1}{2}v_2, \frac{1}{2}v_1)$ when Q is distributed as $F(v_1, v_2)$.

7.5 Testing hypotheses about $\boldsymbol{\mu}$ and $\boldsymbol{\Sigma}$

In Question 1.2.6 and in our comparison in Section 4.3 of the compositional arrays for hongite and kongite we wondered whether there is any essential difference between the two patterns of variability, conjecturing that any difference is more likely to lie between the logratio mean vectors than between the logratio covariance structures. We are now in a position to be more specific about such questions, by testing a simple lattice of hypotheses. For our model we assume that the hongite and kongite compositions of Data 1 and 2 are distributed as $\mathscr{L}^4(\boldsymbol{\mu}_1, \boldsymbol{\Sigma}_1)$ and $\mathscr{L}^4(\boldsymbol{\mu}_2, \boldsymbol{\Sigma}_2)$, respectively. It then seems natural to consider the lattice of Fig. 7.3.

We develop the tests within the general context of D-part compositions with N_1 and N_2 of the two types 1 and 2. Maximum likelihood estimation of $\boldsymbol{\mu}_1, \boldsymbol{\mu}_2, \boldsymbol{\Sigma}_1, \boldsymbol{\Sigma}_2$ under the model and the three hypotheses of the lattice is well documented, for example by Mardia, Kent and Bibby (1979, Section 5.5.3). The primary computations from the logratio compositions y_{1r} $(r = 1, \ldots, N_1)$ and y_{2r} $(r = 1, \ldots, N_2)$ are

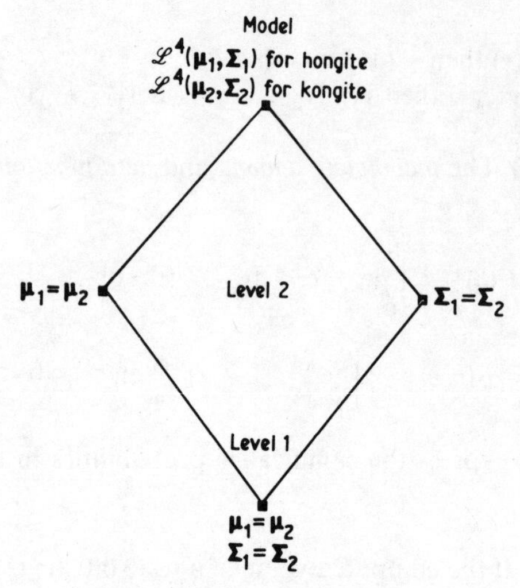

Fig. 7.3. *Lattice of hypotheses for comparison of hongite and kongite compositions.*

the following:

$$\mathbf{m}_1 = N_1^{-1} \sum_{r=1}^{N_1} \mathbf{y}_{1r}, \qquad \mathbf{m}_2 = N_2^{-1} \sum_{r=1}^{N_2} \mathbf{y}_{2r}, \tag{7.17}$$

$$\mathbf{S}_1 = N_1^{-1} \sum_{r=1}^{N_1} (\mathbf{y}_{1r} - \mathbf{m}_1)(\mathbf{y}_{1r} - \mathbf{m}_1)',$$

$$\mathbf{S}_2 = N_2^{-1} \sum_{r=1}^{N_2} (\mathbf{y}_{2r} - \mathbf{m}_2)(\mathbf{y}_{2r} - \mathbf{m}_2)', \tag{7.18}$$

which are the separate sample estimates;

$$\mathbf{S}_p = (N_1 + N_2)^{-1}(N_1\mathbf{S}_1 + N_2\mathbf{S}_2), \tag{7.19}$$

which is the pooled covariance matrix estimate; and

$$\mathbf{m}_c = (N_1 + N_2)^{-1}(N_1\mathbf{m}_1 + N_2\mathbf{m}_2), \tag{7.20}$$

$$\mathbf{S}_c = \mathbf{S}_p + (N_1 + N_2)^{-2} N_1 N_2 (\mathbf{m}_1 - \mathbf{m}_2)(\mathbf{m}_1 - \mathbf{m}_2)', \tag{7.21}$$

which are the combined sample estimates. The approximate generalized likelihood ratio test statistic (7.16) then takes the form

$$N_1 \log(|\hat{\boldsymbol{\Sigma}}_{1h}|/|\hat{\boldsymbol{\Sigma}}_{1m}|) + N_2 \log(|\hat{\boldsymbol{\Sigma}}_{2h}|/|\hat{\boldsymbol{\Sigma}}_{2m}|), \tag{7.22}$$

Table 7.6. *Maximum likelihood estimates and number c of constraints for the lattice of Fig. 7.3*

Model or hypothesis	Estimates of				
	$\boldsymbol{\mu}_1$	$\boldsymbol{\mu}_2$	$\boldsymbol{\Sigma}_1$	$\boldsymbol{\Sigma}_2$	c
Model	$\mathbf{m}_1$	$\mathbf{m}_2$	$\mathbf{S}_1$	$\mathbf{S}_2$	
$\boldsymbol{\mu}_1 = \boldsymbol{\mu}_2,\ \boldsymbol{\Sigma}_1 = \boldsymbol{\Sigma}_2$	$\mathbf{m}_c$	$\mathbf{m}_c$	$\mathbf{S}_c$	$\mathbf{S}_c$	$\frac{1}{2}d(d+3)$
$\boldsymbol{\Sigma}_1 = \boldsymbol{\Sigma}_2$	$\mathbf{m}_1$	$\mathbf{m}_2$	$\mathbf{S}_p$	$\mathbf{S}_p$	$\frac{1}{2}d(d+1)$
$\boldsymbol{\mu}_1 = \boldsymbol{\mu}_2$	$\mathbf{m}_h$	$\mathbf{m}_h$	$\mathbf{S}_{1h}$	$\mathbf{S}_{2h}$	d

where $\mathbf{m}_h, \mathbf{S}_{1h}, \mathbf{S}_{2h}$ are obtained by the following iterative process.

1. $\mathbf{S}_{ih} = \mathbf{S}_i\ (i = 1, 2)$.
2. $\mathbf{m}_h = (N_1\mathbf{S}_{1h}^{-1} + N_2\mathbf{S}_{2h}^{-1})^{-1}(N_1\mathbf{S}_{1h}^{-1}\mathbf{m}_1 + N_2\mathbf{S}_{2h}^{-1}\mathbf{m}_2)$.
3. $\mathbf{S}_{ih} = \mathbf{S}_i + (\mathbf{m}_i - \mathbf{m}_h)(\mathbf{m}_i - \mathbf{m}_h)'$.
4. Repeat steps 2 and 3 until convergence.

to be compared against upper percentage points of $\chi^2(c)$, where the estimates under the model and each hypothesis and the number c of constraints are shown in Table 7.6.

For the hongite and kongite data, $d = 4$, $N_1 = N_2 = 25$, and the relevant computational results are given in Table 7.7. On the basis of the simplicity postulate, our first test is of the hypothesis at level 1, that $\boldsymbol{\mu}_1 = \boldsymbol{\mu}_2$ and $\boldsymbol{\Sigma}_1 = \boldsymbol{\Sigma}_2$, in other words that the two compositional distributions are identical. The computed value of the test statistic (7.22),

$$N_1 \log(|\mathbf{S}_c|/|\mathbf{S}_1|) + N_2 \log(|\mathbf{S}_c|/|\mathbf{S}_2|), \tag{7.23}$$

is 91.7, to be compared against the upper percentage points of the $\chi^2(14)$ distribution. Since the upper 0.05 per cent point of $\chi^2(14)$ is 38.11 we must clearly reject the hypothesis that $\boldsymbol{\mu}_1 = \boldsymbol{\mu}_2, \boldsymbol{\Sigma}_1 = \boldsymbol{\Sigma}_2$. We have described this procedure in traditional terms, to allow a simple comparison with the computation of the significance probability by Property 7.2(a) as

$$1 - \mathscr{I}_{45.9}(7) < 10^{-5}. \tag{7.24}$$

By the rejection of the hypothesis at level 1 we now have a mandate to move to the hypotheses at level 2. We turn first to the hypothesis that $\boldsymbol{\Sigma}_1 = \boldsymbol{\Sigma}_2$. From (7.22) and Table 7.6 the test statistic is

$$N_1 \log(|\mathbf{S}_p|/|\mathbf{S}_1|) + N_2 \log(|\mathbf{S}_p|/|\mathbf{S}_2|), \tag{7.25}$$

Table 7.7. *Maximum likelihood estimates associated with the lattice of Fig. 7.3*

$$\mathbf{m}_1 = (1.715, 0.914, 0.115, 0.167),$$
$$\mathbf{m}_2 = (1.639, 0.754, 0.191, 0.104),$$
$$\mathbf{m}_c = (1.677, 0.834, 0.153, 0.135),$$
$$\mathbf{m}_h = (1.727, 0.957, 0.014, 0.170)$$

$$\mathbf{S}_1 = \begin{bmatrix} 0.133 & 0.254 & -0.214 & 0.116 \\ 0.254 & 0.623 & -0.674 & 0.139 \\ -0.214 & -0.674 & 0.910 & 0.011 \\ 0.116 & 0.139 & 0.011 & 0.180 \end{bmatrix},$$

$$\mathbf{S}_2 = \begin{bmatrix} 0.109 & 0.226 & -0.193 & 0.092 \\ 0.226 & 0.629 & -0.694 & 0.102 \\ -0.193 & -0.694 & 1.008 & 0.087 \\ 0.092 & 0.102 & 0.087 & 0.187 \end{bmatrix},$$

$$\mathbf{S}_c = \begin{bmatrix} 0.127 & 0.252 & -0.209 & 0.109 \\ 0.252 & 0.652 & -0.696 & 0.130 \\ -0.209 & -0.696 & 0.965 & 0.044 \\ 0.109 & 0.130 & 0.044 & 0.187 \end{bmatrix},$$

$$\mathbf{S}_p = \begin{bmatrix} 0.121 & 0.240 & -0.204 & 0.104 \\ 0.240 & 0.626 & -0.684 & 0.120 \\ -0.204 & -0.684 & 0.959 & 0.049 \\ 0.104 & 0.120 & 0.049 & 0.184 \end{bmatrix},$$

$$\mathbf{S}_{1h} = \begin{bmatrix} 0.133 & 0.254 & -0.216 & 0.117 \\ 0.254 & 0.625 & -0.678 & 0.139 \\ -0.216 & -0.678 & 0.920 & 0.011 \\ 0.117 & 0.139 & 0.011 & 0.180 \end{bmatrix},$$

$$\mathbf{S}_{2h} = \begin{bmatrix} 0.116 & 0.244 & -0.209 & 0.098 \\ 0.244 & 0.670 & -0.730 & 0.115 \\ -0.209 & -0.730 & 1.040 & 0.076 \\ 0.098 & 0.115 & 0.076 & 0.192 \end{bmatrix}$$

with computed value 10.6 to be compared against the upper percentage points of $\chi^2(10)$. The significance probability is then

$$1 - \mathscr{I}_{5.3}(5) = 0.39$$

and so we would have to accept that taking the two logistic normal models with a common logratio covariance matrix is certainly acceptable as a working model.

The problem of testing the hypothesis that $\boldsymbol{\mu}_1 = \boldsymbol{\mu}_2$ at level 2 with the possibility of $\boldsymbol{\Sigma}_1$ and $\boldsymbol{\Sigma}_2$ being different is, of course, the multivariate version of the awkward Behrens–Fisher problem, for which no explicit forms for the maximum likelihood estimates exist. Hence the need for the simple iterative method set out in Table 7.6. From (7.22) and Table 7.6 the test statistic is

$$N_1 \log(|\mathbf{S}_{1h}|/|\mathbf{S}_1|) + N_2 \log(|\mathbf{S}_{2h}|/|\mathbf{S}_2|), \tag{7.26}$$

and, from Table 7.7, this has computed value 33.7 to be compared with the upper percentage points of the $\chi^2(4)$ distribution. The significance probability is

$$1 - \mathscr{I}_{16.9}(2) < 10^{-5},$$

and so we must clearly reject the hypothesis that $\boldsymbol{\mu}_1 = \boldsymbol{\mu}_2$.

Thus through the lattice testing procedure we arrive at a working model which assumes that the hongite and kongite compositions follow $\mathscr{L}^4(\boldsymbol{\mu}_1, \boldsymbol{\Sigma})$ and $\mathscr{L}^4(\boldsymbol{\mu}_2, \boldsymbol{\Sigma})$ distributions, with common logratio covariance matrix parameter $\boldsymbol{\Sigma}$ and different logratio mean vector parameters $\boldsymbol{\mu}_1$ and $\boldsymbol{\mu}_2$. The maximum likelihood estimates of $\boldsymbol{\mu}_1, \boldsymbol{\mu}_2$ and $\boldsymbol{\Sigma}$ are $\mathbf{m}_1, \mathbf{m}_2$ and $\mathbf{S}_p$ as given in Table 7.7. We emphasize that $\mathbf{S}_p$ here is the maximum likelihood estimate of $\boldsymbol{\Sigma}$, with divisor $N_1 + N_2$, not $N_1 + N_2 - 2$ as for the usual unbiased estimate of such a pooled covariance matrix. Our practice in future will be to use maximum likelihood estimates in testing with the generalized likelihood ratio tests but to quote the usual unbiased versions when involved in estimation.

The lattice procedure here differs from what appears to be common textbook practice. The recommendation there is often first to test the hypothesis $\boldsymbol{\Sigma}_1 = \boldsymbol{\Sigma}_2$ within the model. This would exactly follow what we have done as our second test, and at level 2. On non-rejection of that hypothesis the common practice is then to regard the hypothesis $\boldsymbol{\Sigma}_1 = \boldsymbol{\Sigma}_2$ as a new accepted model and then within it, rather than the original general model, to test the hypothesis that $\boldsymbol{\mu}_1 = \boldsymbol{\mu}_2$; in our terms the hypothesis at our level 1 that $\boldsymbol{\mu}_1 = \boldsymbol{\mu}_2, \boldsymbol{\Sigma}_1 = \boldsymbol{\Sigma}_2$. This practice leads to the familiar Hotelling T^2 test of equality of means (Mardia, Kent and Bibby, 1979, p. 139), in which, with $\mathbf{m}_1, \mathbf{m}_2$ and $\mathbf{S}_p$ as in Table 7.6,

$$Q = \frac{N_1 N_2 (N_1 + N_2 - d - 1)}{(N_1 + N_2)(N_1 + N_2 - 2)d} (\mathbf{m}_1 - \mathbf{m}_2)' \mathbf{S}_p^{-1} (\mathbf{m}_1 - \mathbf{m}_2) \quad (7.27)$$

is compared with upper percentage points of the $F(d, N_1 + N_2 - d - 1)$ distribution. The computed value q is 11.91 and so, by Property 7.2(b) and its method of evaluation by the incomplete beta function, the significance probability is $< 10^{-5}$, confirming the lattice decision above to reject the hypothesis that $\boldsymbol{\mu}_1 = \boldsymbol{\mu}_2, \boldsymbol{\Sigma}_1 = \boldsymbol{\Sigma}_2$. As we have already explained in Section 7.4, we prefer an insistence on clarifying the hierarchy of hypotheses and on the common sense of the simplicity postulate, both leading to the lattice approach.

The detection of a difference between the mean logratio vectors of hongite and kongite compositions suggests the feasibility of constructing a differential classification system based on the compositions of Data 1 and 2 for application to new, but unclassified, rock samples. We shall investigate this problem, the remaining part of Question 1.2.6, in Section 7.11.

Applying the lattice of Fig. 7.3 to a comparison of the hongite and boxite compositions of Data 1 and 3, rejection starts at the level 1 hypothesis and continues for both level 2 hypotheses, all with significance probabilities $< 10^{-5}$. These established differences thus justify the retention of the different parameters $\boldsymbol{\mu}_1, \boldsymbol{\mu}_2, \boldsymbol{\Sigma}_1, \boldsymbol{\Sigma}_2$ for hongite and boxite compositions.

The tests described above can be extended to more than two types.

7.6 Logratio linear modelling

The extent to which variability in $\mathscr{R}^d$ is explainable by, or dependent on, other factors, concomitants or covariates has given rise to extensive theory and practice involving linear modelling and hypothesis formulation of the multivariate mean. Multivariate analysis of variance, the so-called MANOVA, and multivariate regression fall into this category. We now approach the compositional counterpart by investigating the possible dependence of the 39 Arctic lake sediment compositions of Data 5 on the covariate water depth as a motivating problem.

Let $\mathbf{y} = (y_1, y_2) = \{\log(x_1/x_3), \log(x_2/x_3)\}$ denote a typical logratio composition formed from a sediment (sand, silt, clay) composition $\mathbf{x} = (x_1, x_2, x_3)$ sampled at depth u. The general problem is to model the conditional distribution or density function of $\mathbf{x}$ for given u through the conditional density function $p(\mathbf{y}|u)$ of logratio composition $\mathbf{y}$ for given depth u. Following linear modelling practice in $\mathscr{R}^d$ we suppose this to be bivariate normal with a mean which depends on some function of u which is linear in its parameters, but not necessarily linear in u. For example, we may be ready to assume that a functional form which is quadratic together with a possible logarithmic term will be adequate to describe the variability. This would involve adopting a model in which the distribution of $\mathbf{y}|u$ is $\mathscr{N}^2(\boldsymbol{\alpha} + \boldsymbol{\beta}u + \boldsymbol{\gamma}u^2 + \boldsymbol{\delta}\log u, \boldsymbol{\Sigma})$, where $\boldsymbol{\alpha}, \boldsymbol{\beta}, \boldsymbol{\gamma}, \boldsymbol{\delta}$ are 2×1 vectors, or equivalently $\mathbf{x}|u$ is

$$\mathscr{L}^2(\boldsymbol{\alpha} + \boldsymbol{\beta}u + \boldsymbol{\gamma}u^2 + \boldsymbol{\delta}\log u, \boldsymbol{\Sigma}). \tag{7.28}$$

An alternative and familiar way of expressing this is as follows:

$$y_1 = \alpha_1 + \beta_1 u + \gamma_1 u^2 + \delta_1 \log u + e_1, \tag{7.29}$$

$$y_2 = \alpha_2 + \beta_2 u + \gamma_2 u^2 + \delta_2 \log u + e_2, \tag{7.30}$$

where (e_1, e_2) is $\mathscr{N}^2(\mathbf{0}, \boldsymbol{\Sigma})$, or in a more convenient matrix form

$$[y_1 \quad y_2] = [1 \quad u \quad u^2 \quad \log u]\begin{bmatrix} \alpha_1 & \alpha_2 \\ \beta_1 & \beta_2 \\ \gamma_1 & \gamma_2 \\ \delta_1 & \delta_2 \end{bmatrix} + [e_1 \quad e_2]. \tag{7.31}$$

We can then in standard multivariate format relate the complete 39×2 logratio matrix $\mathbf{Y}$ to the 4×2 parameter matrix $\boldsymbol{\Theta}$ through a premultiplying 39×4 covariate matrix $\mathbf{A}$ with rth row consisting of $[1 \quad u_r \quad u_r^2 \quad \log u_r]$, where u_r is the depth of the rth sediment. The array form of (7.31) is then

$$\mathbf{Y} = \mathbf{A}\boldsymbol{\Theta} + \mathbf{E}, \tag{7.32}$$

where the rows of the error matrix are assumed independent and each row is distributed as $\mathscr{N}^2(\mathbf{0}, \boldsymbol{\Sigma})$.

We can provide a general definition and notation.

Definition 7.4 A set of N independent D-part compositions is said to follow a *logratio linear model* m if the logratio data matrix $\mathbf{Y}$ can be expressed in the form

$$\mathbf{Y} = \mathbf{A}\boldsymbol{\Theta} + \mathbf{E},$$

where the *covariate matrix* $\mathbf{A}$, of order $N \times p_m$ and full rank p_m, is a matrix of known constants, the *parameter matrix* $\boldsymbol{\Theta}$ is of order $p_m \times d$ and the $N \times d$ error matrix $\mathbf{E}$ is assumed to consist of independent row vectors, each distributed as $\mathcal{N}^d(\mathbf{0}, \boldsymbol{\Sigma})$.

Notes

1. It is possible to widen the definition to allow the covariate matrix $\mathbf{A}$ to be of less than full rank but in our applications we shall not require this extension.
2. Since $\mathrm{E}(\mathbf{Y}) = \mathbf{A}\boldsymbol{\Theta}$ or $\mathrm{E}(\mathbf{Y}') = \boldsymbol{\Theta}'\mathbf{A}'$, an alternative definition could have been that each mean logratio vector is expressible as a known linear combination of p_m, d-dimensional parameter vectors, namely the columns of $\boldsymbol{\Theta}'$.
3. The parameters of the model are the 'regression' parameters collected in the matrix $\boldsymbol{\Theta}$, and the 'error' logratio covariance matrix is $\boldsymbol{\Sigma}$.

Estimation of $\boldsymbol{\Theta}$ and $\boldsymbol{\Sigma}$ under the model is standard, either by maximum likelihood under the normality assumption or by multivariate least squares.

Property 7.3 The maximum likelihood estimate $\hat{\boldsymbol{\Theta}}$ of $\boldsymbol{\Theta}$ under the logratio linear model m of Definition 7.4,

$$\mathbf{Y} = \mathbf{A}\boldsymbol{\Theta} + \mathbf{E},$$

is

$$\hat{\boldsymbol{\Theta}}_{\mathrm{m}} = (\mathbf{A}'\mathbf{A})^{-1}\mathbf{A}'\mathbf{Y}$$

and the residual sum of squares and cross products matrix is

$$\begin{aligned} \mathbf{R}_{\mathrm{m}} &= (\mathbf{Y} - \mathbf{A}\hat{\boldsymbol{\Theta}}_{\mathrm{m}})'(\mathbf{Y} - \mathbf{A}\hat{\boldsymbol{\Theta}}_{\mathrm{m}}) \\ &= \mathbf{Y}'\mathbf{Y} - \hat{\boldsymbol{\Theta}}'_{\mathrm{m}}\mathbf{A}'\mathbf{Y} \\ &= \mathbf{Y}'\{\mathbf{I} - \mathbf{A}(\mathbf{A}'\mathbf{A})^{-1}\mathbf{A}'\}\mathbf{Y}. \end{aligned}$$

Moreover an estimate of $\boldsymbol{\Sigma}$, unbiased under model m, is given by

$$\hat{\boldsymbol{\Sigma}}_{\mathrm{m}} = (N - p_{\mathrm{m}})^{-1}\mathbf{R}_{\mathrm{m}}.$$

Note

1. The effects of a permutation P of the parts of each composition are of interest here. What relation does the resulting logratio linear model

$$\mathbf{Y}_P = \mathbf{A}\boldsymbol{\Theta}_P + \mathbf{E}_P \tag{7.33}$$

bear to the original logratio linear model $\mathbf{Y} = \mathbf{A}\boldsymbol{\Theta} + \mathbf{E}$? Since each *row* of $\mathbf{Y}$ is a logratio vector, application of (7.3) gives

$$\mathbf{Y}_P = \mathbf{Y}\mathbf{Q}'_P \tag{7.34}$$

where $\mathbf{Q}_P = \mathbf{FPF'H}^{-1}$. Then since

$$\mathrm{E}(\mathbf{Y}_P) = \mathrm{E}(\mathbf{Y})\mathbf{Q}'_P = \mathbf{A}\boldsymbol{\Theta}\mathbf{Q}'_P$$

we have

$$\boldsymbol{\Theta}_P = \boldsymbol{\Theta}\mathbf{Q}'_P. \tag{7.35}$$

Moreover, the rows of $\mathbf{E}_P$ are still independent and, from (7.6), each row is distributed as $\mathcal{N}^d(\mathbf{0}, \boldsymbol{\Sigma}_P)$, where

$$\boldsymbol{\Sigma}_P = \mathbf{Q}_P\boldsymbol{\Sigma}\mathbf{Q}'_P. \tag{7.36}$$

It is now easy to confirm that the estimation procedure of Property 7.3 applied to the permuted model yields estimates conforming with the above relationships. We have, using (7.34),

$$\hat{\boldsymbol{\Theta}}_P = (\mathbf{A}'\mathbf{A})^{-1}\mathbf{A}'\mathbf{Y}_P = (\mathbf{A}'\mathbf{A})^{-1}\mathbf{A}'\mathbf{Y}\mathbf{Q}'_P = \hat{\boldsymbol{\Theta}}\mathbf{Q}'_P$$

conforming with (7.35): also, dropping the suffix m denoting model, we have

$$\mathbf{R}_P = (\mathbf{Y}_P - \mathbf{A}\hat{\boldsymbol{\Theta}}_P)'(\mathbf{Y}_P - \mathbf{A}\hat{\boldsymbol{\Theta}}_P) = \mathbf{Q}_P\mathbf{R}\mathbf{Q}'_P$$

so that

$$\hat{\boldsymbol{\Sigma}}_P = \mathbf{Q}_P\hat{\boldsymbol{\Sigma}}\mathbf{Q}'_P$$

conforming with (7.36).

Applying Property 7.3 to the Arctic lake sediments of Data 5 we obtain

$$\hat{\boldsymbol{\Theta}}_{\mathrm{m}} = \begin{bmatrix} 8.79 & 2.06 \\ -0.066 & -0.098 \\ 0.00049 & 0.00060 \\ -2.04 & 0.436 \end{bmatrix}, \tag{7.37}$$

$$\mathbf{R}_{\mathrm{m}} = \begin{bmatrix} 45.40 & 24.11 \\ 24.11 & 17.06 \end{bmatrix}. \tag{7.38}$$

7.7 Testing logratio linear hypotheses

Again we use Question 1.3.1 on the possible dependence of Arctic lake sediment compositions on the covariate water depth to motivate the ideas here. If we extend the lattice of Fig. 7.2 with its single characteristic to include the two logratios y_1, y_2 formed from the (sand, silt, clay) composition, then we may consider the lattice of Fig. 7.4. Let us examine the nature of these hypotheses, for example the hypothesis h at level 2 that the dependence is linear on u, that $\gamma_1 = \gamma_2 = \delta_1 = \delta_2 = 0$. Such a hypothesis can be expressed as a trivial

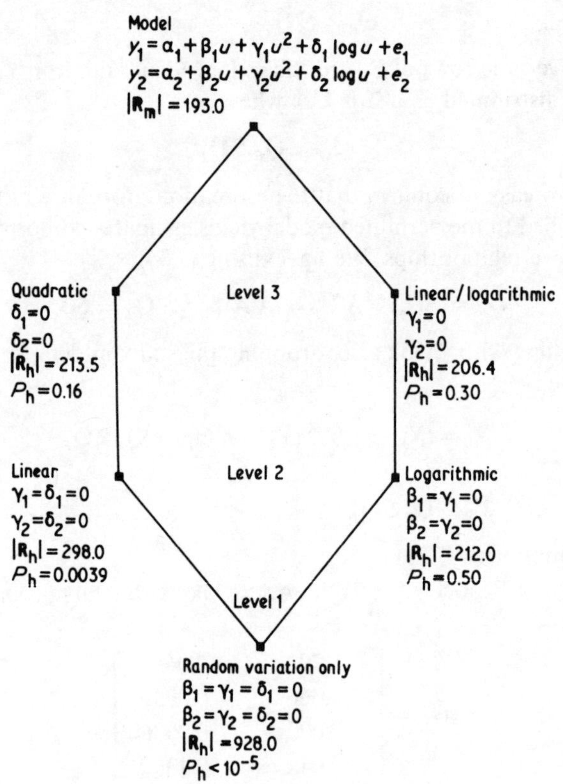

Fig. 7.4. *Lattice of hypotheses for investigation of dependence of Arctic lake sediment compositions on water depth, showing residual determinants* $|\mathbf{R}_h|$ *and significance probabilities* P_h.

reparametrization,

$$\boldsymbol{\Theta} = \begin{bmatrix} 1 & 0 \\ 0 & 1 \\ 0 & 0 \\ 0 & 0 \end{bmatrix} \begin{bmatrix} \alpha_1 & \alpha_2 \\ \beta_1 & \beta_2 \end{bmatrix} \tag{7.39}$$

$$= \mathbf{K}\boldsymbol{\Psi},$$

say, where the new parameter matrix $\boldsymbol{\Psi}$ contains fewer parameters than the original $\boldsymbol{\Theta}$. If we now write $\mathbf{B} = \mathbf{AK}$ so that $\mathbf{B}$ has rth row $[1 \quad u_r]$, we see that this hypothesis h can be expressed in an array form similar to that of the model m, with

$$\mathbf{Y} = \mathbf{B}\boldsymbol{\Psi} + \mathbf{E}, \tag{7.40}$$

with each row of $\mathbf{E}$ assumed distributed independently as $\mathcal{N}^2(\mathbf{0}, \boldsymbol{\Sigma}_\mathrm{h})$. The matrix $\mathbf{B}$, of order 39×2 in our example and in general of order $N \times p_\mathrm{h}$ and full rank p_h, plays the role of covariate matrix. We have thus an estimation problem similar to that of Property 7.3 with

$$\hat{\boldsymbol{\Psi}}_\mathrm{h} = (\mathbf{B}'\mathbf{B})^{-1}\mathbf{B}'\mathbf{Y}, \tag{7.41}$$

so that the corresponding estimate $\hat{\boldsymbol{\Theta}}_\mathrm{h}$ of $\boldsymbol{\Theta}$ under hypothesis h is

$$\hat{\boldsymbol{\Theta}}_\mathrm{h} = \mathbf{K}\hat{\boldsymbol{\Psi}}_\mathrm{h} \tag{7.42}$$

$$= \mathbf{K}(\mathbf{K}'\mathbf{A}'\mathbf{A}\mathbf{K})^{-1}\mathbf{K}'\mathbf{A}'\mathbf{Y}$$

$$= \mathbf{K}(\mathbf{K}'\mathbf{A}'\mathbf{A}\mathbf{K})^{-1}\mathbf{K}'\mathbf{A}'\mathbf{A}\hat{\boldsymbol{\Theta}}_\mathrm{m}. \tag{7.43}$$

Moreover,

$$\mathbf{R}_\mathrm{h} = (\mathbf{Y} - \mathbf{B}\hat{\boldsymbol{\Psi}}_\mathrm{h})'(\mathbf{Y} - \mathbf{B}\hat{\boldsymbol{\Psi}}_\mathrm{h})$$

$$= \mathbf{Y}'\mathbf{Y} - \hat{\boldsymbol{\Psi}}_\mathrm{h}'\mathbf{B}'\mathbf{Y}$$

$$= \mathbf{Y}'\{\mathbf{I} - \mathbf{B}(\mathbf{B}'\mathbf{B})^{-1}\mathbf{B}'\}\mathbf{Y}, \tag{7.44}$$

and an estimate of $\boldsymbol{\Sigma}_\mathrm{h}$, unbiased under hypothesis h, is given by

$$\hat{\boldsymbol{\Sigma}}_\mathrm{h} = (N - p_\mathrm{h})^{-1}\mathbf{R}_\mathrm{h}. \tag{7.45}$$

Applying the above estimation procedure to the Arctic lake sediments we obtain for our selected illustrative hypothesis h the following estimates.

$$\hat{\boldsymbol{\Theta}}_\mathrm{h} = \begin{bmatrix} 2.61 & 1.97 \\ -0.062 & -0.025 \\ 0 & 0 \\ 0 & 0 \end{bmatrix}, \tag{7.46}$$

$$\mathbf{R}_h = \begin{bmatrix} 63.98 & 32.45 \\ 32.45 & 21.12 \end{bmatrix}. \tag{7.47}$$

The approximate form of the generalized likelihood ratio test of a hypothesis h within the logratio linear model m is again standard, comparing

$$N \log(|\mathbf{R}_h|/|\mathbf{R}_m|) \tag{7.48}$$

against upper percentage points of the $\chi^2(k)$ distribution where $k = d(p_m - p_h)$. For some special cases the exact distribution is known, and also for this particular situation improved approximations of the distribution have been obtained. All of these approximations can be expressed in terms of a test statistic which is a simple monotonic function of the generalized likelihood ratio, of the form

$$(c/b)\{(|\mathbf{R}_h|/|\mathbf{R}_m|)^a - 1\} \tag{7.49}$$

which, under the hypothesis h, follows a $F(2b, 2c)$ distribution. Since $2b$ and $2c$ are in some cases not integers we note, from Property 7.1(b), that $1 - (|\mathbf{R}_m|/|\mathbf{R}_h|)^a$ is then $\mathrm{Be}(b, c)$ and so we can obtain the significance probabilities in terms of the incomplete beta function.

Property 7.4 For testing the linear hypothesis h,

$$\mathbf{Y} = \mathbf{B}\boldsymbol{\Psi} + \mathbf{E},$$

where $\mathbf{B}$ is $N \times p_h$ and of full rank p_h, within the model m

$$\mathbf{Y} = \mathbf{A}\boldsymbol{\Theta} + \mathbf{E},$$

where $\mathbf{A}$ is of order $N \times p_m$ and of full rank p_m, let q be the computed value of the test statistic

$$Q = 1 - (|\mathbf{R}_m|/|\mathbf{R}_h|)^a,$$

where the residual matrices $\mathbf{R}_m$, $\mathbf{R}_h$ are as in Property 7.3 and (7.44). Then the significance probability is

$$1 - \mathscr{I}_q(b, c) = \mathscr{I}_{1-q}(c, b),$$

where a, b, c for special cases are given in Table 7.8 and

$$k = p_m - p_h.$$

The result is exact for the first four cases and approximate, for large N, in the final case.

Table 7.8. *Test characteristics*

d	k	N	a	b	c
1	any	$>k+1$	1	$\frac{1}{2}k$	$\frac{1}{2}(N-k-1)$
2	any	$>k+2$	$\frac{1}{2}$	k	$(N-k-2)$
any	1	$>d+1$	1	$\frac{1}{2}d$	$\frac{1}{2}(N-d-1)$
any	2	$>d+2$	$\frac{1}{2}$	d	$N-d-2$
any	any	'large'	$\left(\dfrac{d^2+k^2-5}{d^2k^2-4}\right)$	$\frac{1}{2}dk$	$\frac{1}{2}[\{N-1 - \frac{1}{2}(d+k+1)\}a^{-1} - \frac{1}{2}dk+1]$

Returning now to our example of testing the hypothesis $\gamma_1=\gamma_2=\delta_1=\delta_2=0$ of linearity, we have $p_m=4, p_h=2$ so that $k=2$. From Property 7.4 and (7.44), we compute

$$|\mathbf{R}_m| = 193.0, \quad |\mathbf{R}_h| = 298.0.$$

Moreover, we have the special case $d=2$ so that

$$q = 1-(|\mathbf{R}_m|/|\mathbf{R}_h|)^{1/2} = 0.197,$$

and the significance probability is

$$1-\mathscr{I}_q(b,c) = 1-\mathscr{I}_{0.197}(2,35) = 0.0039.$$

The lattice in Fig. 7.4 shows the values of $|\mathbf{R}|$ and the significance probabilities for each of the five hypotheses. Starting at level 1 we see that the hypothesis of completely random variation with no dependence on depth has a significance probability $<10^{-5}$ and must be rejected. At level 2, while the hypothesis of linear dependence on depth with its significance probability of 0.0039 must also be rejected, we clearly cannot reject the hypothesis of logarithmic dependence on depth. We thus stop testing at this level and adopt as our working model

$$\begin{aligned} y_1 &= \alpha_1+\delta_1\log u+e_1, \\ y_2 &= \alpha_2+\delta_2\log u+e_2, \end{aligned} \tag{7.50}$$

with maximum likelihood estimates

$$\begin{aligned} \hat{\alpha}_1 &= 9.70, \quad & \hat{\delta}_1 &= -2.74, \\ \hat{\alpha}_2 &= 4.80, \quad & \hat{\delta}_2 &= -1.10, \end{aligned} \tag{7.51}$$

and estimate of the error covariance matrix $\boldsymbol{\Sigma}$ given by

$$\hat{\boldsymbol{\Sigma}} = \begin{bmatrix} 1.27 & 0.69 \\ 0.69 & 0.49 \end{bmatrix}. \tag{7.52}$$

In Fig. 7.4 we have for illustrative reasons shown the computations for the complete lattice. In practice it is not necessary to go beyond the level at which a first non-rejection is found. From this full lattice we note that had we reached level 2 we would have found the quadratic and the linear-logarithmic dependence hypotheses also acceptable. Lattice testing prefers the logarithmic dependence because it provides a simpler working model with only four parameters, compared with the six parameters of the quadratic dependence hypothesis.

7.8 Further aspects of logratio linear modelling

With our motivating problem still the question of dependence on water depth of Arctic lake sediment compositions, we now examine a number of aspects of logratio linear modelling.

Validation of working model

In Section 7.3 we found that tests applied to the Arctic lake sediment compositions revealed that they could not reasonably be regarded as having a logistic normal $\mathscr{L}^d(\boldsymbol{\mu}, \boldsymbol{\Sigma})$ distribution. It is important to realize that this finding need not be in conflict with our current modelling and its assumption of a multinormal error vector in the logratio linear model. It is perfectly possible for the conditional distribution $\mathbf{y}|u$ to be normal with the unconditional distribution of $\mathbf{y}$ far from normal. Having reached the working model (7.50), what we must certainly do is examine the validity of the error assumption, by subjecting the bivariate residuals

$$\begin{aligned} \hat{e}_1 &= y_1 - \hat{\alpha}_1 - \hat{\delta}_1 \log u, \\ \hat{e}_2 &= y_2 - \hat{\alpha}_2 - \hat{\delta}_2 \log u, \end{aligned} \tag{7.53}$$

to normality tests. Table 7.9 gives the results of the same battery of tests as was applied to the unadjusted logratio vectors in Table 7.5. Comparison shows that whereas the unconditional distribution is far from logistic normality, the assumption that normality of the error vector in the working logratio linear model (7.50), or equivalently that the conditional distribution of composition on depth is logistic

Table 7.9. *Values of test statistics for logistic normality of residuals of the logratio linear model for Arctic lake sediments*

			Anderson–Darling	Cramer–von Mises	Watson
Marginal	i				
	1		0.477	0.066	0.061
	2		2.621	0.414	0.367
Bivariate	i	j			
	1	2	0.985	0.091	0.098
Radius			2.535	0.417	0.169

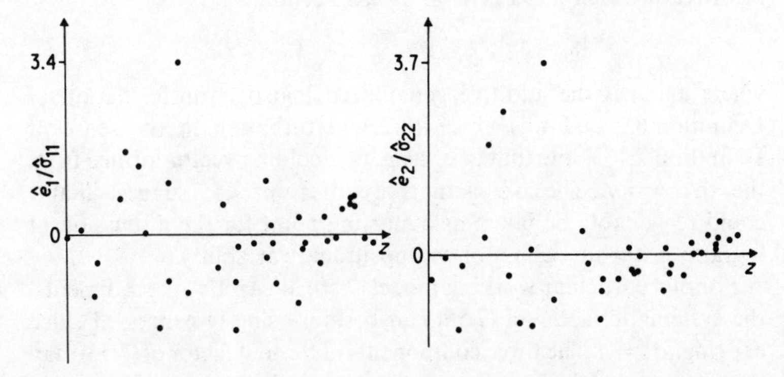

Fig. 7.5. *Plots of standardized residuals $\hat{e}_1/\hat{\sigma}_{11}$ and $\hat{e}_2/\hat{\sigma}_{22}$ against $z = \log u$ for logarithmic dependence of Arctic lake sediment compositions on water depth u.*

normal, is, although not fully validated, at least more reasonable.

Another sensible check on model validity is to examine residual plots. In Fig. 7.5 we show the plots of the standardized residuals $\hat{e}_1/\sqrt{\hat{\sigma}_{11}}$ and $\hat{e}_2/\sqrt{\hat{\sigma}_{22}}$ against $z = \log u$. The only suspect appears to be sediment sample S12 with its standardized residuals of 3.40 and 3.68, possibly arising from its extremely low clay component. A possible course of action here is to investigate the influence of S12 on the analysis by comparing the present analysis with that obtained by its omission. We leave this investigation as an exercise for the reader but shall further study S12 in Section 7.10 when we consider the detection of outliers.

For the general case when we are considering the validity of the multinormality of error in the logratio linear model

$$\mathbf{Y} = \mathbf{A}\boldsymbol{\Theta} + \mathbf{E}$$

we apply our battery of normality tests to the $N \times d$ matrix of residuals

$$\begin{aligned}\hat{\mathbf{E}} &= \mathbf{Y} - \mathbf{A}\hat{\boldsymbol{\Theta}} \\ &= \{\mathbf{I} - \mathbf{A}(\mathbf{A}'\mathbf{A})^{-1}\mathbf{A}'\}\mathbf{Y}.\end{aligned} \tag{7.54}$$

Compositional form of logratio linear model

We can express a logratio linear model in terms of the composition $\mathbf{x}$ by an application of the additive logistic transformation. The form is most readily seen in its general model version:

$$\mathbf{x}_r = \text{agl}(\boldsymbol{\Theta}'\mathbf{a}_r) \circ \mathbf{v} \qquad (r = 1, \ldots, N), \tag{7.55}$$

where agl$(\cdot)$ is the additive generalized logistic transformation of Definition 6.9 and $\mathbf{v}$ is an $\mathscr{L}^d(\mathbf{0}, \boldsymbol{\Sigma})$ perturbation in the sense of Definition 2.13. Note that $\mathbf{a}_r$ is the $p_{\text{m}} \times 1$ column vector formed from the rth row of $\mathbf{A}$. The compactness of expression (7.55) suggests that it could reasonably be taken as a starting point for the discussion of logratio linear modelling of compositional variability.

For our particular working model (7.50) for Arctic lake sediments, the systematic factor of (7.55) can be more simply expressed since $\exp\{\log(u)\} = u$. The three components of the first factor of (7.55) may then be expressed as

$$\mathscr{C}(\varepsilon_1 u^{\delta_1}, \varepsilon_2 u^{\delta_2}, 1)$$

where $\varepsilon_i = \exp(\alpha_i)$ $(i = 1, 2)$.

Constraint form for logratio linear hypothesis

In Section 7.7 we chose to express the linear dependence hypothesis $\gamma_1 = \gamma_2 = \delta_1 = \delta_2 = 0$ for Arctic lake sediments as a reparametrization (7.39) of the original parameter matrix $\boldsymbol{\Theta}$ in terms of a new, lower-order, parameter matrix $\boldsymbol{\Psi}$. By doing so we could carry out the maximum likelihood estimation under the hypothesis by exactly the same routine as under the model. There may be occasions on which it is more convenient to express the hypothesis as placing constraints on $\boldsymbol{\Theta}$ of the form

$$\mathbf{L}\boldsymbol{\Theta} = \mathbf{0}, \tag{7.56}$$

where in general $\mathbf{L}$ is of order $k \times p_{\mathrm{m}}$ and k is related to the orders $N \times p_{\mathrm{m}}$ and $N \times p_{\mathrm{h}}$ of $\mathbf{A}$ and $\mathbf{B}$ by $k = p_{\mathrm{m}} - p_{\mathrm{h}}$, as in Property 7.4. For the above linear dependence hypothesis, $\mathbf{L}\boldsymbol{\Theta} = \mathbf{0}$ would take the form

$$\begin{bmatrix} 0 & 0 & 1 & 0 \\ 0 & 0 & 0 & 1 \end{bmatrix} \begin{bmatrix} \alpha_1 & \alpha_2 \\ \beta_1 & \beta_2 \\ \gamma_1 & \gamma_2 \\ \delta_1 & \delta_2 \end{bmatrix} = \mathbf{0}. \tag{7.57}$$

The maximum likelihood estimate of $\boldsymbol{\Theta}$ under h can then be found, by use of the Lagrangian multiplier technique, to be

$$\begin{aligned} \hat{\boldsymbol{\Theta}}_{\mathrm{h}} &= [(\mathbf{A}'\mathbf{A})^{-1}\mathbf{A}' - (\mathbf{A}'\mathbf{A})^{-1}\mathbf{L}'\{\mathbf{L}(\mathbf{A}'\mathbf{A})^{-1}\mathbf{L}'\}^{-1}\mathbf{L}(\mathbf{A}'\mathbf{A})^{-1}\mathbf{A}']\mathbf{Y} \\ &= [\mathbf{I} - (\mathbf{A}'\mathbf{A})^{-1}\mathbf{L}'\{\mathbf{L}(\mathbf{A}'\mathbf{A})^{-1}\mathbf{L}'\}^{-1}]\hat{\boldsymbol{\Theta}}_{\mathrm{m}}. \end{aligned} \tag{7.58}$$

The corresponding matrix of residual sums of squares and cross-products is then

$$\begin{aligned} \mathbf{R}_{\mathrm{h}} &= \mathbf{Y}'[\mathbf{I} - \mathbf{A}(\mathbf{A}'\mathbf{A})^{-1}\mathbf{A}' \\ &\qquad - \mathbf{A}(\mathbf{A}'\mathbf{A})^{-1}\mathbf{L}'\{\mathbf{L}(\mathbf{A}'\mathbf{A})^{-1}\mathbf{L}'\}^{-1}\mathbf{L}(\mathbf{A}'\mathbf{A})^{-1}\mathbf{A}']\mathbf{Y} \\ &= \mathbf{R}_{\mathrm{m}} - \mathbf{Y}'\mathbf{A}(\mathbf{A}'\mathbf{A})^{-1}\mathbf{L}'\{\mathbf{L}(\mathbf{A}'\mathbf{A})^{-1}\mathbf{L}'\}^{-1}\mathbf{L}(\mathbf{A}'\mathbf{A})^{-1}\mathbf{A}'\mathbf{Y}. \end{aligned} \tag{7.59}$$

Other forms of hypothesis within logratio linear models

Although all the hypotheses within logratio linear models which we shall consider in this monograph are of the forms already discussed, it should be pointed out that these do not exhaust the possibilities. For example, there is a well-established theory for testing hypotheses of the form

$$\mathbf{L}\boldsymbol{\Theta}\mathbf{K} = \mathbf{0} \tag{7.60}$$

within the logratio linear model $\mathbf{Y} = \mathbf{A}\boldsymbol{\Theta} + \mathbf{E}$ through the use of either the generalized likelihood ratio test or the union-intersection principle (Mardia, Kent and Bibby, 1979, Section 6.3). For example, with

$$\mathbf{L} = \begin{bmatrix} 0 & 1 & 0 & 0 \\ 0 & 0 & 1 & 0 \\ 0 & 0 & 0 & 1 \end{bmatrix}, \qquad \mathbf{K} = \begin{bmatrix} 1 \\ -1 \end{bmatrix}$$

we would have the hypothesis that $\beta_1 = \beta_2, \gamma_1 = \gamma_2, \delta_1 = \delta_2$, stating that the dependence on depth is the same for the two logratios

$\log(x_1/x_3)$, $\log(x_2/x_3)$, or equivalently that $\log(x_1/x_2)$ does not depend on depth.

The form (7.60) for a hypothesis is clearly more general than the equivalent reparametrization and constraint forms (7.39) and (7.56) already discussed since the special case $\mathbf{K} = \mathbf{I}$ of (7.60) gives (7.56). Even so, it may not be general enough to include hypotheses of possible practical interest. For example, the hypothesis that $\gamma_1 = \delta_1 = \beta_2 = \gamma_2 = 0$, or that the dependence of sediment composition on water depth takes different forms for the two logratios, linear for y_1 and logarithmic for y_2, cannot be expressed in the form (7.60). For such a hypothesis a convenient test is the Wald (1943) test which is asymptotically equivalent to the generalized likelihood ratio test: see, for example, Aitchison and Silvey (1960) for a simple account of the procedure. For this purpose we would require an estimate of the variances and covariances of the maximum likelihood estimator $\hat{\boldsymbol{\Theta}}_{\mathrm{m}}$ under the model. The standard result here is that if the d columns of $\hat{\boldsymbol{\Theta}}_{\mathrm{m}}$ are formed into an extended vector $\mathrm{vec}(\hat{\boldsymbol{\Theta}}_{\mathrm{m}})$ of order $dp_{\mathrm{m}} \times 1$, in the sediment example $(\hat{\alpha}_1, \hat{\beta}_1, \hat{\gamma}_1, \hat{\delta}_1, \hat{\alpha}_2, \hat{\beta}_2, \hat{\gamma}_2, \hat{\delta}_2)$, then an estimate of the covariance matrix of $\mathrm{vec}(\hat{\boldsymbol{\Theta}}_{\mathrm{m}})$ is given by

$$\hat{\boldsymbol{\Sigma}}_{\mathrm{m}} \otimes (\mathbf{A}'\mathbf{A})^{-1}, \tag{7.61}$$

where $\otimes$ denotes the Kronecker product of two matrices.

7.9 An application of logratio linear modelling

We now consider the application of logratio linear modelling to answer Questions 1.6.1 and 1.6.2 concerning supervisory behaviour with compositions of Data 7. We first transform the 18 compositions to their 3-dimensional logratio vectors, denoting by $\mathbf{y}_{ij}$ the 3×1 logratio composition associated with the ith supervisee and the jth fortnight. In standard multivariate analysis we may consider a model which allows both a supervisee and a time effect

$$\mathbf{y}_{ij} = \boldsymbol{\mu} + \boldsymbol{\alpha}_i + \boldsymbol{\beta}_j + \mathbf{e}_{ij},$$

within which we can test hypotheses such as:

1. Random variability only,

$$\mathbf{y}_{ij} = \boldsymbol{\mu} + \mathbf{e}_{ij}.$$

2. Supervisee effect only

$$\mathbf{y}_{ij} = \boldsymbol{\mu} + \boldsymbol{\alpha}_i + \mathbf{e}_{ij}.$$

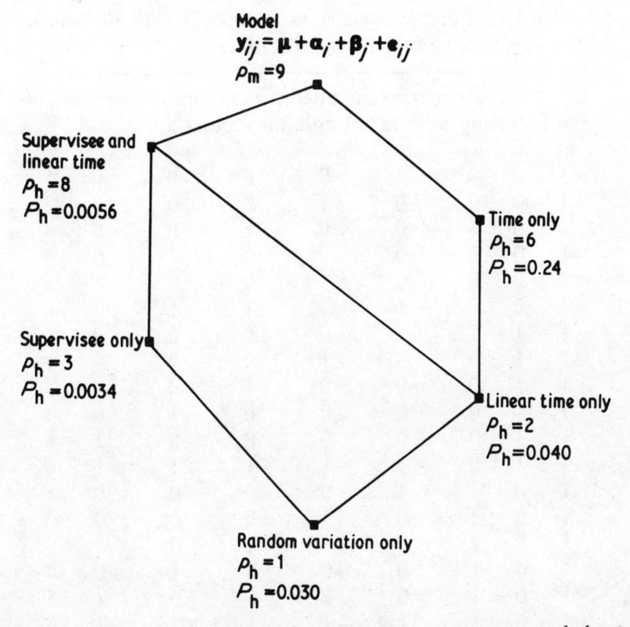

Fig. 7.6. *Lattice of hypotheses for investigating supervisory behaviour.*

3. Time effect only,

$$\mathbf{y}_{ij} = \boldsymbol{\mu} + \boldsymbol{\beta}_j + \mathbf{e}_{ij}.$$

4. Linear time effect only,

$$\mathbf{y}_{ij} = \boldsymbol{\mu} + j\boldsymbol{\beta} + \mathbf{e}_{ij}.$$

5. Supervisee and linear time effect,

$$\mathbf{y}_{ij} = \boldsymbol{\mu} + \boldsymbol{\alpha}_i + j\boldsymbol{\beta} + \mathbf{e}_{ij}.$$

In the above forms we would have to introduce constraints on the parameters to ensure identifiability: to avoid this and the nuisance of working with covariate matrices not of full rank, we prefer to modify the model and hypotheses by reparametrizing so that each is associated with a covariate matrix of full rank. The full lattice for this testing process is shown in Fig. 7.6 and the associated covariate matrices are identified in Table 7.10.

From the significance probabilities recorded in the lattice we arrive

Table 7.10 *Covariate matrices associated with the model and hypotheses of the lattice of Fig. 7.6*

The covariate matrices are defined in terms of selections of the following nine 18×1 column vectors $\mathbf{a}_1, \ldots, \mathbf{a}_9$

$\mathbf{a}_1$	$\mathbf{a}_2$	$\mathbf{a}_3$	$\mathbf{a}_4$	$\mathbf{a}_5$	$\mathbf{a}_6$	$\mathbf{a}_7$	$\mathbf{a}_8$	$\mathbf{a}_9$
1	0	0	1	0	0	0	0	1
1	0	0	0	1	0	0	0	2
1	0	0	0	0	1	0	0	3
1	0	0	0	0	0	1	0	4
1	0	0	0	0	0	0	1	5
1	0	0	−1	−1	−1	−1	−1	6
1	1	0	1	0	0	0	0	1
1	1	0	0	1	0	0	0	2
1	1	0	0	0	1	0	0	3
1	1	0	0	0	0	1	0	4
1	1	0	0	0	0	0	1	5
1	1	0	−1	−1	−1	−1	−1	6
1	0	1	1	0	0	0	0	1
1	0	1	0	1	0	0	0	2
1	0	1	0	0	1	0	0	3
1	0	1	0	0	0	1	0	4
1	0	1	0	0	0	0	1	5
1	0	1	−1	−1	−1	−1	−1	6

Model $\mathbf{a}_1$ $\mathbf{a}_2$ $\mathbf{a}_3$ $\mathbf{a}_4$ $\mathbf{a}_5$ $\mathbf{a}_6$ $\mathbf{a}_7$ $\mathbf{a}_8$

Hypothesis

1. $\mathbf{a}_1$
2. $\mathbf{a}_1$ $\mathbf{a}_2$ $\mathbf{a}_3$
3. $\mathbf{a}_1$ $\mathbf{a}_4$ $\mathbf{a}_5$ $\mathbf{a}_6$ $\mathbf{a}_7$ $\mathbf{a}_8$
4. $\mathbf{a}_1$ $\mathbf{a}_9$
5. $\mathbf{a}_1$ $\mathbf{a}_2$ $\mathbf{a}_3$ $\mathbf{a}_9$

at a working model corresponding to a general time effect only, say

$$\mathbf{y}_{ij} = \boldsymbol{\mu}_j + \mathbf{e}_{ij},$$

with estimates

$$\hat{\boldsymbol{\mu}}_1 = (0.623, -0.135, 2.27)$$
$$\hat{\boldsymbol{\mu}}_2 = (0.191, 0.898, 0.801)$$
$$\hat{\boldsymbol{\mu}}_3 = (-0.009, 1.203, 0.734) \qquad \hat{\boldsymbol{\Sigma}} = \begin{bmatrix} 0.488 & 0.219 & -0.075 \\ 0.219 & 0.420 & -0.262 \\ -0.075 & -0.262 & 0.499 \end{bmatrix}$$
$$\hat{\boldsymbol{\mu}}_4 = (0.643, 0.155, 1.196)$$
$$\hat{\boldsymbol{\mu}}_5 = (0.991, 1.138, 0.769)$$
$$\hat{\boldsymbol{\mu}}_6 = (1.269, -1.387, 1.730)$$

7.10 Predictive distributions, atypicality indices and outliers

Question 1.2.2 asked whether the composition

$$(44.0, 20.4, 13.9, 9.1, 12.6)$$

is reasonably typical of the hongite experience of Data 1. Within other compositional data sets some particular compositions may be suspected of being outliers in the usual statistical sense and may in consequence have an undue influence on our statistical analysis. For example, is kongite composition K5 of Data 2 with its large B and small C components possibly an outlier?

In attempting to compare a composition with the variability experienced in a compositional data set $\mathbf{X}$ we shall find it convenient to encapsulate the experience of $\mathbf{X}$ within an assessment of the compositional distribution or density function. If this distribution can be assumed to be $\mathscr{L}^d(\boldsymbol{\mu}, \boldsymbol{\Sigma})$ with, of course, $\boldsymbol{\mu}$ and $\boldsymbol{\Sigma}$ unknown, then our problem is to try to find some reasonable assessment of the density function $p(\mathbf{x}|\boldsymbol{\mu}, \boldsymbol{\Sigma})$. The unreliability of the so-called estimative method which simply substitutes estimates $\hat{\boldsymbol{\mu}}, \hat{\boldsymbol{\Sigma}}$ for parameters to obtain $p(\mathbf{x}|\hat{\boldsymbol{\mu}}, \hat{\boldsymbol{\Sigma}})$ is well documented (Aitchison and Dunsmore, 1975; Aitchison, 1976): the reason is basically that such a substitution takes no account of the sampling variability of the estimation process. Instead we recommend the use of a predictive distribution which takes account of sampling variability by forming a mixture of all possible $p(\mathbf{x}|\boldsymbol{\mu}, \boldsymbol{\Sigma})$ in which the weighting of particular $(\boldsymbol{\mu}, \boldsymbol{\Sigma})$ is roughly proportional to the plausibility of $(\boldsymbol{\mu}, \boldsymbol{\Sigma})$; see Aitchison and Dunsmore (1975) for precise details.

The effect here is to produce as assessment $p(\mathbf{x}|\mathbf{X})$ of the density function of a future composition $\mathbf{x}$ following the same pattern of variability as the compositions of the array $\mathbf{X}$ what could be termed a logistic-Student density function on $\mathscr{S}^d$. This term is appropriate since the distribution on $\mathscr{S}^d$ is obtained from a multivariate Student distribution on $\mathscr{R}^d$ by an additive generalized logistic transformation in exactly the same way as $\mathscr{L}^d$ is obtained from $\mathscr{N}^d$. We prefer here to express the result in terms of the logratio vector $\mathbf{y}$ of $\mathbf{x}$ and the logratio array $\mathbf{Y}$ formed from $\mathbf{X}$. We follow the definition and notation of Aitchison and Dunsmore (1975, Table 2.2) for the multivariate Student distribution.

Definition 7.5 The distribution of $\mathbf{y} \in \mathscr{R}^d$ is said to be *multivariate Student*, written $\mathrm{St}^d(k, \boldsymbol{\mu}, \boldsymbol{\Sigma})$, where the parameter $\boldsymbol{\mu}$ is a $d \times 1$ vector and the parameter $\boldsymbol{\Sigma}$ is a $d \times d$ non-singular matrix, when its density function is

$$\frac{\Gamma\{\frac{1}{2}(k+1)\}}{\pi^{d/2}\Gamma\{\frac{1}{2}(k-d+1)\}|k\boldsymbol{\Sigma}|^{1/2}} \cdot \frac{1}{\{1+(\mathbf{y}-\boldsymbol{\mu})'(k\boldsymbol{\Sigma})^{-1}(\mathbf{y}-\boldsymbol{\mu})\}^{(k+1)/2}}.$$

The *predictive density function* assessment $p(\mathbf{y}|\mathbf{Y})$ which we shall use for the $\mathcal{N}^d(\boldsymbol{\mu}, \boldsymbol{\Sigma})$ form of $p(\mathbf{y}|\boldsymbol{\mu}, \boldsymbol{\Sigma})$ is then

$$\mathrm{St}^d\{n, \hat{\boldsymbol{\mu}}, (1+N^{-1})\hat{\boldsymbol{\Sigma}}\}, \tag{7.62}$$

where $\hat{\boldsymbol{\mu}}$ and $\hat{\boldsymbol{\Sigma}}$ are given by (7.7)–(7.9) and $n = N-1$.

To return now to the question at the beginning of this section. Is the composition (44.0, 20.4, 13.9, 9.1, 12.6) reasonably typical of the hongite experience or is it essentially an outlier? We can conveniently base our assessment on the following consideration. The predictive distribution assigns a probability density to each possible composition and the smaller the density assigned to a composition, the more it inclines to atypicality. First determine the density associated with the given composition. Then compute the probability, on the basis of the predictive distribution, that a hongite composition has a density greater than the density of the given composition and call this the *atypicality index* of the composition. The atypicality index therefore ranges between 0 and 1 and the closer it is to 1, the more atypical is the given composition. Fortunately, atypicality indices associated with the predictive density function (7.62) are easily computed through the use of incomplete beta functions: see Aitchison and Dunsmore (1975, Section 11.4) for details. We here simply quote the result.

Property 7.5 The atypicality index of a composition $\mathbf{x}$ with respect to a compositional data set $\mathbf{X}$ is given by

$$\mathcal{I}_{q(\mathbf{y})/\{q(\mathbf{y})+n\}}\{\tfrac{1}{2}d, \tfrac{1}{2}(n-d+1)\},$$

where

$$q(\mathbf{y}) = (1+N^{-1})^{-1}(\mathbf{y}-\hat{\boldsymbol{\mu}})'\hat{\boldsymbol{\Sigma}}^{-1}(\mathbf{y}-\hat{\boldsymbol{\mu}}).$$

For the composition (44.0, 20.4, 13.9, 9.1, 12.6) relative to the hongite experience we have $q(\mathbf{y}) = 27.67$ and the atypicality index is 0.997. We must therefore express some doubt as to the specimen being an example of hongite.

A related concept here is that of a predictive region. On the basis of the hongite experience of Data 1, within which region of $\mathscr{S}^4$ may we expect, say, 95 per cent of compositions of new hongite specimens to lie? If we again capture the hongite experience in the predictive density function (7.62), then clearly the required region will consist of every composition with atypicality index at most 0.95. We can easily construct this region in terms of the logratio vector **y** as

$$\{\mathbf{y}: q(\mathbf{y}) \leqslant q\},$$

where

$$\mathscr{I}_{q/(q+n)}\{\tfrac{1}{2}d, \tfrac{1}{2}(n-d+1)\} = 0.95.$$

In the transformed space $\mathscr{R}^d$ a predictive region is thus contained within a familiar ellipsoidal boundary and it is straightforward to translate this into an equivalent, though not ellipsoidal, region in $\mathscr{S}^d$.

Property 7.6 A predictive region of content c for a D-part composition **x** based on the experience of a compositional data set **X** is given by

$$\{\mathbf{x}: q[\mathrm{alr}(\mathbf{x})] \leqslant q\},$$

where $q(\mathbf{y})$ ($\mathbf{y} \in \mathscr{R}^d$) is as defined in Property 7.5 and q is determined by

$$\mathscr{I}_{q/(q+n)}\{\tfrac{1}{2}d, \tfrac{1}{2}(n-d+1)\} = c.$$

For 3-part compositions we can give a graphical representation of a predictive region within the ternary diagram. For example, for the AFM aphyric Skye lava compositions of Data 6, $d = 2$, $N = 23$, $n = 22$ and, for content 0.95,

$$\mathscr{I}_{q/(q+22)}(1, 10.5) = 0.95$$

gives $q = 7.26$. It is then straightforward, though tedious, to convert the elliptical predictive region for logratio **y** into the corresponding predictive region in the AFM triangle by the transformation $\mathbf{x} = \mathrm{agl}(\mathbf{y})$. Figure 7.7 shows the resulting predictive region and its relation to the original data set within the ternary diagram.

To answer the question of whether any of the kongite compositions should be considered as outliers, we can again consider using the concept of atypicality index to compare each composition $\mathbf{x}_r$ ($r = 1, \ldots, N$) with the experience of the data array $\mathbf{X}_{-r}$ formed from **X** by deleting the composition $\mathbf{x}_r$. This is a familiar missing-one-out technique which is easy to apply with adjustment algorithms for

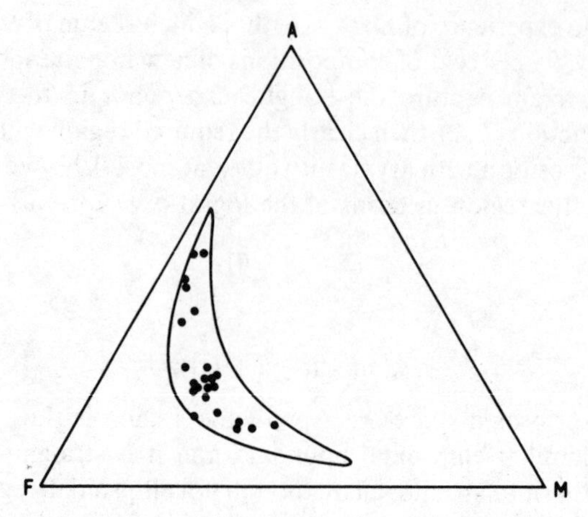

Fig. 7.7. *Predictive region of 0.95 content for AFM compositions of aphyric Skye lavas.*

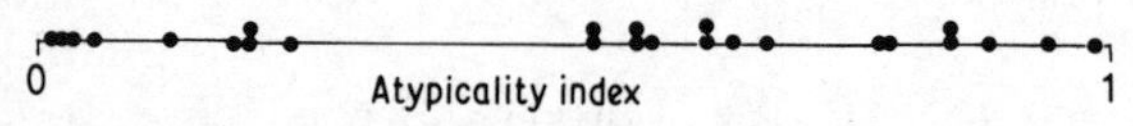

Fig. 7.8. *Atypicality indices for kongite compositions by the missing-one-out technique.*

estimates of mean vectors and covariance matrices along the lines of Effroymson (1960). For the kongite compositions of Data 2 the atypicality indices so assessed by this missing-one-out technique are shown in the (0, 1) interval in Fig. 7.8. Composition K5 with its atypicality index 0.85 cannot be regarded as an outlier. The most atypical kongite composition is K9 with index 0.98 but for a sample size 25 this can hardly be condemned as an outlier.

For the Arctic lake sediment compositions of Data 5, if we apply this missing-one-out technique to the 39 residual vectors $(\hat{e}_1, \hat{e}_2)$ of (7.53) we find that the atypicality index of S12 is 0.9998, confirming our suspicions in Section 7.8 of its outlier status. Such an analysis also raises the question of whether S7 with its atypicality index of 0.9990 may also be an outlier.

7.11 Statistical discrimination

Our finding in Section 7.5 that the hongite and kongite compositions of Data 1, 2 appear to differ in their mean logratio vectors but not in their logratio covariance matrices immediately suggests that standard linear discriminant analysis developed for variation in $\mathscr{R}^d$ may be adapted to obtain a discriminant rule for compositions in $\mathscr{S}^d$. As in previous work, a linear combination in $\mathscr{R}^d$ corresponds to a logcontrast in $\mathscr{S}^d$. In discriminating between two types 1 and 2 of compositions, such as hongite and kongite, we may in standard terms consider allocating a composition $\mathbf{x}$ to type 1 if

$$p(\mathbf{y}|\hat{\boldsymbol{\mu}}_1,\hat{\boldsymbol{\Sigma}})/p(\mathbf{y}|\hat{\boldsymbol{\mu}}_2,\hat{\boldsymbol{\Sigma}}) > 1, \tag{7.63}$$

where $\hat{\boldsymbol{\mu}}_1,\hat{\boldsymbol{\mu}}_2,\hat{\boldsymbol{\Sigma}}$ are $\mathbf{m}_1,\mathbf{m}_2,\mathbf{S}_p$ of Table 7.7. This simplifies to

$$(\hat{\boldsymbol{\mu}}_1-\hat{\boldsymbol{\mu}}_2)'\hat{\boldsymbol{\Sigma}}^{-1}\mathbf{y}-\tfrac{1}{2}\hat{\boldsymbol{\mu}}_1\hat{\boldsymbol{\Sigma}}^{-1}\hat{\boldsymbol{\mu}}_1+\tfrac{1}{2}\hat{\boldsymbol{\mu}}_2\hat{\boldsymbol{\Sigma}}^{-1}\hat{\boldsymbol{\mu}}_2>0, \tag{7.64}$$

where the first term could be readily expressed as a logcontrast of the composition $\mathbf{x}$. If we divide the logcontrast discriminant score in (7.64) by an estimate of its standard error, namely

$$s=\{(\hat{\boldsymbol{\mu}}_1-\hat{\boldsymbol{\mu}}_2)'\hat{\boldsymbol{\Sigma}}^{-1}(\hat{\boldsymbol{\mu}}_1-\hat{\boldsymbol{\mu}}_2)\}^{1/2},$$

we obtain standardized logcontrast discriminant scores. For hongite and kongite compositions these scores will be distributed approximately as $\mathscr{N}(\frac{1}{2}s,1)$ and $\mathscr{N}(-\frac{1}{2}s,1)$, and a rough approximation of the anticipated misclassification rate for each type is $\Phi(-\frac{1}{2}s)$.

For the hongite and kongite compositions the standardized logcontrast discriminant score may be computed as

$$\begin{aligned}&28.6\log x_1+30.0\log x_2+29.7\log x_3\\&\quad+43.7\log x_4-44.5\log x_5-71.6,\end{aligned} \tag{7.65}$$

and $s=1.98$, so that $\Phi(-\frac{1}{2}s)=0.16$. As a first assessment of discriminatory power we show the discriminant scores for the 25 hongite and 25 kongite compositions of Data 1, 2 in Fig. 7.9. To demonstrate that the apparent success is not due to the well-

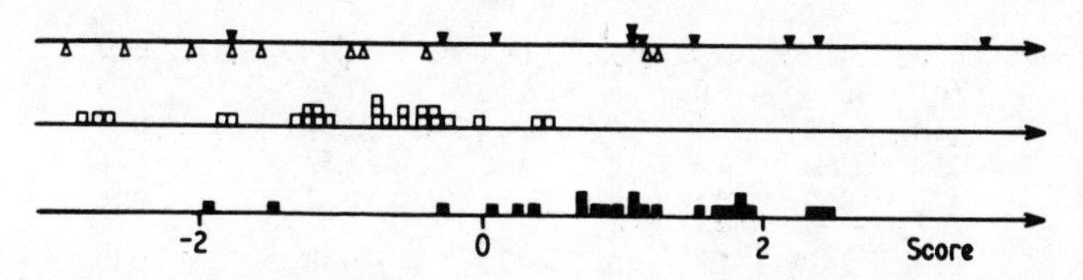

Fig. 7.9. *Standardized logcontrast discriminant scores for the hongite (solid squares) and kongite (open squares) compositions of Data 1 and 2 and the new cases h1–h10 (solid triangles) and k1–k10 (open triangles) of Table 7.11.*

documented favourable bias which such a resubstitution assessment can produce, we compute the scores for the twenty new specimens, ten of each type, recorded in Table 7.11. These are also shown in Fig. 7.9.

Rather than allocate a new case to a particular type, we may prefer simply to assess for it the probabilities of the possible types. From the experience of the compositional data sets $\mathbf{X}_1$ and $\mathbf{X}_2$ for the two types 1 and 2 we can construct as in Section 7.10 assessments of their compositional density functions in the form of predictive density functions. Here, because of the adoption of the assumption of a common logratio covariance matrix, these take the logratio form

$$p_1(\mathbf{y}|\mathbf{Y}_1, \mathbf{Y}_2) = \mathrm{St}^d\{N_1 + N_2 - 2, \hat{\boldsymbol{\mu}}_1, (1 + N_1^{-1})\hat{\boldsymbol{\Sigma}}\},$$
$$p_2(\mathbf{y}|\mathbf{Y}_1, \mathbf{Y}_2) = \mathrm{St}^d\{N_1 + N_2 - 2, \hat{\boldsymbol{\mu}}_2, (1 + N_2^{-1})\hat{\boldsymbol{\Sigma}}\},$$

where $\hat{\boldsymbol{\mu}}_1, \hat{\boldsymbol{\mu}}_2, \hat{\boldsymbol{\Sigma}}$ are the estimates $\mathbf{m}_1, \mathbf{m}_2$ and $\mathbf{S}_p$ of Table 7.7. To assign probabilities to type 1 and 2 for a new case with logratio vector $\mathbf{y}$, Bayes's formula allows us to convert prior odds of type 1 to type 2 in

Table 7.11 *Twenty new compositions of hongite h1–h10 and of kongite k1–k10, their standardized discriminant scores and their predictive probabilities for hongite*

Case no.	Percentages A	B	C	D	E	Standardized discriminant score	Predictive probability for hongite
h1	49.8	28.9	4.2	12.0	5.1	2.41	0.98
h2	43.3	43.3	1.3	6.3	5.8	−1.80	0.05
h3	33.6	8.0	38.0	13.6	6.8	−0.30	0.37
h4	48.7	19.4	12.3	9.5	10.1	1.49	0.94
h5	48.5	26.5	6.7	7.5	10.8	0.06	0.53
h6	38.3	8.1	33.9	12.3	7.4	1.06	0.87
h7	31.6	6.8	43.2	8.0	10.4	1.17	0.89
h8	34.5	10.4	34.8	9.1	11.2	1.07	0.86
h9	43.3	44.6	1.1	4.9	6.0	3.61	0.99
h10	45.5	40.6	1.4	8.3	4.2	2.19	0.97
k1	40.2	48.1	0.8	3.8	7.1	−2.07	0.04
k2	50.0	25.4	6.0	6.3	12.3	−1.78	0.05
k3	15.4	1.8	71.7	4.5	6.6	1.19	0.87
k4	52.4	26.4	4.4	10.3	6.5	−1.59	0.06
k5	47.3	35.7	2.4	9.0	5.6	−0.92	0.16
k6	51.7	24.0	5.9	5.9	12.5	−0.88	0.19
k7	49.9	16.2	13.9	11.4	8.6	−0.41	0.32
k8	42.1	11.9	24.9	11.4	9.7	−2.58	0.01
k9	34.5	6.6	40.3	11.2	7.4	1.16	0.88
k10	54.3	23.7	4.3	12.5	5.2	−3.02	0.01

the following way:

$$\text{posterior odds} = \text{prior odds} \times \frac{p_1(\mathbf{y}|\mathbf{Y}_1, \mathbf{Y}_2)}{p_2(\mathbf{y}|\mathbf{Y}_1, \mathbf{Y}_2)}.$$

If, for illustrative purposes, we assume hongite and kongite specimens to be equally likely to occur, then the prior odds is taken as 1 and we can obtain the posterior odds directly from the ratio of predictive density functions evaluated at $\mathbf{y}$. For the twenty new cases these assessed odds have been converted into probabilities for hongite and are shown in Table 7.11. For each type the proportion of wrong assignments on the basis of these probabilities is 0.2, somewhat in excess of the optimistic assessment of 0.16 above.

Note that we have again used here the predictive density function rather than the estimative density function: for the many advantages of this, see Aitchison, Habbema and Kay (1977).

We now examine Data 20 with its question of classification as posed in Problem 1.5. All 21 compositions are shown in the ternary diagram of Fig. 7.10. We adopt $\mathscr{L}^2(\boldsymbol{\mu}_1, \boldsymbol{\Sigma}_1)$ and $\mathscr{L}^2(\boldsymbol{\mu}_2, \boldsymbol{\Sigma}_2)$ distributions for nearshore and offshore compositions. The statistical problem is again assumed to be the assessment of a reasonable factor for the conversion of prior odds to posterior odds for type. With such a small data set we again adopt a predictive approach to the typing or

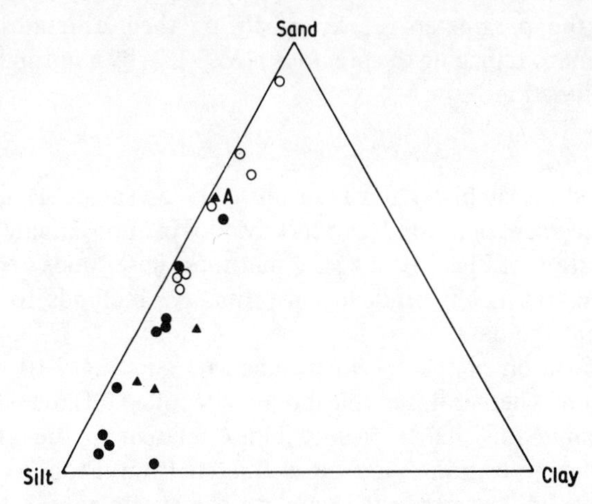

Fig. 7.10. *Ternary diagram for (sand, silt, clay) compositions of 10 offshore (solid circles), 7 nearshore (open circles) sediments and 4 sediment samples of unknown type (solid triangles).*

diagnostic problem, again for the advantages argued, for example, in Aitchison, Habbema and Kay (1977).

The unusual feature of this example, in contrast to the more familiar area of application of predictive discrimination, namely medical diagnosis, is that for the new case we have four replicate observations. For the purposes of predictive diagnosis the predictive problem can then be condensed into obtaining a predictive density function for $\mathbf{M}$ and $\mathbf{V}$, the mean vector and matrix of corrected cross-products of the four (in general N) vectors of logratios. The appropriate theory is summarized in Aitchison and Dunsmore (1975, Table 2.3) and leads to predictive distributions $p(\mathbf{M}, \mathbf{V}|\mathbf{X}_i)$ for $\mathbf{M}$ and $\mathbf{V}$ of Student–Siegel type based on data $\mathbf{X}_i$:

$$\mathrm{StSi}_2\{N_i - 1, \hat{\boldsymbol{\mu}}_i, (N_i^{-1} + N^{-1})\hat{\boldsymbol{\Sigma}}_i; N - 1, (N_i - 1)\hat{\boldsymbol{\Sigma}}_i\},$$

where $N_i, \hat{\boldsymbol{\mu}}_i$ and $\hat{\boldsymbol{\Sigma}}_i$ are respectively the number, the logratio mean vector and covariance matrix of the logratios, of vectors in the data set $\mathbf{X}_i$ for type i. In this assessment we have used the vague prior suggested by Aitchison and Dunsmore (1975) for the reasons given in Aitchison (1976). Straightforward computation then gives $p(\mathbf{M}, \mathbf{V}|\mathbf{X}_1)/p(\mathbf{M}, \mathbf{V}|\mathbf{X}_2) = 0.19$ as the converting factor from prior odds to posterior odds on type 1. Thus if type 1 and 2 are equally likely *a priori*, our evidence leads to odds of 5 to 1 in favour of type 2.

The predictive method can be contrasted with the estimative method (Aitchison, Habbema and Kay, 1977) which would simply replace the parameters $\boldsymbol{\mu}_1, \boldsymbol{\Sigma}_1, \boldsymbol{\mu}_2, \boldsymbol{\Sigma}_2$ by their estimates. This is equivalent to replacing the previous $p(\mathbf{M}, \mathbf{V}|\mathbf{X}_i)$ by a normal-Wishart density function

$$\mathrm{NoWi}_2(\hat{\boldsymbol{\mu}}_i, \hat{\boldsymbol{\Sigma}}_i/N, N - 1, \hat{\boldsymbol{\Sigma}}_i)$$

as defined in Aitchison and Dunsmore (1975, Table 2.1), leading to the replacement of the odds of 5 to 1 by odds of approximately 80 to 1. Examination of Fig. 7.10 suggests that these latter odds are extravagant, illustrating the tendency of estimative methods to read too much into the data.

The position of the specimen labelled A in Fig. 7.10 raises the question of whether it is atypical of the identified offshore standards. To examine this, its atypicality index relative to the other nine offshore compositions may be evaluated from Property 7.5. The atypicality index is only 0.73 and so the specimen can hardly be regarded as atypical.

7.12 Conditional compositional modelling

Data 20 can be further used to illustrate the nature of the conditioning property. Suppose that for an offshore specimen we wish to study the variability of the composition of (sand, clay) for a given silt/clay ratio. We can proceed as follows. First find the predictive distribution for a new complete logratio vector based on $\mathbf{X}_2$. This will be 2-dimensional Student with 9 degrees of freedom. We can then easily derive the predictive conditional distribution of (sand, clay) for given value v of log(silt/clay) as Student with 9 degrees of freedom. In more familiar terms the conditional distribution of log(sand/clay) is

$$\mathrm{St}^1\{9, -3.29 + 1.66v, 2.10 + 0.422(v - 2.71)^2\}$$

in the notation of Definition 7.5. Figure 7.11 shows the considerable differences in this logistic-Student distribution for three silt/clay ratios spanning the range of silt/clay ratios observed in the specimens.

7.13 Bibliographic notes

The fact that the application of the additive logratio transformation to compositions opened up the whole of the standard repertoire of multivariate analysis for the statistical investigation of compositional data was first pointed out by Aitchison and Shen (1980). Most of the

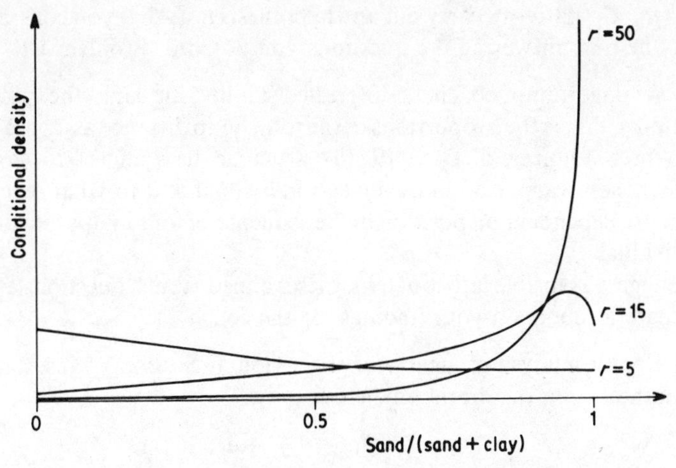

Fig. 7.11. *Conditional density functions of (sand, clay) composition for given value r of silt/clay ratio.*

particular forms of logratio analysis described in this chapter are reported by Aitchison and Shen (1980) and Aitchison (1982).

Problems

7.1 Investigate the validity of a logistic normal form for the variability of
(i) the AFM compositions of the aphyric Skye lavas of Data 6;
(ii) the statistician's activity patterns of Data 10.

7.2 For the subjective diagnostic assessments of Data 17 estimate, separately for clinicians and statisticians, the logratio mean vector and covariance matrix.

Investigate whether there are differences between clinicians and statisticians in the patterns of their diagnostic assessments. Could either clinicians or statisticians as a group be considered as successful in pinpointing the objective diagnostic assessment (0.08, 0.05, 0.87)?

7.3 Subject the two compositional data sets, the 6-part colour-size compositions of Data 14 and 15, to the tests of the lattice of Fig. 7.2. What relevance do your findings have to the questions posed in Section 1.13?

7.4 On the assumption that the two sets of serum protein compositions A1–A14 and B1–B16 of Data 16 follow $\mathscr{L}^3(\boldsymbol{\mu}_1, \boldsymbol{\Sigma}_1)$ and $\mathscr{L}^3(\boldsymbol{\mu}_2, \boldsymbol{\Sigma}_2)$ patterns, carry out any hypothesis tests that you consider may help in answering the questions you posed in Problem 1.1.

7.5 A pogo-jump coach is interested in investigating the jump patterns, that is the proportions of the total jump distance assigned to the three bounces, of Data 19. In particular, he wonders to what extent these vary from individual to individual and to what extent they are dependent on position in the sequence of four jumps for each individual.

Suggest a sensible lattice of hypotheses aimed at such questions and prepare a report on your findings for the coach.

7.6 For the analysis of supervisory behaviour in Section 7.9 insert the hypothesis of a quadratic time effect only,

$$\mathbf{y}_{ij} = \boldsymbol{\mu} + \boldsymbol{\beta}_1 j + \boldsymbol{\beta}_2 j^2 + \mathbf{e}_{ij},$$

at an appropriate position in the lattice of Fig. 7.6. Test this hypothesis and consider how it may affect the conclusions already drawn from investigation of the simpler lattice.

7.7 The statistician's activity patterns of Data 10 are associated with a sequence of alternate weeks. Is there any evidence of a trend in these patterns over time?

7.8 Compute the atypicality indices of the four AFM compositions in Table 7.12 with respect to the aphyric Skye lavas of Data 6.

Table 7.12. *AFM compositions*

Specimen no.	*A*	*F*	*M*
T1	21	64	16
T2	43	47	10
T3	14	42	44
T4	25	60	15

Verify that the positions of the points representing these four compositions relative to the predictive region of Fig. 7.7 conform with the atypicality indices you have computed.

7.9 Devise a system from the serum protein compositions A1–A14 and B1–B16 of Data 16 for the differential diagnosis of diseases A and B. What are the implications of your system for the new patients C1–C6 of Data 16?

7.10 On the assumption of logistic normality of the AFM compositions of Data 6, find the predictive forms for the conditional distributions of the AM subcomposition for the given values 1.5, 5, 10 of the ratio F/M. Sketch the graphs of the corresponding density functions. Do the differences in these correspond to what you might have expected from study of the ternary diagram of Fig. 1.9?

7.11 Reinvestigate the dependence of the Arctic lake sediment compositions of Data 5 on depth excluding sediment S12. Comment on the influence this sediment may have had on the previous analyses, including the logistic normality of the residuals.

CHAPTER 8

Dimension-reducing techniques

Simplicity, simplicity, simplicity! I say, let your affairs be as two or three, and not a hundred or a thousand. Simplify, simplify.

H.D. Thoreau: *Walden*

8.1 Introduction

The fact that many compositions are high-dimensional, for example geochemical compositions consisting of ten or more oxides, and the unfortunate inability of the human eye to see in more than three dimensions have naturally led analysts to consider devices for obtaining lower-dimensional insights into the nature of their data sets. Awareness of the difficulties of interpretation associated with the unit-sum constraint seems to reinforce this wish to reduce the effective dimensionality of the data set.

In handling unconstrained vectors in $\mathscr{R}^d$ we have become familiar with two forms of dimension-reducing techniques. The first is marginal analysis, where we simply select some of the components and look at their marginal distribution in fewer dimensions than d, hoping that we are thereby extracting a substantial part of the total variability. For compositions the counterpart of marginal analysis is subcompositional analysis, the device of examining selected subcompositions, which has a long history, for example in geology, dating back at least to Becke (1897) and the introduction of ternary diagrams. The second is principal component analysis with its attempt to capture most of the variability in a few linear combinations of the components. Although a version of principal component analysis dates back to Pearson (1901), the full development of the methodology in $\mathscr{R}^d$ is of more recent origin (Hotelling, 1933).

We recall the main features of standard principal component analysis for variation in $\mathscr{R}^d$. Attention is confined to standardized linear combinations $\mathbf{a}'\mathbf{y}$ ($\mathbf{y} \in \mathscr{R}^d$), with $\mathbf{a}$ of unit length or $\mathbf{a}'\mathbf{a} = 1$, and the first principal component is the standardized combination $\mathbf{a}'\mathbf{y}$ which has maximum variance. For the purposes of computation the maximum

$$\max\{\text{var}(\mathbf{a}'\mathbf{y}):\quad \mathbf{a}\in\mathcal{R}^d,\quad \mathbf{a}'\mathbf{a}=1\}$$

is usually identified with the maximum eigenvalue λ_1 of the eigenproblem

$$(\boldsymbol{\Sigma}-\lambda\mathbf{I})\mathbf{a}=\mathbf{0},$$

where $\boldsymbol{\Sigma}$ is the covariance matrix of $\mathbf{y}$. The principal component $\mathbf{a}_1'\mathbf{y}$ is obtained from the corresponding eigenvector $\mathbf{a}_1$ standardized so that $\mathbf{a}_1'\mathbf{a}_1=1$.

The complete eigenstructure of $\boldsymbol{\Sigma}$ can indeed be related to further maximization problems. Suppose that the d eigenvalues of $\boldsymbol{\Sigma}$ are ordered as $\lambda_1>\lambda_2>\cdots>\lambda_d$, with corresponding standardized eigenvectors $\mathbf{a}_1,\ldots,\mathbf{a}_d$. Then $\mathbf{a}_i'\mathbf{y}$ is the ith principal component $(i=1,\ldots,d)$ and the principal components have the following properties.

1. Principal components are orthogonal in the sense that $\mathbf{a}_i'\mathbf{a}_j=0$ $(i\neq j)$.
2. Principal components are uncorrelated, with $\text{cov}(\mathbf{a}_i'\mathbf{y},\mathbf{a}_j'\mathbf{y})=\mathbf{a}_i'\boldsymbol{\Sigma}\mathbf{a}_j$ $=0$ $(i\neq j)$.
3. $\lambda_i=\text{var}(\mathbf{a}_i'\mathbf{y})=\mathbf{a}_i'\boldsymbol{\Sigma}\mathbf{a}_i \qquad (i=1,\ldots,d)$.
4. $\lambda_i=\max\{\text{var}(\mathbf{a}'\mathbf{y}):\quad \mathbf{a}'\mathbf{a}=1,\quad \mathbf{a}'\mathbf{a}_j=0\quad (j=1,\ldots,i-1)\}$ $(i=2,\ldots,d)$.
5. $\text{tr}(\boldsymbol{\Sigma})=\lambda_1+\cdots+\lambda_d$ is commonly used as a measure of total variability of the random vector $\mathbf{y}$. The proportion of this total variability which is retained by, or contained in, the first c principal components is then

$$\sum_{i=1}^{c}\lambda_i\Big/\sum_{i=1}^{d}\lambda_i.$$

6. Geometrically the vectors $\mathbf{a}_i\,(i=1,\ldots,d)$ are the direction cosine vectors of a set of mutually orthogonal directions in $\mathcal{R}^d$ arranged in order of diminishing variation of $\mathbf{y}$ along these directions.

Our initial problem is to discover the extent to which these simple and standard procedures and properties in $\mathcal{R}^d$ can be modified for variation in the simplex $\mathcal{S}^d$.

8.2 Crude principal component analysis

It is only in the last two decades that principal component analysis has been explored as a possible dimension-reducing device in the study of patterns of variability of high-dimensional compositional data. Most of these attempts at principal component analysis for compositional data have been directed at geochemical applications, as for example by Butler (1976), Chayes and Trochimczyk (1978), Le Maître (1962, 1968, 1976), Miesch (1980), Roth, Pierce and Huang (1972), Till and Colley (1973), Trochimczyk and Chayes (1977, 1978), Vistelius *et al.* (1970), Webb and Briggs (1966). Moreover, all except the last cited have been based on the crude covariance structure with the most popular method that advocated by Le Maître (1968). Such methods are, of course, subject to all the difficulties of interpretation associated with the use of crude covariance structures. These difficulties are perhaps nowhere more obvious than within the context of principal component analysis and so we take this opportunity of reinforcing our dismissal of crude compositional analysis before taking the more positive approach of providing a more reasonable alternative.

The $D \times D$ crude covariance matrix $\boldsymbol{K} = [\text{cov}(x_i, x_j)]$ of Definition 3.1 satisfies $\boldsymbol{K}\mathbf{j} = \mathbf{0}$ and so has a zero eigenvalue with corresponding standardized eigenvector $D^{-1/2}\mathbf{j}$. The remaining d eigenvalues are, in general, positive, say $\lambda_1 > \cdots > \lambda_d$ with corresponding standardized eigenvectors $\mathbf{a}_1, \ldots, \mathbf{a}_d$. Crude principal component analysis then uses $\mathbf{a}_i'\mathbf{x}$ as the ith crude principal component. Since $\mathbf{a}_i'\mathbf{j} = 0$ the crude principal components are linear contrasts of the components of the composition.

With all the difficulties of interpretation of crude covariances as set out in Section 3.3 it must be clear that any attempt to support the use of crude principal components through their relationship to $\boldsymbol{K}$ will result in failure. The inadequacy of these principal components can, however, be seen more readily in terms of geometrical representations. This inadequacy and its eventual resolution are best discussed against the background of two specific data sets consisting of 3-part compositions and therefore capable of visual representation in ternary diagrams. Figure 8.1 shows the relative proportions of the urinary excretions of the three steroid metabolites of the 37 adults of Data 9; Fig. 8.2 shows the AFM compositions of the 23 aphyric Skye lavas of Data 6.

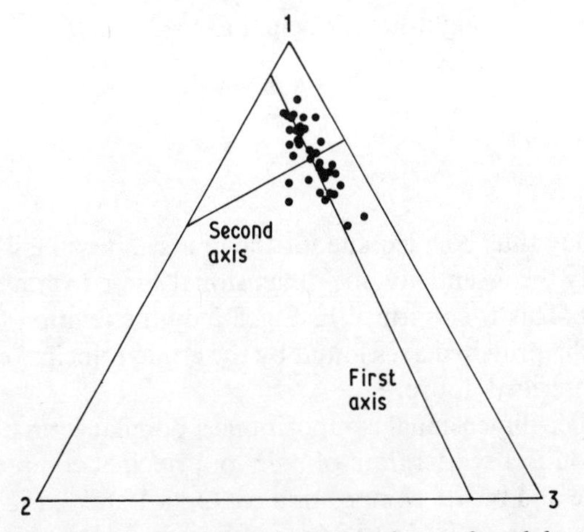

Fig. 8.1. *Ternary diagram and crude principal axes for adult steroid metabolite compositions.*

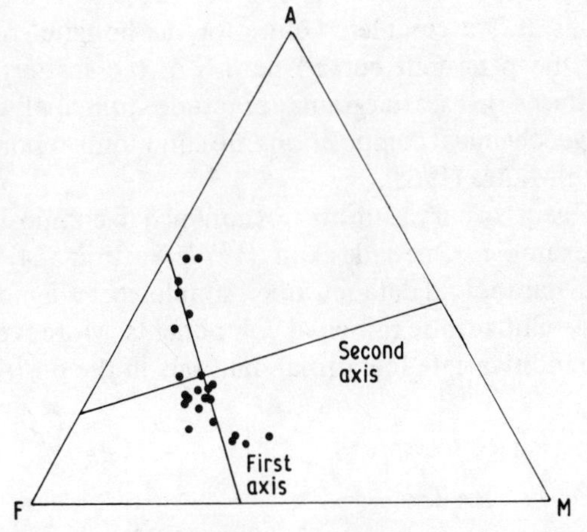

Fig. 8.2. *Ternary diagram and crude principal axes for aphyric Skye lava compositions.*

The difference between these two data sets is striking. The first is decently elliptical, a linear set in the sense of Gnanadesikan (1977, Chapter 2). The second has a decidedly nonlinear pattern but with little variation about a conceptual curved line. In purely geometrical terms crude principal component analysis is a linear reduction technique with straight-line principal axes

$$\mathbf{x} = \bar{\mathbf{x}} + t\mathbf{a}_i,$$

where

$$\bar{\mathbf{x}} = N^{-1} \sum_{r=1}^{N} \mathbf{x}_r.$$

While it may thus be adequate for the first set, it will fail to capture successfully the essentially one-dimensional curved variability of the second set. This is confirmed in Fig. 8.2 by the relation of the data points to the principal axes found by the crude principal component method described above.

For higher-dimensional compositions a popular form of graphical presentation is a scattergram of pairs of principal components. For the hongite and boxite compositions of Data 1 and 3, Fig. 8.3 shows the scattergrams of the first and second crude principal components

$$\{(\mathbf{a}_1' \mathbf{x}_r, \mathbf{a}_2' \mathbf{x}_r): \quad r = 1, \ldots, N\}.$$

Again, while the crude method appears successful for the boxite compositions, it is a complete failure for the hongite compositions because of the persistent curved nature of the scattergram. Such curved patterns in scattergrams of crude principal component analysis of geochemical compositions are commonly reported; see, for example, Le Maître (1968).

Such problems of curvature are not confined to compositional data sets. For example, Gnanadesikan (1977, Section 2.4) uses a 3-dimensional paraboloid data set, albeit simulated, to demonstrate the effectiveness of quadratic principal components. Moreover, it is easy to visualize multivariate lognormal data sets in the positive orthant

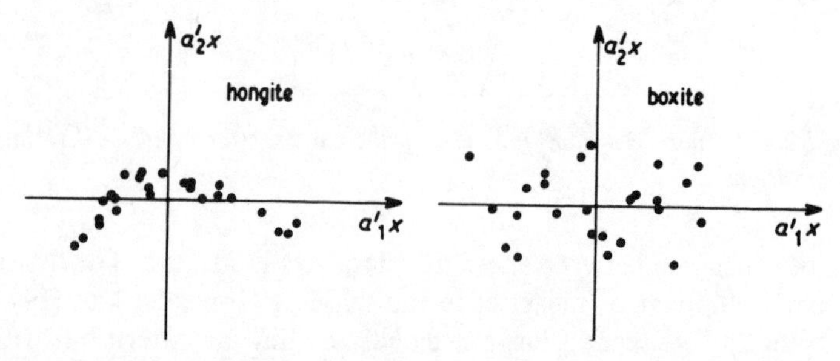

Fig. 8.3. *Scattergram of first and second crude principal components for hongite and boxite compositions.*

with hyperboloid-shaped variability, for which the standard practice of working with logarithmically transformed data would lead to successful principal component analysis.

Webb and Briggs (1966) use the covariance matrix of the ratio vector $\mathbf{x}_{-D}/x_D$ of the composition, and its d eigenvalues and eigenvectors to yield principal components. The implicit exchange of the restrictions of the simplex for the simpler restrictions of the positive orthant through the ratio transformation is certainly a step in a sensible direction. Unfortunately, this method still suffers from the inherent linearity of principal axes and the dependence of the resulting principal components on the choice of common divisor x_D, in other words lack of invariance under the group of permutations of the compositional parts.

To deal with curvature in compositional data we must clearly search for appropriate nonlinear functions of the components from which to build up a suitable principal component system. For the compositional data sets of most practical problems, of dimensions higher than our simple illustrative examples, there is no easy way of detecting whether or not there is curvature, but detailed study of a number of such sets suggests that it would be unwise to ignore the possibility of curvature. Ideally, then, we require a form of principal component analysis which will cope with both linear and nonlinear data patterns. Such a method obviously has to be nonlinear and sufficiently flexible to approximate linearity within its range.

8.3 Logcontrast principal component analysis

For a composition $\mathbf{x}$ in $\mathscr{S}^d$ we saw in Section 4.10 that the counterpart of a linear combination in $\mathscr{R}^d$ is a linear combination of the logratio vector $\mathbf{y}$ or equivalently a logcontrast $\mathbf{a}'\log\mathbf{x}$, where $\mathbf{a}'\mathbf{j}=0$. Let us therefore explore the possibility of devising principal component analysis for compositions by examining the problem of finding the standardized logcontrast (Definition 4.9) of maximum variance. By Property 5.1(a), $\mathrm{var}(\mathbf{a}'\log\mathbf{x})=\mathbf{a}'\boldsymbol{\Gamma}\mathbf{a}$, and so we have to find

$$\max\{\mathbf{a}'\boldsymbol{\Gamma}\mathbf{a}:\mathbf{a}\in\mathscr{R}^D,\quad \mathbf{a}'\mathbf{a}=1,\quad \mathbf{a}'\mathbf{j}=0\}. \tag{8.1}$$

Note that since $\mathbf{a}'\boldsymbol{\Gamma}\mathbf{a}=(-\mathbf{a})'\boldsymbol{\Gamma}(-\mathbf{a})$ we have omitted the standardizing condition $a_1>0$ from the maximizing process: this condition can easily be satisfied by a simple change of sign if necessary. By the

Lagrange multiplier technique (8.1) is equivalent to the maximization of

$$\mathbf{a}'\boldsymbol{\Gamma}\mathbf{a} - \lambda(\mathbf{a}'\mathbf{a} - 1) - 2\mu\mathbf{a}'\mathbf{j} \tag{8.2}$$

with respect to $\mathbf{a} \in \mathscr{R}^D$, $\lambda \in \mathscr{R}^1$, $\mu \in \mathscr{R}^1$. The derivative equations can be expressed as follows:

$$\boldsymbol{\Gamma}\mathbf{a} - \lambda\mathbf{a} - \mu\mathbf{j} = \mathbf{0}, \tag{8.3}$$

$$\mathbf{a}'\mathbf{a} = 1, \tag{8.4}$$

$$\mathbf{a}'\mathbf{j} = 0. \tag{8.5}$$

Premultiplications of (8.3) by $\mathbf{a}'$ and $\mathbf{j}'$ and the use of (8.4) and (8.5) give

$$\mathbf{a}'\boldsymbol{\Gamma}\mathbf{a} - \lambda = 0, \tag{8.6}$$

$$\mathbf{j}'\boldsymbol{\Gamma}\mathbf{a} - D\mu = 0. \tag{8.7}$$

Substitution for λ and μ from (8.6) and (8.7) into (8.3) then gives, after some rearrangement,

$$\{(\mathbf{I} - D^{-1}\mathbf{j}\mathbf{j}')\boldsymbol{\Gamma} - \lambda\mathbf{I}\}\mathbf{a} = \mathbf{0}$$

or

$$(\boldsymbol{\Gamma} - \lambda\mathbf{I})\mathbf{a} = \mathbf{0} \tag{8.8}$$

since $\mathbf{G}\boldsymbol{\Gamma} = \boldsymbol{\Gamma}$ by Property 4.6(b). In (8.8) we have a situation similar to that encountered in the familiar principal component analysis in $\mathscr{R}^d$. If λ_1 is the maximum eigenvalue of (8.8) and $\mathbf{a}_1$ the corresponding standardized eigenvector, then the standardized logcontrast of maximum variance is $\mathbf{a}_1' \log \mathbf{x}$ and by (8.6) this maximum variance is λ_1. In the same way as for $\mathscr{R}^d$ we can extend from the first principal component to a set of d uncorrelated standard principal components with orthogonal principal axes, and so we can use the d positive eigenvalues and their eigenvectors (Property 5.8(c)) to build up principal component analysis for compositions.

Definition 8.1 Let the d positive eigenvalues of the centred logratio covariance matrix $\boldsymbol{\Gamma}$ be labelled in descending order of magnitude, $\lambda_1 > \cdots > \lambda_d$, with corresponding standardized eigenvectors $\mathbf{a}_1, \ldots, \mathbf{a}_d$, satisfying

$$(\boldsymbol{\Gamma} - \lambda_i\mathbf{I})\mathbf{a}_i = \mathbf{0} \qquad (i = 1, \ldots, d).$$

The logcontrast $\mathbf{a}_i' \log \mathbf{x}$ is then termed the ith *logcontrast principal component*.

We can now easily record the main properties of logcontrast principal components using the notation of Definition 8.1.

Property 8.1 The logcontrast principal components $\mathbf{a}_i' \log \mathbf{x}$ $(i = 1, \ldots, d)$ of D-part compositional variation have the following properties.

(a) Logcontrast principal components are orthogonal in the sense of Definition 4.10, with $\mathbf{a}_i'\mathbf{a}_j = 0$ $(i \neq j)$.

(b) Logcontrast principal components are uncorrelated, with $\operatorname{cov}(\mathbf{a}_i' \log \mathbf{x}, \mathbf{a}_j' \log \mathbf{x}) = \mathbf{a}_i' \boldsymbol{\Gamma} \mathbf{a}_j = 0$ $(i \neq j)$.

(c) $\lambda_i = \operatorname{var}(\mathbf{a}_i' \log \mathbf{x}) = \mathbf{a}_i' \boldsymbol{\Gamma} \mathbf{a}_i \qquad (i = 1, \ldots, d)$.

(d) $\lambda_i = \max\{\operatorname{var}(\mathbf{a}' \log \mathbf{x}) : \mathbf{a} \in \mathscr{R}^D, \mathbf{a}'\mathbf{a} = 1, \mathbf{a}'\mathbf{j} = 0, \quad \mathbf{a}'\mathbf{a}_k = 0$ $(k = 1, \ldots, i-1)\} \qquad (i = 2, \ldots, d)$.

(e) $\operatorname{tr}(\boldsymbol{\Gamma}) = \lambda_1 + \cdots + \lambda_d$ may be used as a measure of total variability.

The proportion of this total variability which is retained by, or contained in, the first c logcontrast principal components is then

$$\sum_{i=1}^{c} \lambda_i \Big/ \sum_{i=1}^{d} \lambda_i.$$

Notes

1. From Property 8.1 we see that we have a form of principal component analysis for compositions which completely parallels properties (1)–(5) of principal component analysis in $\mathscr{R}^d$ as recorded in Section 8.1. The single difference is that logcontrasts replace the linear combinations of $\mathscr{R}^d$ theory.
2. The determination of logcontrast principal components makes no more computational demands than for principal components in $\mathscr{R}^d$. All that is required is an algorithm for obtaining the eigenvalues and eigenvectors of a symmetric matrix $\boldsymbol{\Gamma}$.
3. The equivalence of the three eigenproblems of Property 5.8 shows how the logcontrast principal components can be determined from either the variation matrix $\boldsymbol{T}$ or the logratio covariance matrix $\boldsymbol{\Sigma}$.
4. Our derivation of an appropriate principal component analysis throws some light on the lack of invariance of the eigensolutions of the logratio covariance matrix $\boldsymbol{\Sigma}$ under the group of permutations, and the adjustment required to remedy this which is implicit in Property 5.8(b). Principal component analysis in $\mathscr{R}^d$ can be seen as a search for standard linear combinations which show successively, within a mutually orthogonal constraint, the greatest, second

greatest, ..., least variation. In this context it is a search for mutually orthogonal directions subject to maximization requirements, and mathematically it is the selection of an optimum orthogonal transformation. When the covariance matrix is $\boldsymbol{\Sigma}$ this optimum orthogonal transformation $\mathbf{A}$ can be thought of as simultaneously reducing $\boldsymbol{\Sigma}$ to diagonal form and $\mathbf{I}_d$ trivially to $\mathbf{I}_d$:

$$\mathbf{A}'\boldsymbol{\Sigma}\mathbf{A} = \operatorname{diag}(\lambda_1, \ldots, \lambda_d), \qquad \mathbf{A}'\mathbf{I}_d\mathbf{A} = \mathbf{I}_d. \tag{8.9}$$

For our purposes an enlightening view of this is the role of the special covariance matrix $\mathbf{I}_d$. It is the unique $\mathscr{R}^d$ covariance matrix for which there is nothing to be gained by an application of principal component analysis: it is linear combination isotropic in the sense that every standardized linear combination has the same variance. The transformation (8.9) recognizes this.

When we turn to logcontrast principal component analysis through the use of the logratio covariance matrix $\boldsymbol{\Sigma}$ we now recognize the special role that the elementary matrix $\mathbf{H}$ has to play. Apart from a scalar multiple, $\mathbf{H}$ is by Property 5.15(b) the only logratio covariance matrix which is logcontrast isotropic in the sense of Definition 5.3. There is thus nothing to be gained by an application of logcontrast principal component analysis to $\mathbf{H}$. Our required matrix reduction of a general logratio covariance matrix $\boldsymbol{\Sigma}$ to the diagonal form must simultaneously leave $\mathbf{H}$ unaltered. Thus logcontrast principal component analysis seeks a $d \times d$ matrix $\mathbf{B}$ such that

$$\mathbf{B}'\boldsymbol{\Sigma}\mathbf{B} = \operatorname{diag}(\lambda_1, \ldots, \lambda_d), \qquad \mathbf{B}'\mathbf{H}\mathbf{B} = \mathbf{H}, \tag{8.10}$$

and the determination of $\lambda_1, \ldots, \lambda_d$ and $\mathbf{B} = [\mathbf{b}_1, \ldots, \mathbf{b}_d]$ is equivalent to the eigenproblem

$$(\boldsymbol{\Sigma} - \lambda_i \mathbf{H})\mathbf{b}_i = \mathbf{0} \qquad (i = 1, \ldots, d),$$

with the scale of $\mathbf{b}_i$ determined by $\mathbf{b}_i'\mathbf{H}\mathbf{b}_i = 1$.

5. If we apply the argument of note 4 to logcontrast principal component analysis through the use of the centred logratio covariance matrix $\boldsymbol{\Gamma}$ with its logcontrast isotropic form $\mathbf{G}$, we are led to the positive eigenvalues of

$$(\boldsymbol{\Gamma} - \lambda \mathbf{G})\mathbf{a} = \mathbf{0}.$$

At first sight this appears different from (8.8) but the equivalence is immediate on recognition that $\mathbf{a}'\mathbf{j} = 0$ and $\mathbf{G}\mathbf{a} = \mathbf{a}$ are identical conditions on $\mathbf{a}$, by property G11 of Appendix A.

6. Comparison of the equivalent forms of eigenproblems in $\boldsymbol{T}$, $\boldsymbol{\Sigma}$, and $\boldsymbol{\Gamma}$ as set out in Property 5.8 shows that the simplest form of analysis is in terms of $\boldsymbol{\Gamma}$.

7. In Section 2.9 we noted a duality of geometrical representation of a compositional data matrix $\mathbf{X}$, with row vectors as elements of a sample space $\mathscr{S}^d$ and with column vectors as elements of a replicate space $\mathscr{R}_+^N$. This duality can be exploited, along lines similar to those of Gower (1966), to obtain a logcontrast principal *coordinate* analysis, the dual of logcontrast principal *component* analysis for compositional data. This aspect of the duality arises from the fact that covariance between components can be identified with 'angle' between their representative vectors in the replicate space, and the dual concept of similarity of compositions can be identified with negative 'distance' between the compositional representations in the sample space. With the logratio covariance structure of Definition 4.1 for compositions, the required definition of distance between two compositions $\mathbf{x}_1$ and $\mathbf{x}_2$ is

$$\sqrt{\sum_{i=1}^{D} [\log\{x_{1_i}/g(\mathbf{x}_1)\} - \log\{x_{2_i}/g(\mathbf{x}_2)\}]^2}.$$

For practical purposes the duality can be best expressed in terms of the adjusted data matrix

$$\mathbf{U} = \mathbf{G}_N (\log \mathbf{X})\, \mathbf{G}_D,$$

in which the logarithms of the x_{ri} in $\mathbf{X}$ have been centred so that all row sums and column sums are zero. The logcontrast principal components of Definition 8.1 are then obtained as the d solutions of

$$(\mathbf{U}'\mathbf{U} - \lambda \mathbf{I})\mathbf{a} = \mathbf{0}$$

and logcontrast principal coordinates as the N solutions of

$$(\mathbf{U}\mathbf{U}' - \mu \mathbf{I})\mathbf{c} = \mathbf{0}.$$

These solutions are related, with suitable scaling, by

$$\mathbf{c} = \mathbf{U}\mathbf{a},$$

so that the N elements of $\mathbf{c}$ provide the coordinates of a particular principal component $\mathbf{a}' \log \mathbf{x}$ evaluated for each of the compositions of the data set. We shall not pursue this duality further.

8.4 Applications of logcontrast principal component analysis

We adopt here the symmetric version of Definition 8.1 using the centred logratio covariance matrix $\boldsymbol{\Gamma}$ and recall that the only computational tool required is an algorithm for the determination of eigenvalues and eigenvectors of a symmetric matrix. For a given data set we replace $\boldsymbol{\Gamma}$ by its sample estimate $\hat{\boldsymbol{\Gamma}}$.

Table 8.1 gives the estimates $\hat{\boldsymbol{\Gamma}}$ for the steroid metabolite compositions of Data 9 and the aphyric Skye lava compositions of Data 6, the two data sets of Fig. 8.1 and Fig. 8.2, respectively, together with the non-zero eigenvalues and their corresponding eigenvectors. It is

Table 8.1. *Logcontrast principal component analysis for steroid metabolite and aphyric Skye lava compositions*

Data set	*Steroid metabolites*			*Aphyric Skye lavas*		
Covariance matrix $\boldsymbol{\Gamma}$	0.0562	−0.0116	−0.0445	0.2711	0.0133	−0.2844
	−0.0116	0.0651	−0.0534	0.0133	0.0065	−0.0198
	−0.0445	−0.0534	0.0980	−0.2844	−0.0198	0.3042
Eigenvalues	0.148	0.071		0.574	0.0083	
Eigenvectors	0.334	0.745		0.686	0.444	
	0.478	−0.662		0.041	−0.815	
	−0.812	−0.083		−0.727	0.372	

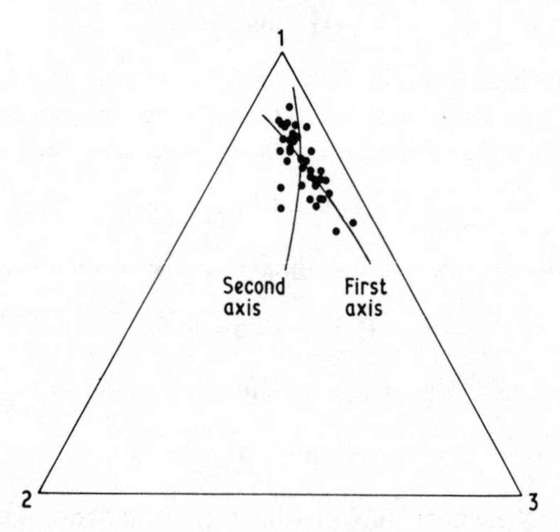

Fig. 8.4. *Ternary diagram and logcontrast principal axes for adult steroid metabolite compositions.*

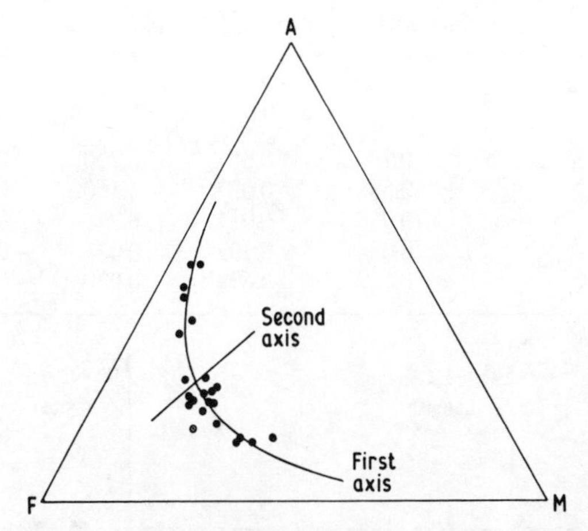

Fig. 8.5. *Ternary diagram and logcontrast principal axes for aphyric Skye lava compositions.*

then a trivial mathematical exercise to convert the principal axes in the **y**-space to their triangular coordinate forms, as shown in Figs 8.4–8.5. It is clear that the logcontrast method faithfully depicts the curved one-dimensional pattern of the aphyric Skye lava compositions (Fig. 8.5), and gives as convincing a near-linear representation of the steroid metabolite compositions (Fig. 8.4) as the specifically linear method shown in Fig. 8.1. From Table 8.1 we can also readily calculate that the proportion of the total variability contained in the first principal component, as $0.148/(0.148 + 0.071) = 0.68$ for the steroid metabolite compositions and as $0.574/(0.574 + 0.0083) = 0.986$ for the aphyric Skye lavas.

For compositions with more than three parts we no longer have the advantage of the ternary diagram to illustrate the marked contrast that can occur between crude and logcontrast principal component analysis. For the hongite compositions of Data 1 the estimate $\hat{\Gamma}$ of the centred logratio covariance matrix has already been computed in Table 4.5. Table 8.2 gives the four positive eigenvalues and their standardized eigenvectors. A standard method of judging the success of a principal component method is to plot the scattergram of the first and second principal components, here the set of points $(\mathbf{a}_1' \log \mathbf{x},\ \mathbf{a}_2' \log \mathbf{x})$ generated by the compositions of the compositional data set. For success such a scattergram should be

Table 8.2. *Eigenvalues and eigenvectors for hongite*

	λ_1	λ_2	λ_3	λ_4
Eigenvalues	1.59	0.091	0.0059	0.00018
Eigenvectors	0.194	0.067	0.790	0.366
	0.588	−0.087	−0.560	0.366
	−0.784	0.011	−0.230	0.364
	0.034	0.707	−0.071	−0.542
	−0.032	−0.699	0.070	−0.533

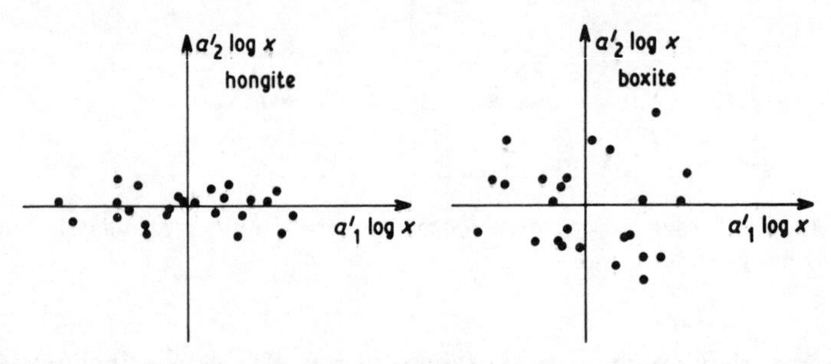

Fig. 8.6. *Scattergram of first and second logcontrast principal components for hongite and boxite compositions.*

reasonably 'elliptical' in nature. Figure 8.6 shows this feature in the first–second logcontrast principal component scattergram for the hongite compositions, in sharp contrast to the persistent curvature in Fig. 8.3. Also shown in Fig. 8.6 is the corresponding logcontrast principal component scattergram for the boxite compositions of Data 3. For this less curved data set the logcontrast method is just as successful as the crude method shown in Fig. 8.3.

8.5 Subcompositional analysis

We are now in a position to re-examine the nature of subcompositional analysis, through a simple relationship that it bears to principal component analysis. We first recall that logcontrast principal component analysis of compositions simply searches out particular forms from among all logcontrasts of the components. Without loss of generality we may consider the C-part, c-dimensional subcomposition

$$\mathbf{X} = (X_1, \ldots, X_C) = \mathscr{C}(x_1, \ldots, x_C)$$

formed from the leading subvector $(x_1, \ldots, x_C)$ of the full composition $(x_1, \ldots, x_D)$. Such a subcomposition is technically a composition with dimension c, smaller than the dimension d of the original composition. It can be completely represented by its logratio vector

$$\mathbf{Y} = \log(\mathbf{X}_{-c}/X_C).$$

Moreover, the logratios $Y_i = \log(X_i/X_C) = \log(x_i/x_C)$ $(i = 1, \ldots, c)$ are obviously logcontrasts, albeit of a specially simple form, of the full composition $\mathbf{x}$, and so belong to the class of functions from which our logcontrast principal components are selected. It therefore follows that the proportion of total compositional variability retained by a subcomposition of dimension c will be at most equal to, and generally less than, that retained by the first c logcontrast principal components.

The measure of variability retained by the subcomposition $\mathbf{X}$ is simply the total variability of $\mathbf{X}$ regarded as a C-part composition or of $\mathbf{Y}$ regarded as a c-dimensional logratio vector. For this particular subcomposition this is

$$\operatorname{tr}(\mathrm{V}[\log\{\mathbf{X}/g(\mathbf{X})\}]), \tag{8.11}$$

where $\mathrm{V}(\cdot)$ denotes covariance matrix, and, more generally, for a C-part subcomposition with $C \times D$ selection matrix $\mathbf{S}$ this is, in the notation of Section 5.7, $\operatorname{tr}(\boldsymbol{\Gamma}_S)$, where $\boldsymbol{\Gamma}_S = \mathbf{G}_C\mathbf{S}\boldsymbol{\Gamma}\mathbf{S}'\mathbf{G}_C$ and $\boldsymbol{\Gamma}$ is the centred logratio covariance matrix of the full composition. Hence the proportion of the total variability of a composition retained by a subcomposition with selection matrix $\mathbf{S}$ is

$$\operatorname{tr}(\boldsymbol{\Gamma}_S)/\operatorname{tr}(\boldsymbol{\Gamma}). \tag{8.12}$$

From Property 5.9 and (5.22) we can, however, find a much simpler computational form which involves only simple summation of subsets of the variation matrix $\boldsymbol{T}$ and so avoids the more awkward construction of $\boldsymbol{\Gamma}_S$. We have

$$\operatorname{tr}(\boldsymbol{\Gamma}) = (2D)^{-1}\mathbf{j}_D'\boldsymbol{T}\mathbf{j}_D, \tag{8.13}$$

$$\operatorname{tr}(\boldsymbol{\Gamma}_S) = (2C)^{-1}\mathbf{j}_C'\boldsymbol{T}_S\mathbf{j}_C = (2C)^{-1}\mathbf{j}_C'\mathbf{S}\boldsymbol{T}\mathbf{S}'\mathbf{j}_C. \tag{8.14}$$

Hence

$$\frac{\operatorname{tr}(\boldsymbol{\Gamma}_S)}{\operatorname{tr}(\boldsymbol{\Gamma})} = \frac{D}{C}\,\frac{\mathbf{j}_C'\mathbf{S}\boldsymbol{T}\mathbf{S}'\mathbf{j}_C}{\mathbf{j}_D'\boldsymbol{T}\mathbf{j}_D}. \tag{8.15}$$

The denominator of the second factor of the right-hand side of (8.15) is the sum of all the elements of $\boldsymbol{T}$; and the numerator is the sum of those elements of $\boldsymbol{T}$ which are in rows and columns associated with the parts of the subcomposition. We can collect these results in the following property, which provides a means of answering Question 1.2.5.

Property 8.2 The proportion of the total variability of a D-part composition retained by a C-part subcomposition is D/C times the ratio of the sum of the elements in the rows and columns of the variation matrix $\boldsymbol{T}$ associated with the subcompositional parts to the sum of all the elements of $\boldsymbol{T}$. Moreover, this proportion cannot be greater than the proportion retained by the first c logcontrast principal components.

Notes

1. Since the variation matrix $\boldsymbol{T}$ is symmetric, the summations could be confined to the superdiagonal triangle or the variance part of the compositional variation array of Definition 4.3. For example, the sum of the ten variance entries of the hongite compositional variation array of Table 4.1 is 8.46; and the sum corresponding to the 3-part subcomposition ACE is $1.53 + 0.14 + 0.95 = 2.62$, formed from the three (row, column) entries (A, C), (A, E), (C, E). The proportion of the total hongite variability retained by the ACE subcomposition is thus, by Property 8.2,

$$\frac{5}{3}\frac{2.62}{8.46} = 0.52.$$

Note that this is appreciably less than the proportion 0.995 of total variability retained by the first two logcontrast principal components of Table 8.2, in conformity with the final statement of Property 8.2.

2. The simplicity of the above results for the covariance specification $\boldsymbol{T}$ reinforces our previous remarks in Section 4.5 about the superiority of the variation array in subcompositional analysis.

8.6 Applications of subcompositional analysis

If the purpose of subcompositional analysis is to retain as much variability as possible for a given number C of parts, then we have to search for subcompositions which maximize (8.12). Since, by Property 8.2, the only computations involved are sums of subsets of the elements of $\boldsymbol{T}$, all we require is an algorithm which identifies these subsets efficiently. This is a very simple task and indeed for low-dimensional compositions inspection is all that is necessary.

Table 8.3. *Percentage retention of total variability of 3-part subcompositions of hongite and boxite*

Hongite		*Boxite*	
Subcomposition	*Percentage*	*Subcomposition*	*Percentage*
ABC	94.5	BCD	77.3
BCD	91.8	ABD	68.0
BCE	90.6	ABC	65.3
ACD	53.7	BCE	63.1
ACE	51.6	BDE	62.1
CDE	44.3	ABE	51.0
BDE	27.3	ACD	35.0
ABE	20.6	CDE	29.5
ABD	17.5	ADE	27.2
ADE	8.0	ACE	21.5
First two principal components	99.5%	First two principal components	81.8%

Table 8.3 shows for the hongite and boxite compositions of Data 1 and 3 the rankings of all ten 3-part subcompositions and the associated proportion (8.12) of total variability retained by each, and compares these with what is achievable by the first two principal components. Note the substantial differences between the proportions retained by the first and last ranked subcompositions for each type and the differences in the orders between hongite and boxite subcompositions. The first ranked 3-part compositions for hongite and boxite capture 94.5 and 77.3 per cent of total variability compared with 99.5 and 81.8 per cent captured by the corresponding first two logcontrast principal components.

It is of interest here to report the results of the above form of subcompositional analysis on four larger geochemical data sets, not publishable here but identified in Table 8.4, to find the top 3-part and 4-part subcompositions. For data sets 1 and 2 this involves the simple computations for $\binom{11}{3} = 165$ submatrices of order 3 and $\binom{11}{4} = 330$ submatrices of order 4; the corresponding numbers for data sets 3 and 4 are 120 and 210 submatrices.

The top three 3-part and 4-part subcompositions, together with the proportion of total variability retained by each and, for comparison purposes, the proportions of total variability retained by the first two

Table 8.4. *The four geochemical data sets analysed*

Data set no.	*Reference*	*Description*	*Number of Specimens*	*Number of Major oxides*
1	Nisbet, Bickle and Martin (1977)	Lavas of the Belingwe Greenstone Belt, Rhodesia: Table 3.	60	11
2	Steiner (1958)	Effusive rocks of the Taupo volcanic association: Table 1, Nos. 1–45 omitting No. 10 which has information missing	44	11
3	Carr (1981)	Igneous rocks of the southern Sydney Basin of New South Wales.	102	10
4	Thompson, Esson and Duncan (1972)	Eocene lavas of the Isle of Skye, Scotland: Table 2 basalts.	32	10

and first three logcontrast principal components respectively, are shown for each of the four data sets in Table 8.5. It is interesting that, on the whole, the best subcompositions perform rather poorly as retainers of variability compared with the corresponding set of principal components. Moreover, the popular (CaO, Na_2O, K_2O) subcomposition displayed in CNK ternary diagrams, as for example in Le Maître (1962), does not appear in Table 8.5, and indeed retains only 62, 28, 23, 29 per cent of total variability for the data sets 1, 2, 3, 4, respectively. The top subcompositions (MgO, K_2O, FeO) and (MgO, K_2O, FeO, Na_2O) for data set 2 are recognizable as forms of (Alkali, F, M) diagrams but retain only 61 and 72 per cent of total variability, compared with the 91 and 95 per cent retained by the corresponding sets of principal components.

Some further insights into subcompositional analysis may be gained by confining attention to the selection of the top 3-part subcomposition. Note that for data set 1 the selection of (MgO, K_2O, Na_2O) is further supported by the appearance of these three oxides in the top three (indeed the top seven) 4-part subcompositions. A similar comment applies to the other three data sets.

Although the dominant role of MgO and K_2O in all of these top subcompositions may appear very reasonable, the presence of TiO_2 as the third oxide for data sets 3 and 4 seems to surprise many

Table 8.5 *3- and 4-part subcompositions retaining the highest percentages of the total variability and the corresponding percentages for the first 2 and 3 logcontrast principal components*

Data set	*3-part subcompositions*	*Percentage retained*	*4-part subcompositions*	*Percentage retained*
1	MgO, K_2O, Na_2O	75.4	MgO, K_2O, Na_2O, P_2O_5	83.3
	FeO, K_2O, Na_2O	63.2	MgO, K_2O, Na_2O, TiO_2	81.4
	SiO_2, K_2O, Na_2O	62.9	MgO, K_2O, Na_2O, CaO	80.8
	Principal components	88.2	Principal components	95.2
2	MgO, K_2O, FeO	60.5	MgO, K_2O, FeO, Na_2O	72.2
	MgO, K_2O, Na_2O	59.3	MgO, K_2O, FeO, SiO_2	70.8
	MgO, K_2O, Fe_2O_3	58.5	MgO, K_2O, FeO, Fe_2O_3	69.0
	Principal components	90.6	Principal components	95.3
3	MgO, K_2O, TiO_2	53.1	MgO, K_2O, TiO_2, MnO	66.4
	MgO, K_2O, P_2O_5	48.9	MgO, K_2O, TiO_2, P_2O_5	62.7
	MgO, K_2O, MnO	47.8	MgO, K_2O, TiO_2, Na_2O	62.2
	Principal components	74.4	Principal components	84.9
4	MgO, K_2O, TiO_2	63.1	MgO, K_2O, TiO_2, P_2O_5	75.4
	MgO, K_2O, P_2O_5	58.5	MgO, K_2O, TiO_2, MnO	74.9
	MgO, K_2O, MnO	52.6	MgO, K_2O, TiO_2, CaO	71.1
	Principal components	82.4	Principal components	95.0

geologists. They argue that TiO_2 contributes only minutely to the total variance of any set of rock analyses. For example, for data set 3 the ten oxides in descending order of the crude variances of their components are:

SiO_2, MgO, CaO, Al_2O_3, FeO, K_2O, Na_2O, TiO_2, MnO, P_2O_5,

with the variance of the eighth-placed TiO_2 just over 2 per cent of that of SiO_2. The surprise is therefore probably ascribable to a conditioned mode of thinking in terms of crude variances and covariances of the components and we now know, from our discussion in Section 3.3 and from the wide literature on the unit-sum problem, that very little of interpretable value comes out of such a mode. If we are to grasp the advantage that the logratio and logcontrast approach to compositional data analysis provides in its removal of unit-sum difficulties, then we must begin to think in terms of a new concept of centred logratio variances

$$\text{var}[\log\{x_i/g(\mathbf{x})\}] \qquad (i = 1, \ldots, D), \tag{8.16}$$

where variation of x_i is measured relative to the variation of the geometric mean of all D components.

The changed viewpoint can make a radical difference to the interpretation of data. For example, for data set 3 the ten oxides in descending order of magnitude of centred logratio variance (8.16) are:

MgO, TiO_2, K_2O, MnO, P_2O_5, CaO, Na_2O, FeO, Al_2O_3, SiO_2,

with TiO_2 now in second position and with a centred logratio variance about eight times that of SiO_2, which is now last in order. Note also that after MgO, TiO_2 and K_2O, the oxides MnO and P_2O_5, small in crude variation, are now contenders for consideration in terms of centred logratio variation.

8.7 Canonical component analysis

In our study in Section 7.11 of the use of compositions to discriminate between hongite and kongite, we arrived at a discriminant (7.65) involving the composition $\mathbf{x}$ in the form of a logcontrast $\mathbf{a}' \log \mathbf{x}$. Since a logcontrast which best distinguishes between two categories must also in some sense be the logcontrast most highly correlated with category, we see that we have a potential dimension-reducing technique for two or more compositional data sets associated with different categories or types. In the parlance of variability in $\mathscr{R}^d$ such an approach constitutes a 'canonical component analysis'. We here confine ourselves to just sufficient detail of its compositional counterpart to allow us to apply the technique.

Suppose that there are C possible categories $1,\ldots,C$ of D-part compositions, where $C \leqslant D$, and that a composition arises from a particular category k with probability π_k $(k = 1,\ldots,C)$. Suppose further that the distribution of compositions of known category k has logratio mean vector $\boldsymbol{\mu}_k$, possibly different for different categories, but logratio covariance matrix $\boldsymbol{\Sigma}$ the same for each category. Then the unconditional logratio mean vector is

$$\boldsymbol{\mu} = \sum_{k=1}^{C} \pi_k \boldsymbol{\mu}_k. \tag{8.17}$$

The eigenproblem

$$\left(\sum_{k=1}^{C} \pi_k (\boldsymbol{\mu}_k - \boldsymbol{\mu})(\boldsymbol{\mu}_k - \boldsymbol{\mu})' - \lambda \boldsymbol{\Sigma} \right) \mathbf{b} = \mathbf{0} \tag{8.18}$$

can then be shown to have $c = C - 1$ positive eigenvalues.

Definition 8.2 Let the c eigenvalues associated with (8.18) be labelled in descending order of magnitude $\lambda_1 > \cdots > \lambda_c$, with corresponding eigenvectors $\mathbf{b}_1,\ldots,\mathbf{b}_c$ standardized by the condition $\operatorname{var}(\mathbf{b}_k'\mathbf{y}) = \mathbf{b}_k' \boldsymbol{\Sigma} \mathbf{b}_k = 1$ $(k = 1,\ldots,c)$. The logcontrast $\mathbf{b}_k'\mathbf{y}$ is then termed the kth *logcontrast canonical component.*

Property 8.3 The logcontrast canonical components $\mathbf{b}_k'\mathbf{y}$ $(k = 1,\ldots,c)$ of D-part compositional variation within C categories have the following properties.

(a) The eigenvalues λ_k and logcontrast canonical components are invariant under the group of permutations of the parts of the compositions.

(b) Logcontrast canonical components are uncorrelated, with $\operatorname{cov}(\mathbf{b}_j'\mathbf{y}, \mathbf{b}_k'\mathbf{y}) = \mathbf{b}_j' \boldsymbol{\Sigma} \mathbf{b}_k = 0$ $(j \neq k)$.

(c) The sequence of logcontrast canonical components is such that the logcontrast which has maximum correlation with category, subject to being uncorrelated with each of $\mathbf{b}_1'\mathbf{y},\ldots,\mathbf{b}_{k-1}'\mathbf{y}$, is $\mathbf{b}_k'\mathbf{y}$ and this maximum correlation is $\lambda_k^{1/2}/(1 + \lambda_k)^{1/2}$.

Notes

1. Unlike the principal component analysis of Section 8.3, the

technique involving the logratio covariance matrix requires no adjustment such as the introduction of the isotropic covariance matrix $\mathbf{H}$ to ensure invariance. The result follows immediately since $\mathbf{y}, \boldsymbol{\mu}, \boldsymbol{\mu}_k$ are related to their permuted forms by Property 5.2(iv) and $\boldsymbol{\Sigma}$ to its permuted form by Property 5.2(v).

2. Another difference from principal component analysis is that the centred logratio covariance specification $\boldsymbol{\Gamma}$ has no advantage over $\boldsymbol{\Sigma}$. In terms of the centred logratio mean vectors γ_k and γ, the equivalent logcontrasts are obtained as $\mathbf{a}_k' \log \mathbf{x}$, where $\mathbf{a}_k$ are the eigenvectors associated with the c positive eigenvalues of

$$\left(\sum_{k=1}^{C} \pi_k(\gamma_k - \gamma)(\gamma_k - \gamma)' - \lambda \boldsymbol{\Gamma}\right)\mathbf{a} = \mathbf{0}, \tag{8.19}$$

standardized by $\mathbf{a}'\boldsymbol{\Gamma}\mathbf{a} = 1$. The $\mathbf{a}$- and $\mathbf{b}$-forms of eigenvector are simply related by

$$\mathbf{b} = \mathbf{a}_{-D}, \quad \mathbf{a} = \mathbf{F}'\mathbf{b}.$$

3. If $C > D$ then there are just $d = D - 1$ positive eigenvalues of (8.18).

In practical applications we have C compositional data sets with N_k compositions $x_{kr}\,(r = 1, \ldots, N_k)$ in the kth data set or kth category $(k = 1, \ldots, C)$. Writing

$$N = \sum_{k=1}^{C} N_k, \qquad \hat{\boldsymbol{\mu}}_k = N_k^{-1} \sum_{r=1}^{N_k} \mathbf{y}_{kr}, \qquad \hat{\boldsymbol{\mu}} = N^{-1} \sum_{k=1}^{C} \sum_{r=1}^{N_k} \mathbf{y}_{kr}, \tag{8.20}$$

we have standard estimates of $\pi_k, \boldsymbol{\mu}_k, \boldsymbol{\mu}$ and $\boldsymbol{\Sigma}$ as $N_k/N, \hat{\boldsymbol{\mu}}_k, \hat{\boldsymbol{\mu}}$ and the pooled logratio covariance matrix

$$\mathbf{W} = (N - C)^{-1} \sum_{k=1}^{C} \sum_{r=1}^{N_k} (\mathbf{y}_{kr} - \hat{\boldsymbol{\mu}}_k)(\mathbf{y}_{kr} - \hat{\boldsymbol{\mu}}_k)'. \tag{8.21}$$

We have adopted the standard canonical component analysis notation of denoting (8.21) by $\mathbf{W}$ for the within-categories covariance matrix. If we also follow convention by writing

$$\mathbf{B} = N^{-1} \sum_{k=1}^{C} N_k(\hat{\boldsymbol{\mu}}_k - \hat{\boldsymbol{\mu}})(\hat{\boldsymbol{\mu}}_k - \hat{\boldsymbol{\mu}})'$$

for the between-categories covariance matrix, the practical version of the eigenproblem (8.18) takes the form

$$(\mathbf{B} - \lambda \mathbf{W})\mathbf{b} = \mathbf{0} \tag{8.22}$$

with standardization of $\mathbf{b}$ through the constraint $\mathbf{b}'\mathbf{W}\mathbf{b} = 1$.

As a simple illustrative example we examine the 30, 3-part fossil pollen compositions of Data 26, with 10 compositions from each of three different locations A, B, C. The estimates of **B** and **W** and the two eigenvalues and corresponding eigenvectors are recorded in Table 8.6. The results of the canonical component analysis can be conveniently shown in the graphical form of canonical component scattergrams where canonical components are computed for each of the original compositions and one canonical component is plotted against another. In our example we have just two canonical

Table 8.6. *Computations for canonical component analysis of fossil pollen compositions*

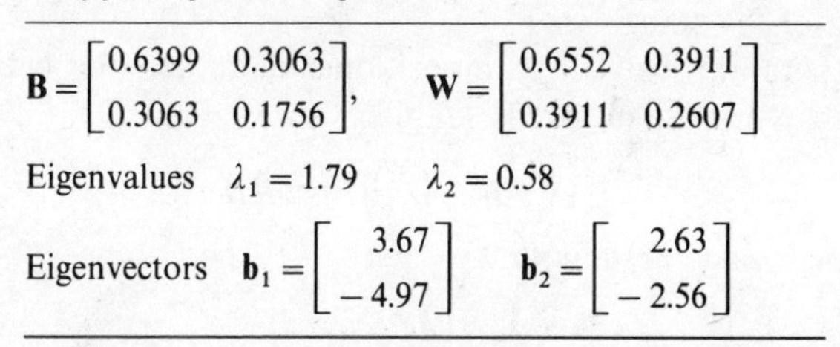

$$\mathbf{B} = \begin{bmatrix} 0.6399 & 0.3063 \\ 0.3063 & 0.1756 \end{bmatrix}, \quad \mathbf{W} = \begin{bmatrix} 0.6552 & 0.3911 \\ 0.3911 & 0.2607 \end{bmatrix}$$

Eigenvalues $\lambda_1 = 1.79$ $\quad \lambda_2 = 0.58$

Eigenvectors $\mathbf{b}_1 = \begin{bmatrix} 3.67 \\ -4.97 \end{bmatrix} \quad \mathbf{b}_2 = \begin{bmatrix} 2.63 \\ -2.56 \end{bmatrix}$

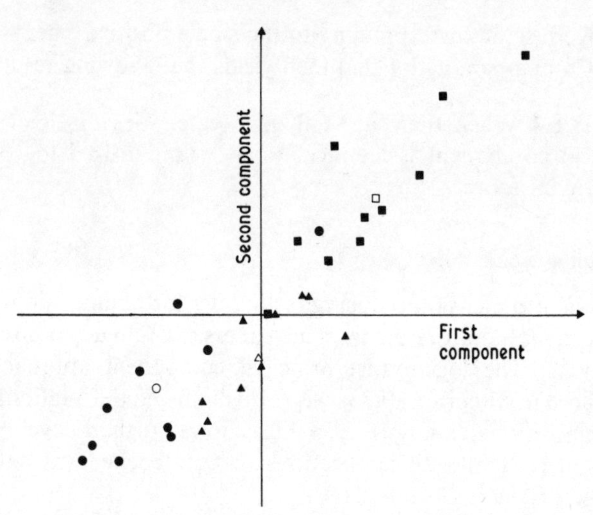

Fig. 8.7. *Scattergram of first and second logcontrast canonical components for fossil pollen compositions, with solid circles, triangles and squares denoting locations A, B, C respectively, and corresponding open symbols denoting mean positions.*

components and the scattergram is shown in Fig. 8.7, where we have conveniently translated the axes so that the origin corresponds to the overall mean logratio vector $\hat{\boldsymbol{\mu}}$. In other words, the points plotted are of the form

$$\{\mathbf{b}_1'(\mathbf{y}-\hat{\boldsymbol{\mu}}), \mathbf{b}_2'(\mathbf{y}-\hat{\boldsymbol{\mu}})\}$$

and we have also indicated the points corresponding to the category means $\hat{\boldsymbol{\mu}}_1, \hat{\boldsymbol{\mu}}_2, \hat{\boldsymbol{\mu}}_3$. The much higher correlation with category of the first canonical component, as compared with the second component, is reflected in Fig. 8.7 by the greater separation of the categories achieved by the first canonical component. This is indeed the dimension-reducing hope of canonical component analysis, that a few of the canonical components will achieve a substantial degree of separation of the categories.

When there are just two compositional data sets, the between-categories covariance matrix $\mathbf{B}$ simplifies to

$$N^{-2}N_1N_2(\hat{\boldsymbol{\mu}}_1-\hat{\boldsymbol{\mu}}_2)(\hat{\boldsymbol{\mu}}_1-\hat{\boldsymbol{\mu}}_2)'$$

and it is then easy to show that

$$\mathbf{b}=\hat{\boldsymbol{\Sigma}}^{-1}(\hat{\boldsymbol{\mu}}_1-\hat{\boldsymbol{\mu}}_2)$$

is an eigenvector corresponding to the single positive eigenvalue of (8.22). Comparison with (7.64) then yields the following result.

Property 8.4 When there are only two categories the logcontrast canonical component is identical to the standardized logcontrast discriminant score.

8.8 Bibliographic notes

Attempts at dimension-reducing techniques for compositional data abound in the literature and the main papers have already been cited in Section 8.2. The logcontrast principal component approach was introduced in Aitchison (1983), who also discussed the relationship to subcompositional analysis. The ideas were further developed in Aitchison (1984b) with applications to the geochemical data sets reported briefly in Section 8.6.

Problems

8.1 For the ABC subcompositions of hongite in Table 1.2 determine both the crude and the logcontrast principal components and compare the adequacy of the corresponding principal axes in the ABC ternary diagram.

8.2 For the 40 past-season fruit compositions of Data 12 compute the logcontrast principal components, show the relation of the principal axes to the compositions in a ternary diagram, and plot the scattergram of first and second principal components. What proportion of the total variability is retained by the first principal component?

8.3 Perform a logcontrast principal component analysis for the compositions of machine operators' shifts in Data 22, and plot the scattergrams of first against second, and of first against third, principal components. What proportions of the total variability are retained by the first principal component and by the first and second principal components?

8.4 By inspection of the variation array for the shift compositions of Data 22, arrange the six 2-part compositions and the four 3-part compositions in decreasing order of variability, and record the proportion of the total compositional variability retained by each. How do these proportions compare with the corresponding proportions already evaluated in Problem 8.3 for the logcontrast principal components?

8.5 For the two compositional data sets consisting of the serum protein compositions A1–A10 and B1–B10 of Data 16, determine the logcontrast canonical component and compare this with the standardized logcontrast discriminant you obtained in Problem 7.9.

8.6 Data 27 provides the 5-part compositions of 25 specimens of the rock type lammite as percentages by weight of the minerals A, B, C, D, E, which are the same as in the hongite and kongite compositions of Data 1 and 2. Carry out a full logcontrast canonical component analysis of the hongite, kongite and lammite compositions, and plot the scattergram of the first and second logcontrast canonical components. Can you assess from the scattergram whether the first two logcontrast canonical components are likely to provide a reasonable means of typing a new specimen of unknown type?

CHAPTER 9

Bases and compositions

E pluribus unum

Adapted from St Augustine: *Confessions*

9.1 Fundamentals

This chapter is concerned with a variety of problems in which compositions are formed from observed bases and interest is in some aspect of the relationship between these compositions and their bases. Two problems of Chapter 1 are of this kind. In Section 1.7 the 4-part monthly budgets of the households are the observed bases (Data 8), and Question 1.7.1 requires us to investigate the extent to which the corresponding compositions, the budget-share patterns, depend on the additive sizes of the bases, namely the total expenditures of the households, and what economic interpretations can be placed on any detected dependence. In Section 1.8 the observed bases are the 3-part urinary excretions of the 37 adults and the 30 children (Data 9). It is clear that the quantities of steroid metabolites excreted by adults are substantially greater than those of children. Questions 1.8.1 and 1.8.2 express interest in determining whether the essential difference between adults and children lies only in the total quantities excreted, the additive sizes of the bases, or whether it may lie also in the relative excretion patterns, the compositions formed from these bases.

At the outset of our discussion of such problems we should recall from Section 2.4 the fundamental many–one aspect of the relationship between bases and compositions. To any D-part basis $\mathbf{w} \in \mathscr{R}_+^D$ there corresponds a unique composition $\mathbf{x} \in \mathscr{S}^d$, determined by

$$\mathbf{x} = \mathscr{C}(\mathbf{w}) = \mathbf{w}/(w_1 + \cdots + w_D), \tag{9.1}$$

where $\mathscr{C}: \mathscr{R}_+^D \rightarrow \mathscr{S}^d$ is the familiar constraining operation of Definition 2.5. To any D-part composition $\mathbf{x} \in \mathscr{S}^d$, however, there correspond many possible bases from which $\mathbf{x}$ could be formed by the constraining operation (9.1). From Property 2.3 any such basis can be expressed as $t\mathbf{x}$, where $t > 0$, and so

$$\mathscr{B}(\mathbf{x}) = \{t\mathbf{x}: t > 0\}, \tag{9.2}$$

the set of all such bases, consists of the straight-line ray in $\mathscr{R}^D_+$ passing through the origin and the composition $\mathbf{x}$.

We note that if $\mathbf{w} = t\mathbf{x}$ then

$$t = w_1 + \cdots + w_D \tag{9.3}$$

is the size of the basis $\mathbf{w}$. As recorded in Property 2.1, any basis $\mathbf{w}$ is thus completely determined by knowledge of its composition $\mathbf{x}$ and its size t, and we give a terminology to the useful one-to-one transformation between the basis space $\mathscr{R}^D_+$ and the composition-size space $\mathscr{S}^d \times \mathscr{R}^1_+$.

Definition 9.1 The *basis to (composition, size) transformation* is defined by

$$\mathscr{R}^D_+ \to \mathscr{S}^d \times \mathscr{R}^1_+ : \mathbf{x} = \mathscr{C}(\mathbf{w}), t = w_1 + \cdots + w_D,$$

with inverse

$$\mathscr{S}^d \times \mathscr{R}^1_+ \to \mathscr{R}^D_+ : \mathbf{w} = t\mathbf{x},$$

and jacobian

$$\operatorname{jac}(\mathbf{w} \mid \mathbf{x}, t) = t^d.$$

9.2 Covariance relationships

One of the motivating factors which led in Chapter 4 to the adoption of logratio transformations in the analysis of compositional data was the simple relationships that could then be identified between the covariance structures of a basis $\mathbf{w}$ and of its composition $\mathbf{x} = \mathscr{C}(\mathbf{w})$. For ease of reference we recall in Table 9.1 the various specifications of covariance structures of bases and of compositions already discussed in Chapter 4.

Since a basis uniquely determines its composition, the compositional covariance structures will be uniquely determined by the basis covariance structure. These determinations are given in element and matrix versions in Table 9.2.

Table 9.1. *Specifications of covariance structures of bases and of compositions*

Name	*Notation*	*Order*	*Definition*
Basis covariance matrix	$\boldsymbol{\Omega} = [\omega_{ij}]$	$D \times D$	$\boldsymbol{\Omega} = \mathrm{V}(\log \mathbf{w})$ $\omega_{ij} = \mathrm{cov}(\log w_i, \log w_j)$
Logratio covariance matrix	$\boldsymbol{\Sigma} = [\sigma_{ij}]$	$d \times d$	$\boldsymbol{\Sigma} = \mathrm{V}\{\log(\mathbf{x}_{-D}/x_D)\}$ $\sigma_{ij} = \mathrm{cov}\{\log(x_i/x_D), \log(x_j/x_D)\}$
Centred logratio covariance matrix	$\boldsymbol{\Gamma} = [\gamma_{ij}]$	$D \times D$	$\boldsymbol{\Gamma} = \mathrm{V}[\log\{\mathbf{x}/g(\mathbf{x})\}]$ $\gamma_{ij} = \mathrm{cov}[\log\{x_i/g(\mathbf{x})\}, \log\{x_j/g(\mathbf{x})\}]$
Variation matrix	$\boldsymbol{T} = [\tau_{ij}]$	$D \times D$	$\tau_{ij} = \mathrm{var}\{\log(x_i/x_j)\}$

Table 9.2. *Specifications of compositional covariance structure derived from a basis covariance matrix* $\boldsymbol{\Omega}$

Element version	*Matrix version*
$\sigma_{ij} = \omega_{ij} - \omega_{iD} - \omega_{jD} + \omega_{DD}$	$\boldsymbol{\Sigma} = \mathbf{F}\boldsymbol{\Omega}\mathbf{F}'$
$\gamma_{ij} = \omega_{ij} - \omega_{i.} - \omega_{j.} + \omega_{..}$	$\boldsymbol{\Gamma} = \mathbf{G}\boldsymbol{\Omega}\mathbf{G}$
$\tau_{ij} = \omega_{ii} + \omega_{jj} - 2\omega_{ij}$	$\boldsymbol{T} = \mathbf{J}\,\mathrm{diag}(\boldsymbol{\Omega}) + \mathrm{diag}(\boldsymbol{\Omega})\mathbf{J} - 2\boldsymbol{\Omega}$ $= \mathbf{j}\,\mathrm{row}(\omega_{11}, \ldots, \omega_{DD})$ $+ \mathrm{col}(\omega_{11}, \ldots, \omega_{DD})\mathbf{j}' - 2\boldsymbol{\Omega}$

A consequence of the many–one relationship of bases to compositions is that $\boldsymbol{\Omega}$ will not be uniquely determined by knowledge of the compositional covariance structure. The extent of this arbitrariness is, however, easily discovered. Since

$$\begin{aligned} \log \mathbf{w}_i &= \log\{w_i/g(\mathbf{w})\} + \log\{g(\mathbf{w})\} \qquad (i = 1, \ldots, D) \\ &= \log\{x_i/g(\mathbf{x})\} + \log\{g(\mathbf{w})\}, \end{aligned} \tag{9.4}$$

we have

$$\omega_{ij} = \gamma_{ij} + \alpha_i + \alpha_j + \beta \qquad (i, j = 1, \ldots, D), \tag{9.5}$$

where

$$\alpha_i = \mathrm{cov}[\log\{x_i/g(\mathbf{x})\}, \log\{g(\mathbf{w})\}], \tag{9.6}$$

$$\beta = \mathrm{var}[\log\{g(\mathbf{w})\}]. \tag{9.7}$$

Table 9.3. *Basis covariance matrix $\boldsymbol{\Omega}$ in relation to specifications of compositional covariance structure*

Element version	*Matrix version*
$\omega_{ij} = \gamma_{ij} + \alpha_i + \alpha_j$	$\boldsymbol{\Omega} = \boldsymbol{\Gamma} + \mathbf{j}\boldsymbol{\alpha}' + \boldsymbol{\alpha}\mathbf{j}'$
$= \sigma_{ij} - \sigma_{i.} - \sigma_{j.} + \sigma_{..} + \alpha_i + \alpha_j$	$= \mathbf{F}'\mathbf{H}^{-1}\boldsymbol{\Sigma}\mathbf{H}^{-1}\mathbf{F} + \mathbf{j}\boldsymbol{\alpha}' + \boldsymbol{\alpha}\mathbf{j}'$
$= -\frac{1}{2}(\tau_{ij} - \tau_{i.} - \tau_{j} + \tau_{..}) + \alpha_i + \alpha_j$	$= -\frac{1}{2}\mathbf{G}\boldsymbol{T}\mathbf{G} + \mathbf{j}\boldsymbol{\alpha}' + \boldsymbol{\alpha}\mathbf{j}'$

The arbitrariness in $\boldsymbol{\Omega}$ is not the combined dimension $D+1$ of $\alpha_1, \ldots, \alpha_D$ and β since the α_i clearly satisfy the condition $\alpha_1 + \cdots + \alpha_D = 0$. We can obtain a convenient expression of the D-dimensionality of the arbitrariness of $\boldsymbol{\Omega}$ by absorbing $\frac{1}{2}\beta$ into each of the α_i, so that they are no longer subject to any constraint. We collect in Table 9.3 the immediate consequences of this for $\boldsymbol{\Omega}$ in terms of each of the various specifications $\boldsymbol{\Sigma}, \boldsymbol{\Gamma}, \boldsymbol{T}$ of the compositional covariance structure, with the $D \times 1$ vector $\boldsymbol{\alpha}$ expressing the arbitrariness in each case.

Two special forms of relationship are readily investigated through the use of these results.

Uncorrelated basis

When the components of a basis $\mathbf{w}$ are uncorrelated, the consequences of the diagonal form, say $\boldsymbol{\Omega} = \text{diag}(\omega_1, \ldots, \omega_D)$, of the basis covariance matrix on the compositional covariance structure are recorded in Table 9.4. Note from the element versions the greater simplicity of the $\boldsymbol{\Sigma}$ and $\boldsymbol{T}$ forms compared with the complications of the $\boldsymbol{\Gamma}$ form. We shall see that a similar feature arises when we study in Chapter 10 a form of compositional independence known as complete subcompositional independence. Note also that the compositional covariance structures associated with an uncorrelated basis agree with the logratio-uncorrelated forms obtained directly in Property 5.11.

Permutation-invariant basis

We use the term 'permutation invariant' to denote a basis whose covariance matrix $\boldsymbol{\Omega}$ remains unchanged under any permutation of the components. Such a matrix must clearly have all its diagonal elements equal and also all its non-diagonal elements equal, so that it must have the form $\alpha\mathbf{I} + \beta\mathbf{J}$. The covariance matrix of a basis

Table 9.4. *Forms of compositional covariance structure corresponding to an uncorrelated basis with* $\boldsymbol{\Omega} = \mathrm{diag}(\omega_1, \ldots, \omega_D)$

Element versions	Matrix versions
$\sigma_{ij} = \begin{cases} \omega_i + \omega_D & (i = j) \\ \omega_D & (i \neq j) \end{cases}$	$\boldsymbol{\Sigma} = \mathrm{diag}(\omega_1, \ldots, \omega_d) + \omega_D \mathbf{J}$
$\gamma_{ij} = \begin{cases} \omega_i - (1/D)(2\omega_i - \omega_.) & (i = j) \\ -(1/D)(\omega_i + \omega_j - \omega_.) & (i \neq j) \end{cases}$	$\boldsymbol{\Gamma} = \mathrm{diag}(\omega_1, \ldots, \omega_D) - D^{-1}(\boldsymbol{\omega}\mathbf{j}' + \mathbf{j}\boldsymbol{\omega}' - \omega_. \mathbf{J})$
$\tau_{ij} = \begin{cases} 0 & (i = j) \\ \omega_i + \omega_j & (i \neq j) \end{cases}$	$\boldsymbol{T} = \boldsymbol{\omega}\mathbf{j}' + \mathbf{j}\boldsymbol{\omega}' - 2\,\mathrm{diag}(\omega_1, \ldots, \omega_D)$

transformed by the permutation matrix $\mathbf{P}$ is then

$$\mathbf{P}(\alpha\mathbf{I} + \beta\mathbf{J})\mathbf{P}' = \alpha\mathbf{I} + \beta\mathbf{J}$$

since $\mathbf{PP}' = \mathbf{I}$ and $\mathbf{PJ} = \mathbf{JP}' = \mathbf{J}$. Thus the general form of permutation-invariant basis covariance matrix is $\alpha\mathbf{I} + \beta\mathbf{J}$, with $\alpha > 0$ and $\alpha + d\beta > 0$ for positive definiteness. The resulting permutation-invariant forms of compositional covariance structure, easily obtained from the relationships of Table 9.2, are given in Table 9.5. We may note here that these permutation-invariant forms of the compositional covariance structure are identical with the isotropic forms already met in Section 5.10. Note that the isotropic basis form occurs when $\beta = 0$. The fact that isotropic compositions arise from other than isotropic bases is just another manifestation of the many–one relationship between bases and compositions.

Our discussion so far has concentrated on the basis covariance structure $\boldsymbol{\Omega}$. Although for many purposes this is a convenient approach, providing simple and evident relationships between bases and compositions, it is, as we shall see, not ideal for many applications. In particular, when interest is in the relationship between the compositional and size aspects of a basis it is more appropriate to consider a basis covariance structure defined in terms of the pattern of variability of $\mathbf{x}$ and t. This is possible since $\mathbf{x}$ and t together determine $\mathbf{w}$ as $t\mathbf{x}$. There is a choice of how we achieve this

Table 9.5. *Permutation-invariant forms of covariance structures*

Specification	*Element version*	*Matrix version*
Basis covariance matrix	$\omega_{ij} = \begin{cases} \alpha + \beta & (i = j) \\ \beta & (i \neq j) \end{cases}$	$\boldsymbol{\Omega} = \alpha\mathbf{I} + \beta\mathbf{J}$
Logratio covariance matrix	$\sigma_{ij} = \begin{cases} 2\alpha & (i = j) \\ \alpha & (i \neq j) \end{cases}$	$\boldsymbol{\Sigma} = \alpha(\mathbf{I} + \mathbf{J}) = \alpha\mathbf{H}$
Centred logratio covariance matrix	$\gamma_{ij} = \begin{cases} (d/D)\alpha & (i = j) \\ -(1/D)\alpha & (i \neq j) \end{cases}$	$\boldsymbol{\Gamma} = \alpha\mathbf{G}$
Variation matrix	$\tau_{ij} = \begin{cases} 0 & (i = j) \\ 2\alpha & (i \neq j) \end{cases}$	$\boldsymbol{T} = 2\alpha(\mathbf{J} - \mathbf{I})$

since we may take any of the three specifications for the covariance structure of the composition $\mathbf{x}$.

If we choose the centred version then we obtain a composition–size form of basis covariance structure defined as follows.

Definition 9.2 For the D-part basis $\mathbf{w}$ let $\mathbf{x} = \mathscr{C}(\mathbf{w})$ and $t = w_1 + \cdots + w_D$ denote its composition and size, and let

$$\mathbf{z} = \log\{\mathbf{w}/g(\mathbf{w})\} = \log\{\mathbf{x}/g(\mathbf{x})\}, \qquad u = \log t.$$

Then the $(D+1) \times (D+1)$ covariance matrix

$$\boldsymbol{\Psi} = \mathrm{V}(\mathbf{z}, u) = \begin{bmatrix} \boldsymbol{\Gamma} & \boldsymbol{\delta} \\ \boldsymbol{\delta}' & \varepsilon \end{bmatrix}$$

is termed the *composition–size covariance matrix* of the basis $\mathbf{w}$.

Note that $\boldsymbol{\Gamma}$ is the compositional centred logratio covariance matrix, ε is the variance of log-size, and since $\mathbf{z}'\mathbf{j} = 0$ the composition–size covariances

$$\delta_i = \mathrm{cov}[\log\{x_i/g(\mathbf{x})\}, \log t]$$

are subject to the condition $\boldsymbol{\delta}'\mathbf{j} = 0$. Together with the known singularity of $\boldsymbol{\Gamma}$ this establishes the singularity of $\boldsymbol{\Psi}$ since

$$\boldsymbol{\Psi}\begin{bmatrix}\mathbf{j}_D \\ 0\end{bmatrix} = \mathbf{0}. \tag{9.8}$$

Again we may consider

$$\operatorname{tr}(\boldsymbol{\Psi}) \tag{9.9}$$

as a suitable measure of total variability of the basis, but note that this measure and that based on $\operatorname{tr}(\boldsymbol{\Omega})$ differ in a way which will be made specific in Property 9.1.

9.3 Principal and canonical component comparisons

In later sections we shall be posing specific distributional questions about the relationship of composition to basis or the extent to which the composition and size of a basis are dependent. In this section we confine our attention to the relative variability of basis and composition as determined by the basis covariance matrices, the logarithmic form $\boldsymbol{\Omega}$ of Definition 4.11 or the composition–size form $\boldsymbol{\Psi}$ of Definition 9.2, and the corresponding compositional covariance structure, specified for convenience here as the centred logratio covariance matrix $\boldsymbol{\Gamma}$ of Definition 4.6. Since a composition is a specially scaled version of a basis, it may be conjectured that in general only a proportion of the total variability of a basis will be retained by its composition. This conjecture is substantiated in the following property.

Property 9.1 If a D-part basis has logarithmic covariance matrix $\boldsymbol{\Omega}$ and composition–size covariance matrix $\boldsymbol{\Psi}$, and its D-part composition has centred logratio covariance matrix $\boldsymbol{\Gamma}$, then

(a) $\operatorname{tr}(\boldsymbol{\Omega}) = \operatorname{tr}(\boldsymbol{\Gamma}) + D\operatorname{var}[\log\{g(\mathbf{w})\}]$,
(b) $\operatorname{tr}(\boldsymbol{\Psi}) = \operatorname{tr}(\boldsymbol{\Gamma}) + \operatorname{var}(\log t)$.

Proof. The proof of (b) is immediate from Definition 9.2. For (a) we have, by the relationships of Table 9.2 and simple properties of the trace operation,

$$\begin{aligned}\operatorname{tr}(\boldsymbol{\Gamma}) &= \operatorname{tr}(\mathbf{G}\boldsymbol{\Omega}\mathbf{G}) = \operatorname{tr}(\mathbf{G}\boldsymbol{\Omega}) \\ &= \operatorname{tr}(\boldsymbol{\Omega}) - D^{-1}\operatorname{tr}(\mathbf{J}\boldsymbol{\Omega}) \\ &= \operatorname{tr}(\boldsymbol{\Omega}) - D^{-1}\mathbf{j}'\boldsymbol{\Omega}\mathbf{j}\end{aligned}$$

$$= \text{tr}(\boldsymbol{\Omega}) - D\,\text{var}(D^{-1}\mathbf{j}'\log\mathbf{w})$$
$$= \text{tr}(\boldsymbol{\Omega}) - D\,\text{var}[\log\{g(\mathbf{w})\}].$$

Notes

1. In general the measure of total variability $\text{tr}(\boldsymbol{\Gamma})$ of a composition is less than either of the measures of total variability, $\text{tr}(\boldsymbol{\Omega})$ and $\text{tr}(\boldsymbol{\Psi})$, of its basis, with equality holding for (a) if and only if the basis has constant geometric mean and for (b) if and only if the basis has constant additive size. In these special cases there is not only equality of the measures of total variability, but identity of the covariance matrices in the sense that $\boldsymbol{\Omega} = \boldsymbol{\Gamma}$ and

$$\boldsymbol{\Psi} = \begin{bmatrix} \boldsymbol{\Gamma} & \mathbf{0} \\ \mathbf{0} & 0 \end{bmatrix}.$$

2. Since in compositional data analysis the natural size of an associated basis is its additive size t, there is much to recommend the use of the composition–size covariance matrix $\boldsymbol{\Psi}$. For example, comparison of the sum of its first D diagonal elements, namely $\text{tr}(\boldsymbol{\Gamma})$, with the final diagonal element $\text{var}(\log t)$ gives an immediate insight into the relative contributions of composition and size to basis variability.

It is easy to illustrate the general relationship of the traces of $\boldsymbol{\Omega}$, $\boldsymbol{\Psi}$ and $\boldsymbol{\Gamma}$ with the sample estimates $\hat{\boldsymbol{\Omega}}$, $\hat{\boldsymbol{\Psi}}$ and $\hat{\boldsymbol{\Gamma}}$ for the 30 children's steroid metabolite bases of Data 9. We have

$$\hat{\boldsymbol{\Omega}} = \begin{bmatrix} 0.1598 & 0.1038 & 0.2173 \\ 0.1038 & 0.2273 & 0.2001 \\ 0.2173 & 0.2001 & 1.0621 \end{bmatrix},$$

$$\hat{\boldsymbol{\Psi}} = \begin{bmatrix} 0.1161 & 0.0433 & -0.1593 & -0.0166 \\ 0.0433 & 0.1501 & -0.1933 & -0.0564 \\ -0.1593 & -0.1933 & 0.3526 & 0.0730 \\ -0.0166 & -0.0564 & 0.0730 & 0.1552 \end{bmatrix},$$

with $\hat{\boldsymbol{\Gamma}}$ the leading 3×3 submatrix of $\hat{\boldsymbol{\Psi}}$. We then have

$$\text{tr}(\hat{\boldsymbol{\Omega}}) = 1.4492, \qquad \text{tr}(\hat{\boldsymbol{\Psi}}) = 0.7740, \qquad \text{tr}(\hat{\boldsymbol{\Gamma}}) = 0.6188,$$
$$\text{var}[\log\{g(\mathbf{w})\}] = 0.2768, \qquad \text{var}(\log t) = 0.1552,$$

and it is easy to verify (a) and (b) of Property 9.1 and the inequalities

$$\text{tr}(\hat{\boldsymbol{\Omega}}) \geqslant \text{tr}(\hat{\boldsymbol{\Gamma}}), \qquad \text{tr}(\hat{\boldsymbol{\Psi}}) \geqslant \text{tr}(\hat{\boldsymbol{\Gamma}}).$$

Even more insight into the relative variabilities of basis and composition is obtained when we compare conformable principal component analyses. For a basis $\mathbf{w} \in \mathscr{R}_+^D$ and logarithmic covariance matrix $\boldsymbol{\Omega}$, the natural counterpart of a linear combination for variability in $\mathscr{R}_+^D$ is a loglinear combination

$$\mathbf{a}' \log \mathbf{w} \tag{9.10}$$

with no restriction on the D-vector $\mathbf{a}$. Moreover, the loglinear principal components are obtained from the standardized eigenvectors of the eigenproblem

$$(\boldsymbol{\Omega} - \mu \mathbf{I})\mathbf{a} = \mathbf{0}.$$

The ordered eigenvalues $\mu_1 \geqslant \mu_2 \geqslant \cdots \geqslant \mu_D$ are equal to the descending variances of the first, second,..., Dth loglinear principal components of the basis.

For a D-part composition $\mathbf{x} \in \mathscr{S}^d$ we are concerned with logcontrasts $\mathbf{a}' \log \mathbf{x}$ where $\mathbf{a}'\mathbf{j} = 0$, which can be regarded as special loglinear combinations of the basis:

$$\mathbf{a}' \log \mathbf{w} \quad (\mathbf{a}'\mathbf{j} = 0), \tag{9.11}$$

and with the positive eigenvalues $\lambda_1 \geqslant \lambda_2 \geqslant \cdots \geqslant \lambda_d$ and their eigenvectors as described in Definition 8.1. From (9.10) and (9.11) we see that when viewed as a succession of variance maximization problems there is just one difference between loglinear principal component analysis for a basis and logcontrast principal component analysis for the corresponding composition, namely the presence of the additional constraint $\mathbf{a}'\mathbf{j} = 0$ in the compositional case. This has an immediate consequence for the relationship of the two sets of eigenvalues $(\mu_1, \ldots, \mu_D)$ and $(\lambda_1, \ldots, \lambda_D)$, where we have added the zero eigenvalue $\lambda_D = 0$ to the compositional eigenvalues to make the result simpler to express. The following property is a special case of a matrix result of Gantmacher (1959, Theorem 14).

Property 9.2 If $\mu_1 \geqslant \mu_2 \geqslant \cdots \geqslant \mu_D$ are the D eigenvalues of the logarithmic covariance matrix $\boldsymbol{\Omega}$ of a D-part basis and $\lambda_1 \geqslant \lambda_2 \geqslant \cdots \geqslant \lambda_d \geqslant \lambda_D = 0$ are the D eigenvalues of the centred logratio covariance matrix $\boldsymbol{\Gamma}$ of the corresponding D-part composition, then

$$\mu_1 \geqslant \lambda_1 \geqslant \mu_2 \geqslant \lambda_2 \geqslant \cdots \geqslant \mu_d \geqslant \lambda_d \geqslant \mu_D \geqslant 0 = \lambda_D.$$

Notes

1. The case of equality of tr($\boldsymbol{\Omega}$) and tr($\boldsymbol{\Gamma}$) in Property 9.1 is easily related to the inequalities of Property 9.2. Equality of the traces corresponds to the eigenvalue equality

$$\begin{aligned}\mu_1 + \cdots + \mu_d + \mu_D &= \lambda_1 + \cdots + \lambda_d \\ &\leqslant \mu_1 + \cdots + \mu_d\end{aligned}$$

by the inequalities of Property 9.2. Thus $\mu_D = 0$, showing that the basis covariance matrix is singular. The consequent equality

$$\mu_1 + \cdots + \mu_d = \lambda_1 + \cdots + \lambda_d,$$

subject to the inequalities of Property 9.2 is only possible when $\mu_1 = \lambda_1, \ldots, \mu_d = \lambda_d$. Thus $\boldsymbol{\Omega}$ and $\boldsymbol{\Gamma}$ have identical eigenvalues.

2. For tr($\boldsymbol{\Gamma}$) = 0 we must have $\lambda_1 = \cdots = \lambda_d = 0$ and so, by the inequalities of Property 9.2, $\mu_2 = \cdots = \mu_D = 0$. Hence $\boldsymbol{\Omega}$ has a single positive eigenvalue, a result corresponding to the fact that basis variability is essentially one-dimensional, confined to the basis geometric mean.

The corresponding comparison in terms of the composition–size covariance matrix $\boldsymbol{\Psi}$ for the basis is even simpler. Basis and compositional principal components then take the respective forms

$$\mathbf{a}' \log \mathbf{x} + b \log t \qquad (\mathbf{a}'\mathbf{j} = 0), \tag{9.12}$$

$$\mathbf{a}' \log \mathbf{x} + b \log t \qquad (\mathbf{a}'\mathbf{j} = 0, b = 0). \tag{9.13}$$

Here there is an even simpler difference in going from the variance maximization problem of principal components for a basis to its composition: the addition of the single, simple constraint $b = 0$. The D basis principal components (9.12) are obtained as the D standardized eigenvectors corresponding to the D positive eigenvalues of

$$(\boldsymbol{\Psi} - \mu \mathbf{I}) \begin{bmatrix} \mathbf{a} \\ b \end{bmatrix} = \mathbf{0}; \tag{9.14}$$

recall from Section 9.2 that $\boldsymbol{\Psi}$ is singular with zero eigenvalue corresponding to the eigenvector $(\mathbf{j}_D', 0)$. There is an eigenvalue separation property for $\boldsymbol{\Psi}$ of exactly the same nature as for $\boldsymbol{\Omega}$.

Property 9.3 If $\mu_1 \geqslant \mu_2 \geqslant \cdots \geqslant \mu_D$ are the D positive eigenvalues of the composition–size covariance matrix $\boldsymbol{\Psi}$ of a D-part basis and $\lambda_1 \geqslant \lambda_2 \geqslant \cdots \geqslant \lambda_d \geqslant \lambda_D = 0$ are the D eigenvalues of the centred

Table 9.6. *Eigenvalues for steroid metabolite compositions*

Covariance matrices	$\boldsymbol{\Omega}$	$\boldsymbol{\Psi}$	$\boldsymbol{\Gamma}$
Eigenvalues	$\mu_1 = 1.16$	$\mu_1 = 0.552$	$\lambda_1 = 0.531$
	$\mu_2 = 0.208$	$\mu_2 = 0.143$	$\lambda_2 = 0.088$
	$\mu_3 = 0.079$	$\mu_3 = 0.079$	$\lambda_3 = 0$

logratio covariance matrix $\boldsymbol{\Gamma}$ of the corresponding D-part composition, then

$$\mu_1 \geqslant \lambda_1 \geqslant \mu_2 \geqslant \lambda_2 \geqslant \cdots \geqslant \mu_d \geqslant \lambda_d \geqslant \mu_D \geqslant 0 = \lambda_D.$$

The property is easily verified for the children's steroid metabolite bases used for illustrative purposes above. The eigenvalues, in the notation of Properties 9.2 and 9.3, are as shown in Table 9.6, conforming with the sequence of inequalities

$$\mu_1 \geqslant \lambda_1 \geqslant \mu_2 \geqslant \lambda_2 \geqslant \mu_3 \geqslant \lambda_3.$$

A simple way to gain some insight into the nature of these eigenvalue separation results is to imagine two statisticians, B and C, engaged in investigation of the variability of bases, B using the original bases but C limited to the compositions constructed from these bases. First it is clear that in general B is in a more favourable position than C; for example, since $\mu_1 \geqslant \lambda_1$, the amount of variability retained by B's first principal component cannot be smaller than that of C's first principal component. But since $\lambda_1 \geqslant \mu_2$, C at least has the assurance that his first principal component is no worse than B's second principal component, which is, however, at least as good in its extraction of variability as C's second principal component, and so on. In this variability extraction contest C can never get ahead of B but is assured that he is never more than one step behind. Although this is the general picture, it is as well to realize that there are two extremes. For example, in terms of composition–size variability, at one extreme all the basis variability may be in the size t in which case $\mu_1 = \text{var}(\log t)$, $\mu_2 = \cdots = \mu_D = \lambda_1 = \cdots = \lambda_D = 0$, and C is in no position to explain any of the variability; at the other extreme all the basis variability may be in the composition $\mathbf{x}$, with $\mu_1 = \lambda_1$, $\mu_2 = \lambda_2, \ldots, \mu_D = \lambda_D$, and C is in no way at a disadvantage.

Completely analogous results hold in the comparison of canonical components obtained from bases and compositions with exactly the same form of separation property. We confine ourselves to the simplest of illustrations, the canonical analysis for the two compositional data sets of Data 9, the bases of steroid metabolite excretions for 37 adults and 30 children. For two data sets there is only one positive eigenvalue for each of the basis logarithmic composition–size and composition canonical analyses. These are computed as 2.33, 1.28, 1.27 respectively, with confirmation of the inequalities $2.33 \geqslant 1.27$ and $1.28 \geqslant 1.27$, of the separation result.

9.4 Distributional relationships

It is clear from our previous discussion that, given the distribution of a basis $\mathbf{w}$, the joint distribution of composition $\mathbf{x}$ and size t is determined, and so also the marginal distribution for $\mathbf{x}$ or for t can be determined. For example, if $f(\mathbf{w})$ is the density function of $\mathbf{w}$ then the joint density function of $(\mathbf{x}, t)$ is, from the basis/composition, size transformation of Definition 9.1,

$$\operatorname{jac}(\mathbf{w}|\mathbf{x}, t) f(t\mathbf{x}) = t^d f(t\mathbf{x}). \tag{9.15}$$

The compositional density function then emerges as the marginal density function

$$\int_0^\infty t^d f(t\mathbf{x}) \mathrm{d}t. \tag{9.16}$$

As a simple illustration consider the Dirichlet composition property of Property 3.3. If the components of the D-part basis $\mathbf{w}$ are independently distributed and of equal-scaled gamma form, say, with w_i having the $\mathrm{Ga}(\alpha_i, \beta)$ form of Definition 3.3, then $t^d f(t\mathbf{x})$ can be expressed in the following product form:

$$\frac{\Gamma(\alpha_1 + \cdots + \alpha_D)}{\Gamma(\alpha_1) \cdots \Gamma(\alpha_D)} x_1^{\alpha_1 - 1} \cdots x_D^{\alpha_D - 1}$$
$$\times \frac{\beta^{\alpha_1 + \cdots + \alpha_D} t^{\alpha_1 + \cdots + \alpha_D - 1} \exp(-\beta t)}{\Gamma(\alpha_1 + \cdots + \alpha_D)} \qquad (\mathbf{x} \in \mathscr{S}^d, t > 0),$$

showing that composition $\mathbf{x}$ and size t are independently distributed as $\mathscr{D}^d(\boldsymbol{\alpha})$ and $\mathrm{Ga}(\alpha_1 + \cdots + \alpha_D, \beta)$. There is therefore in this case no need to perform the integration at (9.16).

If we remove the requirement of equality of scale parameters, with w_i distributed as $\mathrm{Ga}(\alpha_i, \beta_i)$ $(i = 1, \ldots, D)$, then the above factorization is no longer possible, but straightforward integration leads to a generalized Dirichlet density function for the compositional distribution:

$$\frac{\Gamma(\alpha_1+\cdots+\alpha_D)}{\Gamma(\alpha_1)\cdots\Gamma(\alpha_D)}\frac{\beta_1^{\alpha_1}x_1^{\alpha_1-1}\cdots\beta_D^{\alpha_D}x_D^{\alpha_D-1}}{(\beta_1x_1+\cdots+\beta_Dx_D)^{\alpha_1+\cdots+\alpha_D}} \quad (\mathbf{x}\in\mathscr{S}^d). \quad (9.17)$$

We have already seen in Property 6.1 that linear transformation theory for multivariate normal distributions shows that a basis $\mathbf{w}$ which has a multivariate lognormal form $\Lambda^D(\boldsymbol{\zeta},\boldsymbol{\Omega})$ leads to an additive logistic normal form $\mathscr{L}^d(\mathbf{F}\boldsymbol{\zeta},\mathbf{F}\boldsymbol{\Omega}\mathbf{F}')$ for the composition $\mathbf{x}$. The joint distribution for $(\mathbf{x},t)$ or the marginal distribution for t is much more complicated. This suggests that when interest in bases is in the distributional relationships between their compositions and sizes it may be more profitable to follow the line of argument in Section 9.2 and model directly the joint distribution of $(\mathbf{x},t)$. Here if $\mathbf{y}=\log(\mathbf{x}_{-D}/x_D)$ and $u=\log t$ we may consider whether some satisfactory analysis will emerge from the assumption that

$$\begin{bmatrix}\mathbf{y}\\ u\end{bmatrix} \text{ is } \mathscr{N}^D\left\{\begin{bmatrix}\boldsymbol{\mu}\\ \nu\end{bmatrix},\begin{bmatrix}\boldsymbol{\Sigma} & \boldsymbol{\kappa}\\ \boldsymbol{\kappa}' & \varepsilon\end{bmatrix}\right\}. \quad (9.18)$$

We shall see that this approach pays rich dividends.

9.5 Compositional invariance

In our initial examination of the household budgets of Data 8 we asked in Question 1.7.1 whether the budget pattern of expenditure may be independent of total expenditure and, if not, whether we can satisfactorily describe the nature of the dependence of the pattern on total expenditure. Here each vector $\mathbf{w}$ of four commodity expenditures is a 4-part basis, the budget pattern $\mathbf{x}$ is its composition $\mathscr{C}(\mathbf{w})$ and total expenditure t is the basis size $w_1+\cdots+w_4$. Thus our concern is in whether the basis variability is such that composition is independent of size and, if not, the nature of the conditional distribution of $\mathbf{x}$ for given t.

Again each row of Data 9 consists of a 3-part basis $\mathbf{w}$ of steroid metabolite excretions, and Question 1.8.1 asked whether composition or excretion pattern $\mathbf{x}=\mathscr{C}(\mathbf{w})$ is independent of basis size or total excretion $t=\mathbf{w}'\mathbf{j}$ for adults or for children and, if not, Question 1.8.2 asked whether the dependence of pattern on total excretion is the same for adults and children.

This section is concerned with such questions of the relation of composition $\mathbf{x}$ to size t. We start with an obvious definition.

Definition 9.3 A D-part basis $\mathbf{w}$ is said to be *compositionally invariant* if its composition $\mathbf{x} = \mathscr{C}(\mathbf{w})$ is statistically independent of its size $t = w_1 + \cdots + w_D$.

We immediately have the following strong property associated with Dirichlet compositions.

Property 9.4 A basis of independent, equally scaled, gamma components is compositionally invariant, and its composition has a Dirichlet distribution.

Proof. This is immediate from Properties 3.3 and 3.4 of the Dirichlet class of distributions.

Naturally in any practical investigation of compositional invariance we require our modelling to allow the possibility of the basis not possessing this property. Moreover, since compositional invariance is essentially a property of a basis in relation to its composition, it is tempting to start such modelling with a rich parametric class of distributions for the basis. The most appealing one is the class of multivariate lognormal distributions $\Lambda^D(\zeta, \boldsymbol{\Omega})$ of Definition 6.3 for the basis. Unfortunately the following result due to Mosimann (1975b) shatters hopes of this being a useful starting point.

Property 9.5 If a D-part basis has a $\Lambda^D(\zeta, \boldsymbol{\Omega})$ distribution and has compositional invariance, then the basis covariance matrix $\boldsymbol{\Omega}$ is a scalar multiple of $\mathbf{J}_D$ so that the basis distribution is singular with rank 1.

Note

1. For such a degenerate basis, variation is one-dimensional in $\mathscr{R}^D_+$, along a single ray through the origin, and the associated composition is constant. In other words, size t varies but composition $\mathbf{x}$ is fixed.

Compositional invariance is obviously a perfectly sensible scientific hypothesis to consider, so that if it cannot be posed within the framework of a model which demands multivariate lognormality of the basis we must look to other means of modelling. There is no law of nature which dictates that multivariate positive variability must follow lognormality, although it is sometimes convenient to use such a model; in the present case, because of the degeneracy property, it is obviously pointless to insist on a multivariate lognormal model. Since the question at issue is the dependence or otherwise of composition **x** and size t, we concentrate our modelling skills on the joint distribution of $(\mathbf{x}, t)$ or one of the conditional distributions of **x** given t or of t given **x**, rather than on the basis distribution. Probably the simplest formulation is in terms of the familiar logratio composition $\mathbf{y} = \log(\mathbf{x}_{-D}/x_D)$ and of $\log t$. Depending on the particular aim of the investigation of compositional invariance, we then have three simple and standard hypothesis testing possibilities.

1. Assume that the joint distribution of $(\mathbf{y}, \log t)$ is of $\mathcal{N}^D$ form as in (9.18) with compositional invariance corresponding to a hypothesis of covariance pattern, namely that $\boldsymbol{\kappa} = \mathrm{cov}(\mathbf{y}, \log t) = \mathbf{0}$.
2. Assume that the conditional distribution of $\mathbf{y}|t$ or $\mathbf{y}|\log t$ is of $\mathcal{N}^d$ form with mean $\boldsymbol{\alpha} + \boldsymbol{\beta} \log t$ with compositional invariance corresponding to a hypothesis of multivariate regression, namely that the regression coefficient vector $\boldsymbol{\beta} = \mathbf{0}$.
3. Assume that the conditional distribution of $\log t|\mathbf{y}$ is of $\mathcal{N}^1$ form with mean $\gamma + \boldsymbol{\delta}'\mathbf{y}$ with compositional invariance corresponding to a hypothesis of univariate, or multiple, regression, namely that the regression coefficient vector $\boldsymbol{\delta} = \mathbf{0}$.

These are essentially equivalent representations of the compositional invariance hypothesis: for example, the generalized likelihood ratio principle yields exactly the same critical region for each approach. Which of the three approaches is convenient will largely depend on what other aspects of the data are of interest. There is little doubt that the most familiar version, and the simplest computationally, is the multiple regression form (3). Suppose, however, that in the event of our rejection of the hypothesis of compositional invariance we wish to investigate how composition depends on size, rather than how size depends on composition. Then it would be sensible to adopt approach (2) because we shall already have determined the dependence of **y** on t in the process of testing for compositional invariance. Since this is the only type of dependence which will concern us in this monograph, we shall confine attention to (2). We should note, however, that the other, equally tractable, versions may be more appropriate for other situations. We also emphasize that we should not read into any dependence detected any causal mechanism: from the basis data all that we can hope to infer is how composition and size are jointly or conditionally distributed.

A test for compositional invariance

If we approach the testing of compositional invariance through a formulation of the conditional distribution of logratio composition **y** on basis size t then we are simply involved in the process of loglinear compositional modelling as described in Sections 7.6–7.8. In terms of the $N \times d$ logratio array **Y** we can express the model in the familiar notation of Definition 7.4 as

$$\mathbf{Y} = \mathbf{A}\boldsymbol{\Theta} + \mathbf{E}, \tag{9.19}$$

where **A** is $N \times 2$ with rth row $[1 \quad \log t_r]$ and

$$\boldsymbol{\Theta} = \begin{bmatrix} \alpha_1 \cdots \alpha_d \\ \beta_1 \cdots \beta_d \end{bmatrix}. \tag{9.20}$$

The hypothesis of compositional invariance is that $\beta_1 = \cdots = \beta_d = 0$, expressible either in constraint form

$$[0 \quad 1]\boldsymbol{\Theta} = \mathbf{0}, \tag{9.21}$$

or, in our further development here, in reparametrization form analogous to (7.40), as

$$\mathbf{Y} = \mathbf{B}\boldsymbol{\Psi} + \mathbf{E}, \tag{9.22}$$

where $\mathbf{B} = \mathbf{j}_N$ is the $N \times 1$ vector of units and

$$\boldsymbol{\Psi} = [\alpha_1 \cdots \alpha_d]. \tag{9.23}$$

No further details need be provided since the test is of standard multivariate form already described in Property 7.4.

9.6 An application to household budget analysis

In the literature of consumer demand analysis there have been only a few attempts to incorporate compositional analysis directly into the analysis of household budgets. This technique has many advantages and provides opportunities for new forms of investigation. Suppose that **w** is a record of household expenditures on D mutually exclusive and exhaustive commodity groups so that $t = w_1 + \cdots + w_D$ is total expenditure and $\mathbf{x} = \mathscr{C}(\mathbf{w})$ is the proportional pattern of allocation to the groups.

Loglinear modelling as suggested in Section 9.5 with $p(\mathbf{x}|t)$ of $\mathscr{L}^d(\boldsymbol{\alpha} + \boldsymbol{\beta} \log t, \boldsymbol{\Sigma})$ form has interesting consequences. First, the sometimes troublesome budget constraint or Engel aggregation condition (Brown and Deaton, 1972, p. 1163), that for each household total expenditure should equal the sum of all commodity expenditures, is automatically satisfied. Secondly, the hypothesis of composition invariance, $\boldsymbol{\beta} = \mathbf{0}$, has a direct interpretation in terms of the income elasticities $\varepsilon_i = \partial \log w_i / \partial \log t$ of demand $(i = 1, \ldots, D)$, if for the moment and for simplicity we identify household total expenditure with household income. In expectation terms

$$\beta_i = \varepsilon_i - \varepsilon_D \qquad (i = 1, \ldots, d) \tag{9.24}$$

so that compositional invariance, not surprisingly, corresponds to equality of all D income elasticities. Thirdly, whether or not there is compositional invariance, the modelling can clearly be extended to a full consumer demand analysis by the incorporation of commodity prices and other covariates such as household type and household composition into the mean parameter of the logistic normal distribution. Indeed, such an extension can be shown to be identical with the Houthakker (1960) indirect addilog model of consumer demand (Brown and Deaton, 1972, equation 115).

In the above discussion we have identified household total expenditure t with household income s. This is not an essential feature of the modelling since we could approach it through the conditioning $p(s, t, \mathbf{x}) = p(s)p(t|s)p(\mathbf{x}|s, t)$ with perhaps the reasonable assumption that, for given total expenditure t, the pattern $\mathbf{x}$ is independent of income s, leading to the above focus on $p(\mathbf{x}|t)$.

As a simple start to our analysis of the household budgets of Data 8, let us first apply tests of compositional invariance separately to the 20 single male households and to the 20 single female households. For each of these tests, in the notation of Section 9.5 and Property 7.4 we have $d = 3$, $k = 1$, $N = 20$ so that $a = 1$, $b = 3/2$, $c = 8$. Table 9.7 gives the vector estimates $\hat{\boldsymbol{\alpha}}, \hat{\boldsymbol{\beta}}$ of $\boldsymbol{\alpha}, \boldsymbol{\beta}$ under the model, the residual matrix $\mathbf{R}_{\text{m}}$ under the model, the values of $|\mathbf{R}_{\text{m}}|$ and $|\mathbf{R}_{\text{h}}|$, the residual determinants under the model and the hypothesis, and hence the values of q, the computed test statistic of Property 7.4. From these the significance probabilities are less than 10^{-5} for both single male and single female households. Thus for each set of bases, the household

Table 9.7. *Computations for testing compositional invariance of household budgets of single males and single females*

	Males			Females		
$\hat{\boldsymbol{\alpha}}$ $\hat{\boldsymbol{\beta}}$	5.10	−0.57		5.52	−0.58	
	10.02	−1.28		8.70	−1.32	
	−6.44	0.79		−6.81	1.02	
$\mathbf{R}_m$	1.99	1.25	−0.48	1.94	1.19	−0.63
	1.25	1.90	−0.30	1.19	1.92	−0.40
	−0.48	−0.30	0.63	−0.63	−0.40	0.81
$\lvert\mathbf{R}_m\rvert$	1.16			1.41		
$\lvert\mathbf{R}_h\rvert$	24.12			24.74		
q	0.952			0.943		

expenditure vectors of single males and of single females, there is strong evidence against the hypothesis of compositional invariance: in other words, the patterns of expenditures do appear to depend on total expenditure.

From (9.24) we see that, although the D 'income' elasticities ε_i $(i = 1, \ldots, D)$ are not determined by the d regression coefficients β_i $(i = 1, \ldots, d)$, they can at least be placed in order of magnitude. The commodity groups arranged in increasing order of elasticity, that is in conventional economic terminology from necessity to increasing luxury groups, are (for each sex)

1. Foodstuffs, including alcohol and tobacco.
2. Housing, including fuel and light.
3. Services, including transport and vehicles.
4. Other goods, including clothing, footwear and durable goods.

The fact that the ordering is the same for males and females raises the question of whether the dependence of pattern on total expenditure is really different for females and males. This suggests that it might have been more fruitful to consider a whole lattice of hypotheses along the lines of Fig. 9.1, with hypotheses expressed in terms of the parameters of the model

$$\begin{aligned} \mathbf{y} &= \boldsymbol{\alpha}_M + \boldsymbol{\beta}_M \log t + \text{error}, && \text{for males;} \\ \mathbf{y} &= \boldsymbol{\alpha}_F + \boldsymbol{\beta}_F \log t + \text{error}, && \text{for females.} \end{aligned} \tag{9.25}$$

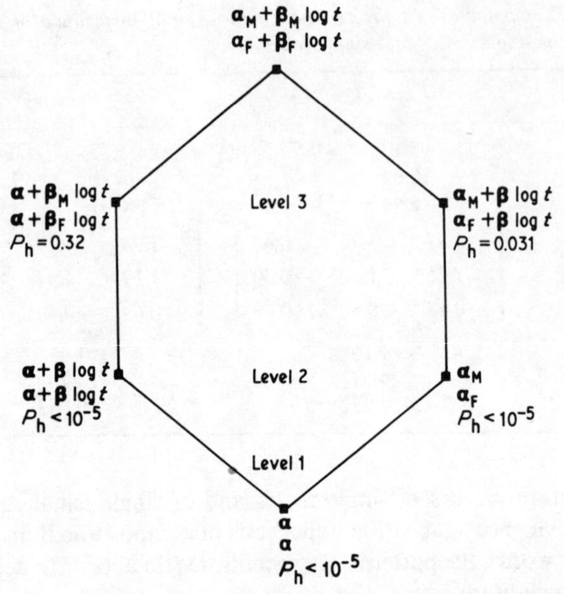

Fig. 9.1. *Lattice of hypotheses for comparison of household budgets of single males and single females.*

Note that the separate compositional invariance hypotheses tested above are the hypotheses $\boldsymbol{\beta}_M = \mathbf{0}$ and $\boldsymbol{\beta}_F = \mathbf{0}$ at level 2. All the hypotheses on the lattice can be tested within the standard framework of Property 7.4. We omit the details here but show on the lattice the significance probabilities associated with each hypothesis, noting that we can move up the lattice by rejection until level 3 when we would fail to reject the hypothesis that $\boldsymbol{\alpha}_M = \boldsymbol{\alpha}_F$.

Before leaving this example we point out that it would be straightforward to introduce some concomitant feature, such as age u, on which the pattern may depend, and test hypotheses within the model

$$\begin{aligned} \mathbf{y} &= \boldsymbol{\alpha}_M + \boldsymbol{\beta}_M \log t + \boldsymbol{\gamma}_M u + \text{error}, \qquad \text{for males,} \\ \mathbf{y} &= \boldsymbol{\alpha}_F + \boldsymbol{\beta}_F \log t + \boldsymbol{\gamma}_F u + \text{error}, \qquad \text{for females.} \end{aligned} \tag{9.26}$$

9.7 An application to clinical biochemistry

We now turn to Data 9, with its 3-part bases (w_1, w_2, w_3) of urinary excretions of steroid metabolites, 37 for adults and 30 for children. To investigate possible compositional invariance properties we follow the procedure of Section 9.6 and transform to obtain the logratio composition $\mathbf{y} = (y_1, y_2)$, where $y_1 = \log(w_1/w_3)$, $y_2 = \log(w_2/w_3)$, and the logarithm of size, $z = \log t$, for each vector. Investigation of possible differences in the patterns of steroid metabolite excretions between adults and children can then be made within the multivariate linear model

$$\begin{aligned} \mathbf{y} &= \boldsymbol{\alpha}_{\mathrm{A}} + \boldsymbol{\beta}_{\mathrm{A}} z + \text{error}, \qquad \text{for adults,} \\ \mathbf{y} &= \boldsymbol{\alpha}_{\mathrm{C}} + \boldsymbol{\beta}_{\mathrm{C}} z + \text{error}, \qquad \text{for children,} \end{aligned} \tag{9.27}$$

where the $\boldsymbol{\alpha}$ and $\boldsymbol{\beta}$ parameters are 2-dimensional vectors.

We could again set out a lattice of hypotheses within model (9.27) in a way similar to that of Fig. 9.1, starting with the hypothesis at level 1 that there are no differences between adults and children and that their excretion patterns show compositional invariance. The outcome of such a procedure is that there is no evidence of dependence of composition on size for either adults or children. There is, however, a significant difference between the mean steroid metabolite compositions of adults and children so that it is still doubtful whether children can be usefully subjected to any diagnostic system devised for adults.

9.8 Reappraisal of an early shape and size analysis

Fisher (1947), in an expository article on the methodology of analysis of covariance, analysed a set of data consisting of heart and total body weights of 47 female and 97 male domestic cats. These are reproduced in Data 28. He had observed that the ratio of average heart weight to average total body weight is almost identical in females and males, and the purpose of his analysis was to investigate whether this close agreement masked some real difference between the sexes. He used the model

$$\begin{aligned} w_{\mathrm{H}} &= \alpha_{\mathrm{F}} + \beta_{\mathrm{F}} w_{\mathrm{B}} + \text{error}, \\ w_{\mathrm{H}} &= \alpha_{\mathrm{M}} + \beta_{\mathrm{M}} w_{\mathrm{B}} + \text{error}, \end{aligned} \tag{9.28}$$

for females and males, where w_{H} and w_{B} denote heart and total body weights. In his testing he found that both β_{F} and β_{M} are significantly different from zero and also that there is a strong probability that α_{F} is different from zero. The estimates of the parameters are

$$\hat{\alpha}_{\mathrm{F}} = 0.00298, \hat{\beta}_{\mathrm{F}} = 0.00264, \hat{\alpha}_{\mathrm{M}} = -0.00118, \hat{\beta}_{\mathrm{M}} = 0.00431.$$

Fisher thus concluded that there is a difference in the heart to body weight relation between female and male cats; more specifically, that female heart weights probably increase less than proportionately to body weight, whereas in males, where the heart weight appears to increase more than proportionately to the body weight, the difference is not significant with these data. Fisher finishes with the comment that what at first sight appears to be close agreement between the sexes has been revealed by analysis of covariance to have concealed differences in the relation of heart to total body weight between the sexes.

Unfortunately, Fisher's modelling, and therefore his conclusion, are seriously open to question. The residuals of the above regressions display appreciable positive skewness. This might have been anticipated from simple examination of the marginal frequency distributions of both heart and body weight which show considerable positive skewness. This feature suggests that it would be more appropriate to model linearly in terms of some transformation of the heart and body weights. Thus before we reformulate the problem in terms of compositional analysis we report the results of using the model m:

$$\begin{aligned} \log w_{\mathrm{H}} &= \alpha_{\mathrm{F}} + \beta_{\mathrm{F}} \log w_{\mathrm{B}} + \text{error}, \\ \log w_{\mathrm{H}} &= \alpha_{\mathrm{M}} + \beta_{\mathrm{M}} \log w_{\mathrm{B}} + \text{error}. \end{aligned} \tag{9.29}$$

This is indeed a now popular form of modelling in biometric allometry; see, for example, Sprent (1972). In the investigation of the possibility of proportional relationship and sex homogeneity of the relationship, it seems sensible as a first step to test the overall hypothesis h that $\beta_{\mathrm{F}} = \beta_{\mathrm{M}} = 1$, $\alpha_{\mathrm{F}} = \alpha_{\mathrm{M}}$. The estimates of the parameters under the model m are

$$\hat{\alpha}_{\mathrm{F}} = -5.295, \hat{\beta}_{\mathrm{F}} = 0.699, \hat{\alpha}_{\mathrm{M}} = -5.662, \hat{\beta}_{\mathrm{M}} = 1.099,$$

and under the hypothesis h are

$$\hat{\alpha}_{\mathrm{F}} = \hat{\alpha}_{\mathrm{M}} = -5.555, \quad \hat{\beta}_{\mathrm{F}} = \hat{\beta}_{\mathrm{M}} = 1.$$

The residual sums of squares under m and under h are

$$\mathrm{RSS}_{\mathrm{m}} = 2.5037, \quad \mathrm{RSS}_{\mathrm{h}} = 2.5823,$$

so that the standard form of test statistic for testing a linear hypothesis within a linear model takes the value

$$\frac{(\text{RSS}_h - \text{RSS}_m)/3}{\text{RSS}_m/140} = 1.47$$

to be compared against upper $F(3, 140)$ values, 2.65 at 5 per cent. There is thus no evidence against this hypothesis and so this logarithmic modelling would regard the conclusion that the relationship between heart weight w_H and body weight w_B is for both sexes approximately

$$\log(w_H/w_B) = -5.555 + \text{error}$$

or

$$w_H = 0.00387 w_B \times \text{error}, \tag{9.30}$$

as tenable. Moreover, the residuals in this case follow very closely a normal pattern.

If we approach this problem from the viewpoint of compositional data analysis then we recognize heart weight w_H and residual body weight $w_R = w_B - w_H$ as components of a basis (w_H, w_R). From this basis we can then form its composition

$$(x_H, x_R) = (w_H, w_R)/(w_H + w_R),$$

which represents 'shape', and its size

$$w_H + w_R = w_B,$$

which is simply total body weight. A standard question in shape and size analysis is whether shape is independent of size which, in terms of compositional data analysis, is whether the basis has compositional invariance. Our considerations of Section 9.5 suggest that we may attempt to investigate this problem through the logratio form of the composition by adopting the regression model m:

$$\begin{aligned} \log(w_H/w_R) &= \alpha_F + \beta_F \log w_B + \text{error}, \\ \log(w_H/w_R) &= \alpha_M + \beta_M \log w_B + \text{error}, \end{aligned} \tag{9.31}$$

with $\beta_F = 0$ and $\beta_M = 0$ corresponding to lack of dependence of shape on size for females and males respectively. Since we are also interested in possible sex differences it is again natural to take as a starting point the overall hypothesis h that $\beta_F = \beta_M = 0$, $\alpha_F = \alpha_M$. The estimates of the parameters under the model m are then

$$\hat{\alpha}_F = -5.291, \hat{\beta}_F = -0.302, \hat{\alpha}_M = -5.658, \hat{\beta}_M = 0.100,$$

and under the hypothesis h

$$\hat{\alpha}_F = -5.552, \hat{\beta}_F = 0, \hat{\alpha}_M = -5.552, \hat{\beta}_M = 0.$$

Again the standard test fails to reject the hypothesis h and so we would be prepared to consider the relationship as represented by

$$w_H = 0.00388 w_R \times \text{error}. \tag{9.32}$$

This result and that of the logarithmic regression analysis are essentially in agreement, the difference arising from the presence of w_B on the right-hand side of (9.29) instead of the current w_R. Since w_R is only slightly less than w_B, we would expect the estimated ratio 0.00388 in the present analysis to be marginally larger than the estimated ratio 0.00387 for the previous analysis. Support for the above conclusion can be convincingly demonstrated by the plots in Fig. 9.2 of the cumulative frequency distribution of $-\log(w_H/w_R)$ on arithmetic

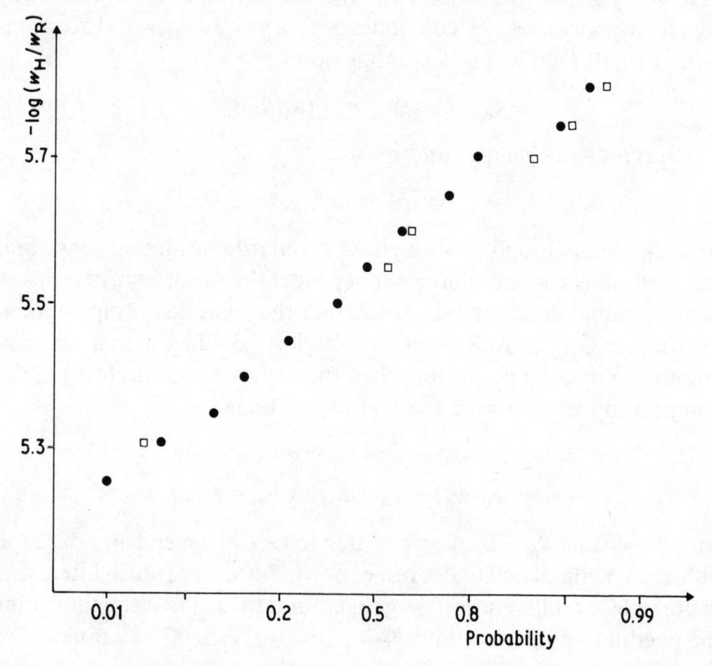

Fig. 9.2. *Cumulative frequency distribution of* $-\log(w_H/w_R)$ *on arithmetic probability paper. Males, solid circles; females, open squares, where different from males.*

probability paper, leading to coincident straight lines for females and males.

9.9 Bibliographic notes

The first attempt to relate properties of bases and compositions is implicit in the correlation analysis of Pearson (1897). Unfortunately this attempt, like so many other subsequent attempts, failed because of its concentration on crude covariance structures. A main line of attack started with Chayes and Kruskal (1966) who, in pursuit of a definition of pseudo-independence for compositions, introduced imaginary bases and their approximate crude covariance relationship with the compositions of primary interest. The unsatisfactory nature of this approach and the many-to-one relationship of bases to compositions was emphasized by Kork (1977). The introduction of logratio covariance structures for compositions by Aitchison (1981a) enabled the expression of exact relationships between covariance structures of bases and compositions, and made explicit the nature of the many-to-one relationship.

The relationships between principal and canonical component analyses of bases and compositions are natural developments of the logcontrast principal component analysis of Aitchison (1983) and the eigenvalue separation results of Gantmacher (1959). Similar results are also found in shape and size analysis in biometric allometry: see Darroch and Mosimann (1985) for a detailed account of such studies. The composition–size forms of such analyses are introduced in this monograph.

The relationship of a basis of independent equally scaled gamma components and its independent Dirichlet-distributed composition and gamma-distributed size is an old-established result of distribution theory. The definition by Aitchison and Shen (1980) of the logistic normal class of distributions on the simplex allowed an expression of the simple connection between a lognormal basis and its logistic normal composition.

The hypothesis of compositional invariance has appeared under a variety of guises: as the Lukacs condition in a characterization of the Dirichlet distribution (Mosimann, 1962); as additive isometry in the analysis of biological shape and size (Mosimann, 1970, 1975a, b); as proportional invariance in the study of F-independence (Darroch and James, 1974). The earliest attempt at devising a test of compositional invariance seems to be in the approach of Fisher (1947) to the domestic cat data, reanalysed in Section 9.8. The composition–size form for the distribution of a basis was introduced by Aitchison (1982) to allow straightforward testing of compositional invariance by providing for the description of basis variability a simple alternative to the lognormal class of distributions which, by the result of Mosimann (1975b), can only support compositional invariance in a degenerate form.

The budget-share approach to household expenditure analysis seems to have been first considered by Working (1943); see also Houthakker (1960), Brown and Deaton (1972), Leser (1963), Deaton (1978), Deaton and Muellbauer (1980), Aitchison (1982).

Problems

9.1 For the 4-part bases of single male household expenditures of Data 8 estimate the basis logarithmic covariance matrix $\boldsymbol{\Omega}$, the composition–size covariance matrix $\boldsymbol{\Psi}$, the logratio covariance matrix $\boldsymbol{\Sigma}$, the centred logratio covariance matrix $\boldsymbol{\Gamma}$ and the variation matrix $\boldsymbol{T}$.

Verify the relationships of Table 9.2 between the basis and the compositional specifications of covariance structure.

Verify also the relationships of Property 9.1.

9.2 Carry out principal component analyses of the bases and compositions associated with the glacial tills of Data 18, omitting tills with zero component and using both the logarithmic covariance matrix estimate $\hat{\boldsymbol{\Omega}}$ and the composition–size $\hat{\boldsymbol{\Psi}}$ for the bases.

Verify the eigenvalue separation results in Properties 9.2 and 9.3.

9.3 Starting from a D-part basis $\mathbf{w}$ of $\Lambda^D(\zeta, \boldsymbol{\Omega})$ form, derive the joint density function of composition–size $(\mathbf{x}, t)$ as in (9.15). By carrying out the integration in (9.16) show that the marginal density function of composition $\mathbf{x}$ is of $\mathscr{L}^d$ form.

9.4 Test the hypothesis of compositional invariance for the 92 glacial tills of Data 18.

9.5 Treating the yat, yee, sam of the pogo jumps of Data 19 as bases, investigate the hypothesis of compositional invariance for each of the seven finalists.

Set up a lattice of hypotheses which will allow not only

investigation of compositional invariance but also allow comparisons to be made between finalists. Do your conclusions from analysis of this lattice enable you to answer the questions you posed in Problem 1.4?

9.6 From 75 heart X-rays, of which 20 were from normotensive and 55 from hypertensive males, measurements of heart width w_H and total thorax width w_T were recorded as shown in Data 29. Regarding $(w_H, w_T - w_H)$ as a basis of size w_T investigate the nature of compositional invariance and explore the possibilities of differences between normotensives and hypertensives within your analysis.

9.7 Data 30 gives the household expenditures by 20 single men on five mutually exclusive and exhaustive commodity groups over the period of one month. Investigate the nature of the income elasticities of demand along the lines of Section 9.6.

CHAPTER 10

Subcompositions and partitions

Independence, like honour, is a rocky island without a beach.

Napoleon Bonaparte: *Sayings of Napoleon*

10.1 Introduction

In this chapter we introduce a number of useful concepts of independence for compositions, provide a means of testing hypotheses of independence within various classes of logistic normal models, and where necessary derive convenient means of describing certain forms of dependence. In such models logratio compositions of one kind or another are assumed to have a multivariate normal distribution and so no distinction need then be made between zero correlation and independence. For example, the concept of a logratio-uncorrelated composition in Definition 5.1 will be seen in the next section to be equivalent to complete subcompositional independence within additive logistic normal modelling. Since the null distributions of the test statistics are derived under the assumption of underlying normality and are therefore conceived as testing independence, in this chapter we shall usually express our ideas in terms of independence.

10.2 Complete subcompositional independence

The unit-sum constraint on the components of a D-part composition compels us not only to reconsider definitions of covariance structure but also to reformulate our ideas of independence. Clearly the components $x_1, \dots, x_D$ cannot be independent since $x_1 + \cdots + x_D = 1$. Can we, however, find some strong form of independence for patterns of variability in the simplex $\mathscr{S}^d$ to match the independence of components in $\mathscr{R}^d$? A useful starting point is to consider the imposition of independence conditions on subcompositions which have played such a central role in compositional theory and practice so far. The following definition places a very demanding set of constraints on the distribution of a composition.

Definition 10.1 A composition has *complete subcompositional independence* if the subcompositions formed from any partition of the composition form an independent set.

The definition has immediate relevance to Dirichlet and logistic normal compositions.

Property 10.1 Every Dirichlet composition has complete subcompositional independence.

Proof. This is immediate from Property 3.4 for the Dirichlet class. An obvious consequence of Property 10.1 is that the Dirichlet class will be useless in describing patterns of variability not having this strong form of independence.

Property 10.2 For additive logistic normal compositions the properties of complete subcompositional independence and of being logratio uncorrelated are equivalent.

Proof. Under additive logistic normality any finite set of logratios of components has a multivariate normal distribution so that zero correlation and independence become equivalent properties.

From Property 5.11 we have an immediate characterization of complete subcompositional independence for additive logistic normal compositions in terms of special covariance patterns. Since the $\boldsymbol{\Gamma}$ form is too cumbersome for practical use we confine attention to the element forms for $\boldsymbol{\Sigma}$ and $\boldsymbol{T}$ in the following result.

Property 10.3 A D-part composition with a $\mathscr{L}^d(\boldsymbol{\mu}, \boldsymbol{\Sigma})$ distribution has complete subcompositional independence if

$$\sigma_{ij} = \begin{cases} \alpha_i + \alpha_D & (i = j), \\ \alpha_D & (i \neq j), \end{cases}$$

or, equivalently,

$$\tau_{ij} = \begin{cases} 0 & (i = j), \\ \alpha_i + \alpha_j & (i \neq j). \end{cases}$$

Notes

1. For 3-part compositions the covariance structure can always be expressed in the special form of Property 10.3 so that all 3-part compositions have complete subcompositional independence. This can be seen to be the case from Definition 10.1 since, for every partition of a 3-part composition, the highest-dimensional subcomposition that can emerge is of two parts and this is trivially independent of any 1-part subcomposition which is simply the constant 1. Thus the concept of complete subcompositional independence is of no practical importance for 2- or 3-part compositions.
2. For $D > 3$ the special covariance pattern for complete subcompositional independence has only D parameters compared with the $\frac{1}{2}dD$ of the full covariance structure. Complete subcompositional independence therefore imposes $\frac{1}{2}D(D-3)$ constraints on a covariance structure and so is quite a restrictive condition to place on a distribution.
3. Property 10.1 shows that there is no possibility of discussing any departure from complete subcompositional independence within the class of Dirichlet compositions. The additive logistic normal class is different: within it we may accommodate both compositions which have, and compositions which do not have, complete subcompositional independence. Thus it is an excellent framework within which to test the hypothesis of complete subcompositional independence.
4. As indicated in Section 9.9, early attempts to arrive at a strong form of compositional independence within the context of the crude covariance structure were conceived in terms of an imaginary basis of independent components. The concept was bound to fail because of the two difficulties: the many–one relationship of bases to compositions, and the lack of a simple, exact relationship between basis and compositional covariance structures. Although we have seen above that the strong form of independence we have named complete subcompositional independence can be defined in terms of compositional variability alone, it does bear a simple relationship to independence in any real basis from which the composition is formed. If a basis $\mathbf{w}$ has independent components, then $\boldsymbol{\Omega} = \operatorname{diag}(\alpha_1, \ldots, \alpha_D)$ so that, from Table 9.2, the composition $\mathbf{x} = \mathscr{C}(\mathbf{w})$ has logratio covariance matrix

$$\boldsymbol{\Sigma} = \mathbf{F}\boldsymbol{\Omega}\mathbf{F}' = \operatorname{diag}(\alpha_1, \ldots, \alpha_d) + \alpha_D \mathbf{J}_d$$

and so, under additive logistic normality, $\mathbf{x}$ has complete subcompositional independence. The converse is not true because of the many–one relationship of bases to compositions: it is clear from Table 9.3 that $\boldsymbol{\Omega}$ constructed from the above special form of $\boldsymbol{\Sigma}$ is not necessarily diagonal.

In any investigation of complete subcompositional independence it is convenient to incorporate a test of the more stringent hypothesis that the covariance structure is isotropic (Definition 5.5), which is equivalent, by Property 5.15, to complete subcompositional independence with d additional constraints, namely the equality of $\alpha_1, \ldots, \alpha_D$.

No exact tests of these hypotheses have been found but the Wilks (1938) asymptotic generalized likelihood ratio test of Section 7.4 can be used in the following forms.

Property 10.4 Let $\hat{\boldsymbol{\Sigma}}_{\mathrm{m}}, \hat{\boldsymbol{\Sigma}}_{\mathrm{h}}$ and $\hat{\boldsymbol{\Sigma}}_{\mathrm{i}}$ denote the maximum likelihood estimates of the logratio covariance matrix $\boldsymbol{\Sigma}$ under the general $\mathscr{L}^d(\boldsymbol{\mu}, \boldsymbol{\Sigma})$ model, under the hypothesis of complete subcompositional independence, and under the isotropic hypothesis, respectively.

(a) A test for complete subcompositional independence compares

$$N\{\log|\hat{\boldsymbol{\Sigma}}_{\mathrm{h}}| - \log|\hat{\boldsymbol{\Sigma}}_{\mathrm{m}}|\}$$

against upper percentage points of the

$$\chi^2\{\tfrac{1}{2}D(D-3)\}$$

distribution.

(b) A test for isotropic covariance structure compares

$$N\{\log|\hat{\boldsymbol{\Sigma}}_{\mathrm{i}}| - \log|\hat{\boldsymbol{\Sigma}}_{\mathrm{m}}|\}$$

against upper percentage points of the

$$\chi^2\{\tfrac{1}{2}(D+1)(D-2)\}$$

distribution.

Notes

1. The estimate $\hat{\boldsymbol{\Sigma}}_{\mathrm{m}}$ under the model is straightforwardly obtained as the sample logratio covariance matrix.
2. Under the isotropic hypothesis $\boldsymbol{\Sigma}$, by Definition 5.5, takes the form $\alpha\mathbf{H}$ and the maximum likelihood estimate $\hat{\alpha}$ of α can be simply related

to the measure of total variability in any one of its equivalent forms of Property 5.9. We have

$$\hat{\alpha} = (2dD)^{-1}\mathbf{j}'\hat{\boldsymbol{T}}_{\mathrm{m}}\mathbf{j} = d^{-1}\mathrm{tr}(\mathbf{H}^{-1}\hat{\boldsymbol{\Sigma}}_{\mathrm{m}}) = d^{-1}\mathrm{tr}(\hat{\boldsymbol{\Gamma}}_{\mathrm{m}}), \qquad (10.1)$$

where the subscript m denotes estimation under the model or equivalently sample covariance estimation as in Section 4.3. We shall shortly see that the variation matrix $\hat{\boldsymbol{T}}_{\mathrm{m}}$ is the simplest form of covariance structure for the problem of computing $\hat{\boldsymbol{\Sigma}}_{\mathrm{h}}$ and so it is reasonable to use the form involving the sum of the elements of $\hat{\boldsymbol{T}}_{\mathrm{m}}$ in the computation of $\hat{\alpha}$. Note also that property H5 of Appendix A gives

$$\log|\hat{\boldsymbol{\Sigma}}_{\mathrm{i}}| = d\log\hat{\alpha} + \log D.$$

3. No explicit form for $\hat{\boldsymbol{\Sigma}}_{\mathrm{h}}$ is obtainable so that estimation here requires an iterative process. We recommend the Nash (1979) modification of the Marquardt (1963) mixture of Newton–Raphson and steepest-ascent methods. For this purpose a form of loglikelihood which acknowledges the symmetry in $\alpha_1, \ldots, \alpha_D$ is advisable. Since the expression of complete subcompositional independence is simpler in terms of the variation matrix $\boldsymbol{T}$ than the centred logratio covariance matrix $\boldsymbol{\Gamma}$, this indicates that we should use the $\mathscr{L}^d(\boldsymbol{\mu}, \boldsymbol{\Sigma})$ density function in its $\boldsymbol{T}$ version (6.4) in constructing the loglikelihood. At first sight the form $\mathbf{F}'(\mathbf{F}\boldsymbol{T}\mathbf{F}')^{-1}\mathbf{F}$ seems an obstacle to simplification but the relationship (4.25) that $\boldsymbol{\Sigma} = -\frac{1}{2}\mathbf{F}\boldsymbol{T}\mathbf{F}'$ allows us to exploit the simpler form in Property 5.13(b) for $\boldsymbol{\Sigma}^{-1}$ under complete subcompositional independence to obtain

$$\mathbf{F}'(\mathbf{F}\boldsymbol{T}\mathbf{F}')^{-1}\mathbf{F} = -\tfrac{1}{2}(\mathrm{diag}\,\boldsymbol{\beta} - B^{-1}\boldsymbol{\beta}\boldsymbol{\beta}'),$$

where $\boldsymbol{\beta} = (\alpha_1^{-1}, \ldots, \alpha_D^{-1})$ and $B = \beta_1 + \cdots + \beta_D$. With the special form for $|\boldsymbol{\Sigma}|$ in Property 5.13(a) we can then express the loglikelihood, already maximized with repect to the uninteresting mean parameter $\boldsymbol{\mu}$ or $\boldsymbol{\xi}$, as

$$\begin{aligned} l &= \tfrac{1}{2}N\left[\sum_{i=1}^{D}\log\beta_i - \log B + \tfrac{1}{2}\mathrm{tr}\{(\mathrm{diag}\,\boldsymbol{\beta} - B^{-1}\boldsymbol{\beta}\boldsymbol{\beta}')\hat{\boldsymbol{T}}\}\right] \\ &= \tfrac{1}{2}N\left\{\sum_{i=1}^{D}\log\beta_i - \log B - (2B)^{-1}\boldsymbol{\beta}'\hat{\boldsymbol{T}}\boldsymbol{\beta}\right\}, \end{aligned} \qquad (10.2)$$

since $\mathrm{tr}\{(\mathrm{diag}\,\boldsymbol{\beta})\hat{\boldsymbol{T}}\} = 0$ and $\mathrm{tr}(\boldsymbol{\beta}\boldsymbol{\beta}'\hat{\boldsymbol{T}}) = \mathrm{tr}(\boldsymbol{\beta}'\hat{\boldsymbol{T}}\boldsymbol{\beta}) = \boldsymbol{\beta}'\hat{\boldsymbol{T}}\boldsymbol{\beta}$.

The simple form of (10.2) suggests that maximization will be simpler in terms of $\boldsymbol{\beta}$ rather than $\boldsymbol{\alpha}$ and we can easily obtain the

Newton–Raphson factors. The vector $\mathbf{l}_1(\boldsymbol{\beta})$ of first derivatives with ith element $\partial l/\partial\beta_i$ can be expressed as

$$\mathbf{l}_1(\boldsymbol{\beta}) = \tfrac{1}{2}N\{\boldsymbol{\alpha} - B^{-1}\mathbf{j}_D - B^{-1}\hat{\boldsymbol{T}}\boldsymbol{\beta} + \tfrac{1}{2}B^{-2}(\boldsymbol{\beta}'\hat{\boldsymbol{T}}\boldsymbol{\beta})\mathbf{j}_D\}. \quad (10.3)$$

Using the easily established results that, under complete subcompositional independence,

$$\mathrm{E}(\hat{\boldsymbol{T}}\boldsymbol{\beta}) = B\boldsymbol{\alpha} + (d-1)\mathbf{j}_D, \qquad \mathrm{E}(\boldsymbol{\beta}'\hat{\boldsymbol{T}}\boldsymbol{\beta}) = 2dB, \quad (10.4)$$

we can express the matrix $\mathbf{L}_2(\boldsymbol{\beta})$ of expectations of second derivatives with (i,j)th element $\mathrm{E}(-\partial^2 l/\partial\beta_i\partial\beta_j)$ as

$$\mathbf{L}_2(\boldsymbol{\beta}) = \tfrac{1}{2}NB^{-2}[\mathrm{diag}\{(B-\beta_1)^2/\beta_1^2, \ldots, (B-\beta_D)^2/\beta_D^2\} + \mathbf{J}_D - \mathbf{I}_D]. \quad (10.5)$$

We can now describe the computational process for obtaining the estimate $\hat{\boldsymbol{\Sigma}}_{\mathrm{h}}$ for the logratio covariance matrix under complete subcompositional independence.

Modified Marquardt iteration for $\hat{\boldsymbol{\Sigma}}_{\mathrm{h}}$.

At each iterative step we obtain an *updated* estimate $\boldsymbol{\beta}_{(1)}$ of $\boldsymbol{\beta}$ from the *current* estimate $\boldsymbol{\beta}_{(0)}$ by a computation of the form

$$\boldsymbol{\beta}_{(1)} = \boldsymbol{\beta}_{(0)} + \{\mathbf{M}(\boldsymbol{\beta}_{(0)})\}^{-1}\mathbf{l}_1(\boldsymbol{\beta}_{(0)}). \quad (10.6)$$

Here

$$\mathbf{M}(\boldsymbol{\beta}) = \mathbf{L}_2(\boldsymbol{\beta}) + k\mathbf{I}_D,$$

where k is a factor which may change from iteration to iteration. When $k=0$ the process is standard Newton–Raphson; when k is large relative to $\mathbf{L}_2(\boldsymbol{\beta})$ then the process is essentially steepest ascent. Each time we apply (10.6) we must ensure that the updated $\boldsymbol{\beta}_{(1)}$ satisfies two criteria.

1. $\boldsymbol{\beta}_{(1)}$ must be feasible in the sense that the corresponding $\hat{\boldsymbol{\Sigma}}_{\mathrm{h}}$ is positive definite. This means testing whether $\boldsymbol{\beta}_{(1)}$, or rather its associated $\boldsymbol{\alpha}_{(1)}$, satisfies the condition of Property 5.13(c).
2. The updated loglikelihood $l(\boldsymbol{\beta}_{(1)})$ calculated for $\boldsymbol{\beta}_{(1)}$ must be greater than the current loglikelihood $l(\boldsymbol{\beta}_{(0)})$.

In any event of failure to meet either of criteria (1) and (2) the factor k is increased by multiplication by $f > 1$, and (10.5) is attempted again. In the event of success in meeting both criteria the factor k is decreased by division by $f > 1$. Convergence is conveniently determined on the basis of a sufficiently small increase in the relative loglikelihood, that is when

$$\{l(\boldsymbol{\beta}_{(1)}) - l(\boldsymbol{\beta}_{(0)})\}/|l(\boldsymbol{\beta}_{(0)})| < \varepsilon, \tag{10.7}$$

a preassigned convergence criterion.

All that remains is to suggest practical initial values for $\boldsymbol{\beta}_{(0)}$ and k, and a reasonable value for f. Our own experience suggests the following.

To obtain an initial $\boldsymbol{\beta}_{(0)}$ first compute

$$\boldsymbol{\alpha}_{(0)} = (d-1)^{-1}\{\hat{\boldsymbol{T}}\mathbf{j} - (2d)^{-1}\mathbf{j}'\hat{\boldsymbol{T}}\mathbf{j}\}. \tag{10.8}$$

If $\boldsymbol{\alpha}_{(0)}$ satisfies the feasibility criterion then proceed with $\boldsymbol{\beta}_{(0)}$ consisting of the reciprocals of $\boldsymbol{\alpha}_{(0)}$. If the feasibility criterion is not satisfied then compute the estimate $\hat{\alpha}$ given by (10.1) for the isotropic parameter α and then set

$$\boldsymbol{\beta}_{(0)} = \hat{\alpha}^{-1}\mathbf{j}_D. \tag{10.9}$$

A convenient initial Marquardt factor k can now be computed as

$$k = \operatorname{tr}\{\mathbf{L}_2(\boldsymbol{\beta}_{(0)})\}/(100D), \tag{10.10}$$

and for the step adjustment factor we suggest $f = 10$.

The hongite and boxite compositions of Data 1 and 3 provide a convenient contrast to illustrate testing the hypotheses of complete subcompositional independence and of isotropic covariance structure. At the initialization stage for the hongite compositions, the initial estimate given by (10.8) is not feasible and so recourse has to be made to the isotropic estimate (10.9); for the boxite compositions (10.8) provides a feasible initial estimate. Table 10.1 gives the results of the iterative process.

For hongite the hypotheses of isotropic covariance structure and complete subcompositional independence are both soundly rejected with significance probabilities less than 10^{-5}. This might have been anticipated by simple inspection of the hongite covariance structure: for example, in Table 7.1 the non-diagonal elements of the hongite logratio covariance matrix are obviously not all equal, a requirement for complete subcompositional independence.

For the boxite compositions the conclusions are quite different. Although again the hypothesis of isotropic covariance structure has to be firmly rejected with a significance probability of less than 10^{-5}, complete subcompositional independence remains a very tenable hypothesis since the significance probability is as high as 0.65.

Table 10.1. *Computations for the investigation of complete subcompositional independence of hongite and boxite compositions*

Model or hypothesis	*Loglikelihood*	
	Hongite	*Boxite*
Model	125.9	95.5
Complete subcompositional independence	17.3	93.8
Isotropic	−27.1	78.3
	Estimates	
Parameter	*Hongite*	*Boxite*
α_1	0.00	0.24
α_2	0.26	1.34
α_3	1.53	0.38
α_4	0.083	0.49
α_5	0.14	0.076

10.3 Partitions of order 1

After the successful development of a testing procedure for complete subcompositional independence, we can now turn our attention to other concepts of independence for patterns of compositional variability. The simplest next step is to consider various forms of independence when the questions of interest relate to a single division $(x_1, \ldots, x_c | x_C, \ldots, x_D)$ of the composition, where $C = c + 1$, and to the corresponding transformation to the partition $(\mathbf{t}; \mathbf{s}_1, \mathbf{s}_2)$ of order 1 of Definitions 2.11 and 2.12, with

$$\mathbf{s}_1 = \mathscr{C}(x_1, \ldots, x_c), \qquad \mathbf{s}_2 = \mathscr{C}(x_C, \ldots, x_D),$$

the first and second subcompositions, and

$$\mathbf{t} = (x_1 + \cdots + x_c, x_C + \cdots + x_D),$$

the amalgamation recording the division of the original unit between the parts of the two subcompositions. With three ingredients $\mathbf{s}_1, \mathbf{s}_2, \mathbf{t}$, there are clearly many ways in which we can formulate hypotheses of independence and it is convenient to discuss these against the background of a particular compositional data set. We address ourselves to Questions 1.9.1 and 1.9.2.

Consider a single division of the statistician's day of Data 10 into the four parts relating to work:

(1) teaching, (2) consultation, (3) administration, (4) research,

and the remaining two parts relating to leisure:

(5) other wakeful activities, (6) sleep.

For this division $(x_1, x_2, x_3, x_4 | x_5, x_6)$, the amalgamation **t** describes the division of the day into total work $t_1 = x_1 + x_2 + x_3 + x_4$ and total leisure $t_2 = x_5 + x_6$; the first and second subcompositions $\mathbf{s}_1$ and $\mathbf{s}_2$ describe the patterns of work and of leisure within the times allocated to these major divisions of the day. In the investigation of such a partition we may well wish to start at the most stringent independence hypothesis possible, namely that there is no dependence at all between the three aspects of the day, the work and leisure patterns and the proportions of the day allocated to work and leisure. The appropriate definition of this form of independence can now be given.

Definition 10.2 A partition $(\mathbf{t}; \mathbf{s}_1, \mathbf{s}_2)$ of order 1 of a D-part composition $\mathbf{x}$ has *partition independence* if $\mathbf{t}, \mathbf{s}_1, \mathbf{s}_2$ are independent.

Before we consider the problem of testing the hypothesis of partition independence it is convenient to consider three other forms of independence which also involve all three aspects $\mathbf{t}, \mathbf{s}_1, \mathbf{s}_2$, but in special groupings. Suppose that we wish to investigate whether the work and leisure patterns depend on the proportions of the day allocated to work and leisure. We may then be interested in whether $(\mathbf{s}_1, \mathbf{s}_2)$ depends on **t**, and so in the following form of independence.

Definition 10.3 A partition $(\mathbf{t}; \mathbf{s}_1, \mathbf{s}_2)$ of order 1 of a D-part composition $\mathbf{x}$ has *subcompositional invariance* if $(\mathbf{s}_1, \mathbf{s}_2)$ is independent of **t**.

Suppose that each of the statistician's days was recorded over a 24-hour period starting at 6 a.m. We may then be interested in the extent to which the statistician's leisure pattern depends on the length and pattern of the work day. Perhaps the statistician's ratio of sleep to other leisure activity may be affected by the period available for leisure and even by the kind of work day; for example, if he has had a work day with a long and exciting research component he may find sleep difficult. In asking such questions we are interested in whether or not $\mathbf{s}_2$, the second subcomposition or 'composition on the right', is dependent on $(\mathbf{s}_1, \mathbf{t})$ or equivalently on $\mathbf{x}^{(c)}$, the actual components on the left. This leads naturally to the following definition.

Definition 10.4 A partition $(\mathbf{t}; \mathbf{s}_1, \mathbf{s}_2)$ of order 1 of a D-part composition $\mathbf{x}$ is *neutral on the right* if $\mathbf{s}_2$ is independent of $(\mathbf{s}_1, \mathbf{t})$, or equivalently if $\mathbf{s}_2$ is independent of $\mathbf{x}^{(c)}$.

On the other hand, if the recorded days run from 6 p.m. then interest may focus on the extent to which the work pattern is affected by the amounts of preceding leisure and sleep. In other words, to what extent may $\mathbf{s}_1$, the composition on the left, be dependent on $(\mathbf{s}_2, \mathbf{t})$ or $\mathbf{x}_{(c)}$? This leads us to the following definition.

Definition 10.5 A partition $(\mathbf{t}; \mathbf{s}_1, \mathbf{s}_2)$ of order 1 of a D-part composition $\mathbf{x}$ is *neutral on the left* if $\mathbf{s}_1$ is independent of $(\mathbf{s}_2, \mathbf{t})$, or equivalently if $\mathbf{s}_1$ is independent of $\mathbf{x}_{(c)}$.

Tests of all the four hypotheses of Definitions 10.2–10.5 can easily be obtained within the framework of partitioned logistic normal distributions introduced in Section 6.14. We first make logratio transformations

$$\mathbf{y}_1 = \text{alr}(\mathbf{s}_1), \quad \mathbf{y}_2 = \text{alr}(\mathbf{s}_2), \quad \mathbf{z} = \text{alr}(\mathbf{t}) = \log(t_1/t_2). \qquad (10.11)$$

We then take as our model for the vector $(\mathbf{y}_1, \mathbf{y}_2, \mathbf{z})$ a $\mathcal{N}^d(\boldsymbol{\mu}, \boldsymbol{\Sigma})$ distribution, where the mean vector $\boldsymbol{\mu}$ and covariance matrix $\boldsymbol{\Sigma}$ may be expressed, in partitioned forms conformable with the dimensions $c-1, d-c, 1$ of $\mathbf{y}_1, \mathbf{y}_2, \mathbf{z}$ as

$$\boldsymbol{\mu} = \begin{bmatrix} \boldsymbol{\mu}_1 \\ \boldsymbol{\mu}_2 \\ \boldsymbol{\mu}_3 \end{bmatrix}, \qquad \boldsymbol{\Sigma} = \begin{bmatrix} \boldsymbol{\Sigma}_{11} & \boldsymbol{\Sigma}_{12} & \boldsymbol{\Sigma}_{1z} \\ \boldsymbol{\Sigma}_{21} & \boldsymbol{\Sigma}_{22} & \boldsymbol{\Sigma}_{2z} \\ \boldsymbol{\Sigma}_{z1} & \boldsymbol{\Sigma}_{z2} & \boldsymbol{\Sigma}_{zz} \end{bmatrix}. \qquad (10.12)$$

Within this model we can immediately characterize the four forms of independence in terms of special forms of the logratio covariance matrix.

Property 10.5 For a partition $(\mathbf{t}; \mathbf{s}_1, \mathbf{s}_2)$ of order 1 of a D-part composition $\mathbf{x}$ the following forms of the covariance matrix $\boldsymbol{\Sigma}$ of the logratio vector $(\mathbf{y}_1, \mathbf{y}_2, \mathbf{z})$ characterize the independence hypotheses of

Definitions 10.2–10.5.

(a) Partition independence
$$\begin{bmatrix} \boldsymbol{\Sigma}_{11} & \mathbf{0} & \mathbf{0} \\ \mathbf{0} & \boldsymbol{\Sigma}_{22} & \mathbf{0} \\ \mathbf{0} & \mathbf{0} & \boldsymbol{\Sigma}_{zz} \end{bmatrix}$$

(b) Subcompositional invariance
$$\begin{bmatrix} \boldsymbol{\Sigma}_{11} & \boldsymbol{\Sigma}_{12} & \mathbf{0} \\ \boldsymbol{\Sigma}_{21} & \boldsymbol{\Sigma}_{22} & \mathbf{0} \\ \mathbf{0} & \mathbf{0} & \boldsymbol{\Sigma}_{zz} \end{bmatrix}$$

(c) Neutrality on the right
$$\begin{bmatrix} \boldsymbol{\Sigma}_{11} & \mathbf{0} & \boldsymbol{\Sigma}_{1z} \\ \mathbf{0} & \boldsymbol{\Sigma}_{22} & \mathbf{0} \\ \boldsymbol{\Sigma}_{z1} & \mathbf{0} & \boldsymbol{\Sigma}_{zz} \end{bmatrix}$$

(d) Neutrality on the left
$$\begin{bmatrix} \boldsymbol{\Sigma}_{11} & \mathbf{0} & \mathbf{0} \\ \mathbf{0} & \boldsymbol{\Sigma}_{22} & \boldsymbol{\Sigma}_{2z} \\ \mathbf{0} & \boldsymbol{\Sigma}_{z2} & \boldsymbol{\Sigma}_{zz} \end{bmatrix}$$

From these special forms of covariance matrix we have simple implication relationships.

Property 10.6
(a) Partition independence implies subcompositional invariance, neutrality on the right, neutrality on the left.
(b) Any two of the hypotheses of subcompositional invariance, neutrality on the right, neutrality on the left, imply partition independence.

Implication (a) means that the four hypotheses may be arranged in a lattice as in Fig. 10.1 with partition independence at level 1 and subcompositional invariance, neutrality on the right, neutrality on the left at a higher level. Testing within such a lattice can be carried out with the Wilks (1938) generalized likelihood ratio test procedure.

Property 10.7 For a compositional data set let $\hat{\boldsymbol{\Sigma}}_{\mathrm{m}}$ be the sample covariance matrix of the logratio vectors $(\mathbf{y}_1, \mathbf{y}_2, \mathbf{z})$. For any hypothesis h of Definitions 10.2–10.5 let $\hat{\boldsymbol{\Sigma}}_{\mathrm{h}}$ denote $\hat{\boldsymbol{\Sigma}}_{\mathrm{m}}$ with submatrices set equal to zero according to the associated hypothesis pattern of Property 10.5. To test the hypothesis h within the partitioned logistic normal model m, compare

$$N\{\log|\hat{\boldsymbol{\Sigma}}_{\mathrm{h}}| - \log|\hat{\boldsymbol{\Sigma}}_{\mathrm{m}}|\}$$

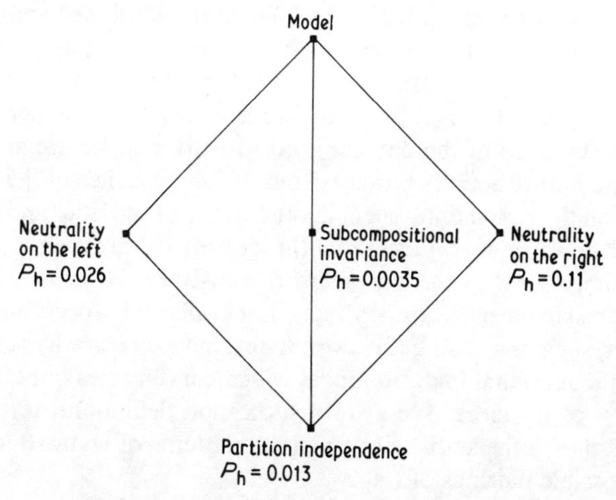

Fig. 10.1. *Lattice of independence hypotheses for partition of order 1 of statistician's activity patterns.*

against upper percentage points $\chi^2(k)$, where for
(a) partition independence, $k = dc - c^2 + c - 1$,
(b) subcompositional invariance, $k = d - 1$,
(c) neutrality on the right, $k = c(d - c)$,
(d) neutrality on the left, $k = (c - 1)(d - c + 1)$.

For the work–leisure partition of the statistician's day we have $D = 6$, $d = 5$, $c = 4$, $C = 5$, and

$$\hat{\boldsymbol{\Sigma}}_m = 10^{-2} \times \begin{bmatrix} 2.734 & 0.545 & -0.572 & 1.405 & 0.158 \\ 0.545 & 4.383 & 1.732 & 0.233 & -0.311 \\ -0.572 & 1.732 & 3.868 & -1.097 & 0.002 \\ 1.405 & 0.233 & -1.097 & 9.587 & 0.426 \\ 0.158 & -0.311 & 0.002 & 0.426 & 0.121 \end{bmatrix},$$

$$|\hat{\boldsymbol{\Sigma}}_m| = 10^{-8} \times 1.628.$$

We start at level 1 with the hypothesis of partition independence, for which $|\hat{\boldsymbol{\Sigma}}_h| = 10^{-8} \times 3.952$. The computed value of the test statistic of Property 10.7 is then 17.74, to be compared against upper percentage points of $\chi^2(7)$. The significance probability is 0.013, and so we reject the hypothesis of partition independence and move to the next level. The test procedures are similar at this level and we record only the significance probabilities in Fig. 10.1. At this level the hypotheses of subcompositional invariance and of neutrality on the left are rejected. There is no evidence against the hypothesis of neutrality on the right. Thus we conclude that there is evidence of a dependence between the ratio of the day assigned to work and leisure and the work and leisure activity patterns, and of a dependence of the work pattern on the actual times spent in other wakeful activities and sleep. There is, however, no evidence that the leisure pattern depends on the actual times spent in the various work activities.

If interest in the partition $(\mathbf{t}; \mathbf{s}_1, \mathbf{s}_2)$ is not in all three aspects but only in a pair, such as $\mathbf{s}_1$ and $\mathbf{t}$, then we can test independence hypotheses within the marginal logratio model which concentrates only on the variation of $\mathbf{y}_1$ and $\mathbf{z}$. We give in succession definitions, tests and applications to the statistician's activity patterns of Data 10 for the three possible pairings of $\mathbf{t}, \mathbf{s}_1, \mathbf{s}_2$.

Definition 10.6 A partition $(\mathbf{t}; \mathbf{s}_1, \mathbf{s}_2)$ of order 1 of a D-part composition has *first subcompositional invariance* if $\mathbf{s}_1$ is independent of $\mathbf{t}$.

Note that first subcompositional invariance is equivalent to composition invariance, as described in Definition 9.3, of the subvector $\mathbf{x}^{(c)}$ regarded as a basis. A test of first subcompositional invariance could thus be carried out within the loglinear modelling procedure as set out in Section 9.5. Since our earlier tests in this section have been presented in terms of the partitioned logratio covariance matrix (10.12), we prefer to describe an equivalent test of first subcompositional invariance along similar lines.

Property 10.8 A test for first subcompositional invariance of Definition 10.6 compares

$$N\left\{\log(|\hat{\boldsymbol{\Sigma}}_{11}||\hat{\boldsymbol{\Sigma}}_{zz}|) - \log\begin{vmatrix}\hat{\boldsymbol{\Sigma}}_{11} & \hat{\boldsymbol{\Sigma}}_{1z}\\ \hat{\boldsymbol{\Sigma}}_{z1} & \hat{\boldsymbol{\Sigma}}_{zz}\end{vmatrix}\right\}$$

against upper percentage points of the $\chi^2(c-1)$ distribution.

Applied to the work–leisure partition of the statistician's day this gives a significance probability of 0.018 so that we would have to conclude that the work pattern does depend on the proportion of time allocated to work.

Second subcompositional invariance can be defined and tested in a similar, obvious way and we need not go into detail here. The significance probability is 0.064, so that although the significance probability is quite small we would hardly conclude that the leisure pattern depends on the proportion of time allocated to leisure. Note that these findings on first and second subcompositional invariance, taken together, are consistent with our previous rejection of subcompositional invariance.

The third form of pairwise independence is concerned with $\mathbf{s}_1$ and $\mathbf{s}_2$.

Definition 10.7 A partition $(\mathbf{t}; \mathbf{s}_1, \mathbf{s}_2)$ of order 1 of a D-part composition has *subcompositional independence* if $\mathbf{s}_1$ and $\mathbf{s}_2$ are independent.

Property 10.9 A test for subcompositional independence of Definition 10.7 compares

$$N\left\{\log(|\hat{\boldsymbol{\Sigma}}_{11}||\hat{\boldsymbol{\Sigma}}_{22}|) - \log\begin{vmatrix}\hat{\boldsymbol{\Sigma}}_{11} & \hat{\boldsymbol{\Sigma}}_{12}\\ \hat{\boldsymbol{\Sigma}}_{21} & \hat{\boldsymbol{\Sigma}}_{22}\end{vmatrix}\right\}$$

against upper percentage points of the $\chi^2\{(c-1)(d-c)\}$ distribution.

When applied to the work and leisure patterns of the statistician's day this test gives a significance probability of 0.56 and so we would have to conclude that there is no evidence of dependence between the two patterns.

10.4 Ordered sequences of partitions

There are a number of situations where a specific ordering of the D parts of a composition has been made and already embodied in $\mathbf{x}$, and where interest is in considering independence properties at a sequence of levels of partitions of order 1.

This is particularly so if we can visualize the available unit as being assigned to the different parts in sequence. For example, in relation to the compositions of the academic statistician's activity patterns of Data 10, the first call on the statistician's time may be to fulfil his teaching duties; out of what remains the next call may be to provide consultation for other university departments and teaching hospitals; and so on, through administration, research, and other wakeful activities such as eating and washing, until the residue is assigned to sleep. Another possible example is in the household budgets of Data 8 if expenditure in the foodstuffs category takes priority, followed by housing, services, and finally other goods.

With any such imagined sequence of allocations of components $x_1, \ldots, x_D$ to the parts $1, \ldots, D$ of the composition there is a question of obvious interest. When we have reached level c with $x_1, \ldots, x_c$ already allocated to parts $1, \ldots, c$, is the allocation of what remains independent of $\mathbf{x}^{(c)} = (x_1, \ldots, x_c)$; in other words, is $\mathscr{C}(\mathbf{x}_{(c)})$ independent of $\mathbf{x}^{(c)}$? Reference to Definition 10.4 will show that this is neutrality on the right and, for a fixed c, this hypothesis could be tested within the class of partitioned logistic normal distributions by the methods of Section 10.3. We have seen in Section 10.3 that neutrality on the right is a tenable hypothesis when $c = 4$, but is this also the case for other values of c? The new feature in the present situation is that we wish to carry out tests of neutrality on the right for a sequence of values of c and we cannot achieve this within the fixed partition modelling of Section 10.3. We require a more flexible modelling approach which will allow consideration of partitions at all levels c. The answer is to be found in the multiplicative logistic and logratio transformations of Definition 6.5 and the resulting class of multiplicative logistic normal distributions $\mathscr{M}^d(\boldsymbol{\mu}, \boldsymbol{\Sigma})$ of Definition 6.6.

First we crystallize some of the fundamental concepts in formal definitions.

Definition 10.8 The *partition at level c* of the D-part composition $\mathbf{x}$ is simply the partition of order 1 formed from the single division $(x_1, \ldots, x_c | x_C, \ldots, x_D)$.

Definition 10.9 For a D-part composition $\mathbf{x}$ let h be an independence property, such as neutrality on the right or subcompositional invariance, associated with a partition of order 1. Let h_k denote the independence property at level k. Then the composition is said to have the *independence property up to level c*, written h^c, if h_k holds for $k = 1, \ldots, c$.

Definition 10.10 A D-part composition $\mathbf{x}$ is said to have the *complete independence property* h if h^d holds, in other words if h_k holds at all possible levels.

Note that h_d may be trivially true for some forms of independence: for example, for neutrality on the right the second subcomposition associated with the partition at level d consists of one part and is therefore equal to the constant 1 and so is trivially independent of $(x_1, \ldots, x_d)$.

Although in the above definitions h may be any independence property associated with a partition of order 1, we shall concentrate on the concept of neutrality on the right and its application to the examples cited above. We now turn to the multiplicative logistic transformation of Definition 6.5 and the associated $\mathcal{M}^d(\boldsymbol{\mu}, \boldsymbol{\Sigma})$ class of Definition 6.6, and their role in testing for neutrality on the right at different levels. The importance of the multiplicative logratio transformation $\mathbf{y} = \mathrm{mlr}(\mathbf{x})$ from $\mathcal{S}^d$ to $\mathcal{R}^d$ and its inverse, the multiplicative logistic transformation $\mathbf{x} = \mathrm{mlg}(\mathbf{y})$ from $\mathcal{R}^d$ to $\mathcal{S}^d$ is, by Note 1 to Definition 6.5, that, *for any* c, $\mathbf{x}^{(c)}$ is uniquely determined by $\mathbf{y}^{(c)}$, and $\mathscr{C}(\mathbf{x}_{(c)})$ is uniquely determined by $\mathbf{y}_{(c)}$. This means that, with this single transformation and its associated transformed normal class, we have the framework of a single class $\mathcal{M}^d(\boldsymbol{\mu}, \boldsymbol{\Sigma})$ of distributions within which we may test neutrality on the right at any fixed level c, up to level c, or indeed complete neutrality. First we express the independence properties in terms of special patterns of the multiplicative logratio covariance matrix

$$\boldsymbol{\Sigma} = \mathrm{V}\{\mathrm{mlr}(\mathbf{x})\}. \tag{10.13}$$

Property 10.10 Let h_c, h^c and h^d denote neutrality at level c, neutrality up to level c, and complete neutrality on the right. The hypotheses h_c, h^c and h^d correspond to the following patterns for the multiplicative logratio covariance matrix $\boldsymbol{\Sigma} = [\sigma_{ij}]$:

$$\begin{bmatrix} \boldsymbol{\Sigma}_{11} & \mathbf{0} \\ \mathbf{0} & \boldsymbol{\Sigma}_{22} \end{bmatrix}, \quad \begin{bmatrix} \mathrm{diag}(\sigma_{11}, \ldots, \sigma_{cc}) & \mathbf{0} \\ \mathbf{0} & \boldsymbol{\Sigma}_{22} \end{bmatrix}, \quad \mathrm{diag}(\sigma_{11}, \ldots, \sigma_{dd}),$$

where $\boldsymbol{\Sigma}_{11}$ is of order $c \times c$. The numbers of constraints imposed by the three hypotheses are $c(d-c)$, $c\{d - \frac{1}{2}(c+1)\}$ and $\frac{1}{2}d(d-1)$.

The corresponding tests, again in the Wilks (1938) form, use the now familiar test statistic.

Property 10.11 Let $\hat{\boldsymbol{\Sigma}}_{\mathrm{m}}$ denote the estimated multiplicative logratio covariance matrix, the sample covariance matrix of the d-dimensional vectors $\mathbf{y}_r = \mathrm{mlr}(\mathbf{x}_r)$ $(r = 1, \ldots, N)$. The estimate $\hat{\boldsymbol{\Sigma}}_{\mathrm{h}}$ under any of the hypotheses of Property 10.10 is then obtained by using the appropriate patterns of Property 10.10. Tests of these hypotheses then compare

$$N\{\log|\hat{\boldsymbol{\Sigma}}_{\mathrm{h}}| - \log|\hat{\boldsymbol{\Sigma}}_{\mathrm{m}}|\}$$

against upper percentage points of the $\chi^2(k)$ distribution, where for
(a) h_c, $k = c(d-c)$,
(b) h^c, $k = c\{d - \frac{1}{2}(c+1)\}$,
(c) h^d, $k = \frac{1}{2}d(d-1)$.

Since h^c implies h^{c-1} and h^c also implies h_c, we can arrange any set of hypotheses conveniently in a lattice. Figure 10.2 shows the entire lattice for investigation of all hypotheses of neutrality on the right for the 6-part compositions of the statistician's day. The estimated multiplicative logratio covariance matrix under the $\mathcal{M}^5(\boldsymbol{\mu}, \boldsymbol{\Sigma})$ model is

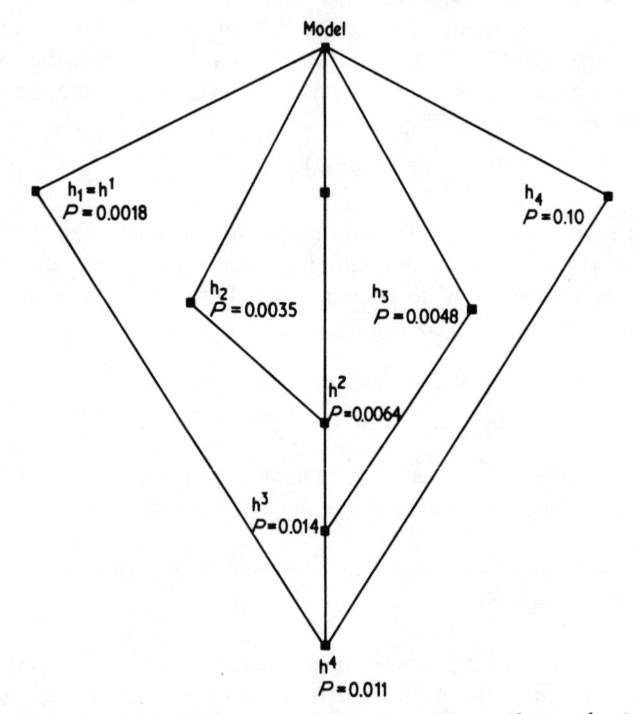

Fig. 10.2. *Lattice for investigation of hypotheses of neutrality on the right for statistician's activity patterns.*

$$\hat{\Sigma}_{\rm m} = 10^{-2} \times \begin{bmatrix} 4.216 & 0.361 & -0.974 & 2.154 & 2.720 \\ 0.361 & 3.158 & 0.035 & 0.679 & 1.761 \\ -0.974 & 0.035 & 2.513 & 0.689 & 0.538 \\ 2.154 & 0.679 & 0.689 & 3.672 & 1.959 \\ 2.702 & 1.761 & 0.538 & 1.959 & 9.507 \end{bmatrix},$$

and from this the test statistic values of Property 10.11 are readily computed. The significance values are shown in Fig. 10.2. Note that h_4 is the hypothesis of neutrality on the right for the work–leisure partition which we investigated in Section 10.3. Although the investigation here is within a different model, we have a confirmation of our previous finding that at this particular level neutrality on the right is a tenable hypothesis. Starting at level 1 with the hypothesis h^4 of complete neutrality on the right, we see that we would certainly reject h^4 and indeed each $h^3, h^2, h^1 = h_1$, in succession. The hypothesis of neutrality h_4 on the right at level 4 is the only hypothesis of the lattice which cannot be rejected.

For the household expenditure compositions of Data 8 the full lattice for single male households is shown in Fig. 10.3, together with the corresponding significance values, for the ordering (foodstuffs,

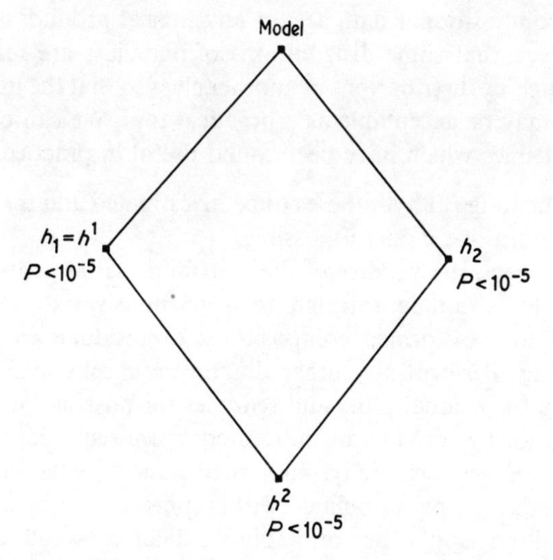

Fig. 10.3. *Lattice for investigation of hypotheses of neutrality on the right for budget patterns of single male households.*

housing, services, other goods) in order of increasing income elasticity. All three hypotheses of neutrality are firmly rejected.

10.5 Caveat

In the analysis of the statistician's activity patterns in Sections 10.3 and 10.4 we have tested the hypothesis that the leisure pattern is independent of the actual components of the working parts of the day within two different models. In the first, in Section 10.3, we considered the fixed division $(x_1, x_2, x_3, x_4 | x_5, x_6)$ and the resultant partition $(\mathbf{t}; \mathbf{s}_1, \mathbf{s}_2)$ and tested the independence of $(\mathbf{s}_1, \mathbf{t})$ and $\mathbf{s}_2$, neutrality on the right, within the partitioned logistic normal model. In the second, in Section 10.4, we considered the multiplicative logratio transformation $\mathbf{y} = \text{mlr}(\mathbf{x})$ and tested the independence of $\mathbf{y}^{(4)}$ and $\mathbf{y}_{(4)} = y_5$, neutrality at level 4, within the multiplicative logistic normal model. We were fortunate to obtain similar significance probabilities and therefore consistent conclusions.

The reader should be warned that such consistency will not always be the case. The partitioned and multiplicative logistic normal classes are separate in the sense that it is not possible to find two distributions, one from each class, which are arbitrarily close to each other. It is thus not possible to defend the use of both classes for the same compositional data set on any logical grounds. We may find, however, that some distributions of one class are reasonably approximated by distributions of another class so that the use of dual modelling may be acceptable as a practical tool. We can offer only broad guidelines which have been found useful in practice.

1. Study the range of hypotheses to be investigated and try to frame them within one model if possible.
2. Always check the validity of the distributional assumption of a model, for example through tests of multivariate normality applied to transformed compositions or residuals in loglinear modelling. Perform any other diagnostic checks available, for example by residual plots and searches for possible outliers.
3. In situations where two or more models are required, check the validity of each carefully. In circumstances where the same hypothesis may be formulated within more than one model and the resulting significance probabilities differ substantially, try to judge which model has the greater validity, but be cautious in the drawing of conclusions.

10.6 Partitions of higher order

For partitions of higher order there is obviously a multiplicity of hypotheses and which are of interest will depend on the circumstances. We confine ourselves to a simple extension of the analysis of Section 10.4, making a second division of the work parts. We form a partition $(\mathbf{t}; \mathbf{s}_1, \mathbf{s}_2, \mathbf{s}_3)$ of order 2 from the division

(teaching, consultation | administration,
research | other wakeful activities, sleep),

so that $\mathbf{t}$ is now a 3-part amalgamation (t_1, t_2, t_3) and $\mathbf{s}_1, \mathbf{s}_2, \mathbf{s}_3$ are all 2-part subcompositions. In addressing ourselves to any form of independence involving these we may be able to model in terms of the subcompositional logratios

$$y_1 = \log(x_1/x_2), \qquad y_2 = \log(x_3/x_4), \qquad y_3 = \log(x_5/x_6), \tag{10.14}$$

and amalgamation logratios

$$z_1 = \log(t_1/t_3), \qquad z_2 = \log(t_2/t_3) \tag{10.15}$$

or

$$z_1 = \log\{t_1/(1 - t_1)\}, \qquad z_2 = \log\{t_2/(1 - t_1 - t_2)\}. \tag{10.16}$$

For example, the extension of the hypothesis of partition independence would involve testing whether y_1, y_2, y_3 and (z_1, z_2) are independent, and it is straightforward to devise a test for this involving the logratio covariance matrix. As a simple illustrative example we examine the hypothesis of subcompositional invariance, asking whether the ratios of teaching to consultation, administration to research, other wakeful activities to sleep are independent of the way in which the day is divided into the three categories of the amalgamation.

Let $\hat{\boldsymbol{\Sigma}}_{\mathrm{m}}$ denote the estimate of the 5×5 covariance matrix of $(\mathbf{y}, \mathbf{z})$, the logratios (10.14) and (10.15), and let

$$\hat{\boldsymbol{\Sigma}}_{\mathrm{m}} = \begin{bmatrix} \hat{\boldsymbol{\Sigma}}_{yy} & \hat{\boldsymbol{\Sigma}}_{yz} \\ \hat{\boldsymbol{\Sigma}}_{zy} & \hat{\boldsymbol{\Sigma}}_{zz} \end{bmatrix}$$

be the 3:2 partition of this estimate. Then to test subcompositional invariance we compare

$$N\{\log(|\hat{\boldsymbol{\Sigma}}_{yy}||\hat{\boldsymbol{\Sigma}}_{zz}|) - \log|\hat{\boldsymbol{\Sigma}}_{\mathrm{m}}|)\}$$

against upper percentage points of the $\chi^2(6)$ distribution. For the statistician's activity patterns

$$\hat{\Sigma}_m = 10^{-1} \times \begin{bmatrix} 0.634 & -0.242 & 0.123 & 0.136 & 0.035 \\ -0.242 & 0.407 & -0.116 & -0.206 & -0.024 \\ 0.123 & -0.116 & 1.001 & 0.294 & 0.149 \\ 0.136 & -0.206 & 0.294 & 0.306 & 0.131 \\ 0.035 & -0.024 & 0.149 & 0.131 & 0.243 \end{bmatrix}$$

and the significance probability for the hypothesis of subcompositional invariance is then 0.010. This is again consistent with our earlier rejection of first and second subcompositional invariance in Section 10.3.

10.7 Bibliographic notes

The concept of complete subcompositional independence was introduced by Aitchison (1982) as a means of providing a strong independence property for compositions without reference to imagined bases, and a test within the additive logistic normal class of distributions was simultaneously developed. The concept of neutrality was introduced by Connor and Mosimann (1969) and has been the source of a number of developments by Darroch (1969), Darroch and James (1974), Darroch and Ratcliff (1970, 1971, 1978), Hohn and Nuhfer (1980), James (1972, 1975, 1981), James and Mosimann (1980), Mosimann (1975a, b), Snow (1975). Much of the statistical analysis of neutrality was hampered because until recently no parametric class of distributions on the simplex rich enough to accommodate both neutrality and non-neutrality had been found. Tests of the different forms of neutrality were developed by Aitchison (1981b). All the other forms of independence associated with partitions discussed in this chapter were introduced by Aitchison (1982).

Problems

10.1 Investigate the hypothesis of complete subcompositional independence for each of the following data sets:
(a) the coxite compositions of Data 4,
(b) the lammite compositions of Data 27,
(c) the shift compositions of Data 22.

10.2 For the variation matrix $\boldsymbol{T}$ of a D-part composition with complete subcompositional independence show that the row sums τ_{i+} $(i = 1, \ldots, D)$ and the sum τ_{++} of all the elements are given by

$$\tau_{i+} = (d-1)\alpha_i + \alpha_+$$
$$\tau_{++} = 2d\alpha_+,$$

in the notation of Property 10.3, and hence that

$$\alpha_i = (d-1)^{-1}\{\tau_{i+} - (2d)^{-1}\tau_{++}\}.$$

For a compositional data set with complete subcompositional independence this suggests that

$$\hat{\alpha}_i = (d-1)^{-1}\{\hat{\tau}_{i+} - (2d)^{-1}\hat{\tau}_{++}\}$$

should give reasonable estimates of α_i $(i = 1, \ldots, D)$. Apply this estimation procedure to the variation matrices of the hongite and boxite compositions of Data 1 and 3. Do your results give any indication of the conclusions reached in Section 10.2, that the hypothesis of complete subcompositional independence is tenable for the boxite, but not for the hongite, compositions?

10.3 Data 31 presents, in the same format as Data 10, the activity patterns of another statistician. Investigate, along the lines of Sections 10.3–10.6, any hypotheses which you consider of interest.

10.4 Consider the partition of the serum protein compositions of patients A1–A10 of Data 16 into albumin and globulin parts. Test all the hypotheses of Definitions 10.2–10.7 in relation to this partition and state your conclusions as precisely as possible.

10.5 Reconsider the questions you formulated in Problem 1.7 about the productive and non-productive parts of the shifts of Data 22. Can they be reformulated as hypotheses associated with a partition of order 1 of the shift composition? If so, what inferences can you make from the compositional data set? Are these conclusions consistent with your findings in Problem 10.1(c)?

10.6 For the ordering A, B, C, D, E of the parts of the hongite and boxite compositions of Data 1 and 3 investigate the complete lattice of hypotheses of neutrality at level c and neutrality up to level c $(c = 1, 2, 3, 4)$.

CHAPTER 11

Irregular compositional data

This is the zero moment of consciousness. Stuck. No answer. Honked. Kaput. It's a miserable experience emotionally. You're losing time. You're incompetent. You don't know what you're doing. You should be ashamed of yourself. You should take the machine to a real *mechanic who knows how to figure these things out.*

Robert Pirsig: *Zen and the Art of Motorcycle Maintenance*

11.1 Introduction

The compositional data analysis which we have so far developed faces substantial problems under three different conditions.

1. Suppose that for some compositions the sum of the components is not unity. Clearly there is some form of imprecision or measurement error, of which we have to take account in our analysis. We consider a number of aspects of this imprecision in Sections 11.2–11.4.
2. Suppose that some components are recorded as zero. Since we cannot take the logarithm of zero, none of the logratio composition techniques which have been so essential to our progress can be applied without modification. We study this problem in Sections 11.5–11.6.
3. Suppose that for some of the compositions some of the components are missing. For example, at some stage of the experimental process something may have prevented the completion of the determining procedure for one or more of the components. We are then faced essentially with a problem of missing data, and the question of whether the compositional nature of the data makes the techniques of handling this difficulty different from the corresponding problem for vector data in $\mathscr{R}^d$. We examine this problem in Section 11.7.

11.2 Modelling imprecision in compositions

The presence of measurement error in compositional data is often made obvious by the fact that the components do not sum exactly to a unit. For example, the major oxide percentage composition of a Skye lava is recorded by Thompson, Esson and Duncan (1972) as

SiO_2	Al_2O_3	Fe_2O_3	MgO	CaO	Na_2O	K_2O	TiO_2	P_2O_5	MnO
47.84	17.47	12.97	6.66	10.70	2.83	0.51	1.25	0.17	0.26

with a percentage total of 100.66. Even when the components of a composition sum exactly to a unit there may still be imprecision present. The components may already have been subjected to the constraining operation $\mathscr{C}(\cdot)$ by their reporter or indeed the measurement process, despite imprecision, may automatically lead to the reporting of unit-sum vectors. For example, in determining the proportions of a geographical area of different types of vegetation we may estimate the areas associated with the different types and then divide by their total. Moreover, it is sometimes impossible to distinguish between measurement error and heterogeneity of the material being sampled. For example, a single blood sample was divided into eight aliquots and serum protein compositions determined for each aliquot, as recorded in Data 32. The variability in these compositions could be due to measurement error or sampling heterogeneity, or both.

In some applications it may not matter what the source of the imprecision is. In other applications a primary objective of experimentation may be to locate the source of imprecision; for example, Data 33 records the results of an observer error study in which three analysts were each given an aliquot from each of seven blood samples and required to determine serum protein composition.

We now attempt to provide a form of modelling within which to discuss imprecision and its effects on various forms of compositional data analysis.

There are many ways in which the modelling of measurement error may be approached, and it must be admitted that we have adopted forms which lead to tractable analyses without any real evidence that they are appropriate to the particular situations. For example, all our modelling involves multiplicative rather than additive errors. In most applications that we have encountered the data seem insufficient to discriminate between such models.

In the single determination of a composition let us suppose that the true D-part composition is ξ, satisfying the constraint

$$\xi_1 + \cdots + \xi_D = 1,$$

and that the observed vector is $\mathbf{w}$, not necessarily satisfying the constraint that the sum of its components is unity. The immediate question is what function of $\mathbf{w}$ should be used as an estimator of ξ. It would be possible to investigate this problem from viewpoints such as least squares and maximum likelihood under various assumptions, but many of these can lead to awkward features such as the possibility of negative components or extremely complicated functions. Here we simply adopt what appears to be common practice, namely to apply the constraining operator of Definition 2.5 to obtain the composition $\mathbf{x} = \mathscr{C}(\mathbf{w})$ from the observed vector $\mathbf{w}$ as estimator of the true composition ξ and to follow through the consequences of this reasonable procedure. Such a procedure always leads in a very direct way to a vector in the appropriate simplex, whereas as mentioned above other principles of estimation do not.

In adopting this approach we are viewing measurement error as made up of two stages illustrated in Fig. 11.1:

1. The extent to which $\mathbf{w}$ fails to satisfy the constraint, that is the departure of the sum $t = w_1 + \cdots + w_D$ from the constant 1.

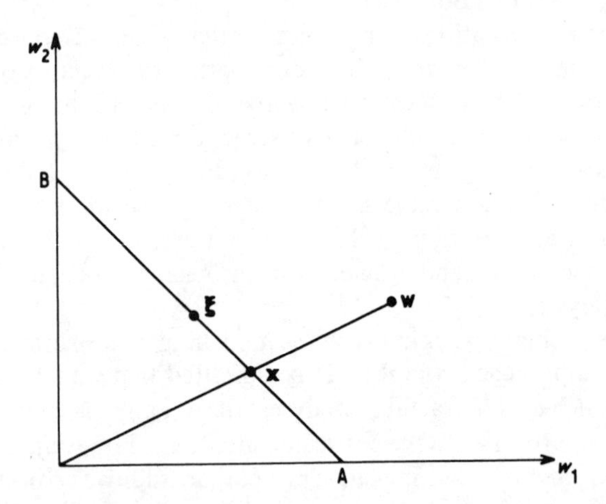

Fig. 11.1. *Representation of the 2-stage measurement error process for a 2-part composition. The line segment AB represents the simplex $\mathscr{L}^1$.*

2. The extent by which $\mathbf{x}$ differs from the true composition $\boldsymbol{\xi}$.

Stage 1 can be trivially modelled as

$$t = 1 \times u, \tag{11.1}$$

where u is a positive multiplicative error. The model that we adopt for stage 2 is then

$$\mathbf{x} = \boldsymbol{\xi} \circ \mathbf{v}, \tag{11.2}$$

where $\mathbf{v}$ can itself be regarded as a compositional measurement error perturbing the true composition $\boldsymbol{\xi}$. When it is necessary to make distributional assumptions about $\mathbf{v}$ we shall adopt additive logistic normal distributions of the form $\mathscr{L}^d(\mathbf{0}, \boldsymbol{\Sigma})$. Note that in such modelling the discrepancy between $\mathbf{w}$ and $\mathbf{x}$ is unimportant from the viewpoint of the analysis of compositional data.

Suppose now that there are N replicate determinations of a single true composition $\boldsymbol{\xi}$ in the form of vectors $\mathbf{w}_1, \ldots, \mathbf{w}_N$, with $\mathbf{w}_r$ having components $(w_{r1}, \ldots, w_{rD})$ $(r = 1, \ldots, N)$. What then is the natural extension of the estimation problem considered above? What composition $\boldsymbol{\xi}$ should be constructed from the observed vectors? The answer that emerges from logistic normal modelling takes a particularly simple form. With independent and identically distributed logistic normal perturbations the maximum likelihood estimator of the logratio composition $\boldsymbol{\eta} = \text{alr}(\boldsymbol{\xi})$ is given by

$$\begin{aligned} \hat{\boldsymbol{\eta}} = \log(\hat{\boldsymbol{\xi}}_{-D}/\hat{\xi}_D) &= N^{-1} \sum_{r=1}^{N} \log(\mathbf{w}_{r,-D}/w_{rD}) \\ &= N^{-1} \sum_{r=1}^{N} \log(\mathbf{x}_{r,-D}/x_{rD}), \end{aligned} \tag{11.3}$$

where

$$\mathbf{x}_r = \mathscr{C}(\mathbf{w}_r) \qquad (r = 1, \ldots, N)$$

are the compositions formed from the observed vectors. Simplifying (11.3) and remembering that $\xi_1 + \cdots + \xi_D = 1$, we readily find that

$$\hat{\boldsymbol{\xi}} = \mathscr{C}(g_1, \ldots, g_D), \tag{11.4}$$

where g_i is the geometric mean of the N observations $w_{1i}, \ldots, w_{Ni}$ on the ith component, or equivalently, the corresponding proportions $x_{1i}, \ldots, x_{Ni}$ in the N compositions $\mathbf{x}_1, \ldots, \mathbf{x}_N$. For the case of a single replicate this result is, of course, equivalent to the use of $\mathbf{x}$, as described earlier.

To illustrate this estimation procedure we apply it to the serum compositions of Data 32. To estimate the true composition we first compute the geometric means of the columns of the table as

$$0.1110, \quad 0.3156, \quad 0.1994, \quad 0.3700$$

and from these, by application of the constraining operator, arrive at the estimate ξ of the true composition,

$$\xi = (0.112, 0.317, 0.200, 0.371).$$

For this particular pattern of measurement error the estimated composition does not differ, to the accuracy recorded, from the arithmetic average of the eight compositional vectors. This will not in general be the case when, for example, the relative error of measurement is large, as the following simple example illustrates. Three subjects, each presented with an aerial photograph, were asked to estimate the proportions of three different types of vegetation. The results are shown in Table 11.1 and in Fig. 11.2, which is the relevant part of a ternary diagram. There is a clear distinction between the arithmetic average composition and the estimated composition (11.4) based on the composition formed from the geometric means of components. To obtain a view of the reliability of this estimate we can construct in Fig. 11.2 a 95 per cent predictive region as described in Property 7.6. The largeness of this region indicates the great imprecision in the subjective determinations of the compositions.

Table 11.1. *Vegetation compositions and their analysis*

	Vegetation types		
Composition no.	*1*	2	3
1	0.03	0.94	0.03
2	0.04	0.85	0.11
3	0.12	0.86	0.02
Arithmetic average	0.063	0.883	0.053
Estimated composition (11.4)	0.054	0.905	0.041

Sample logratio covariance matrix

$$\begin{bmatrix} 2.015 & 1.133 \\ 1.133 & 0.834 \end{bmatrix}$$

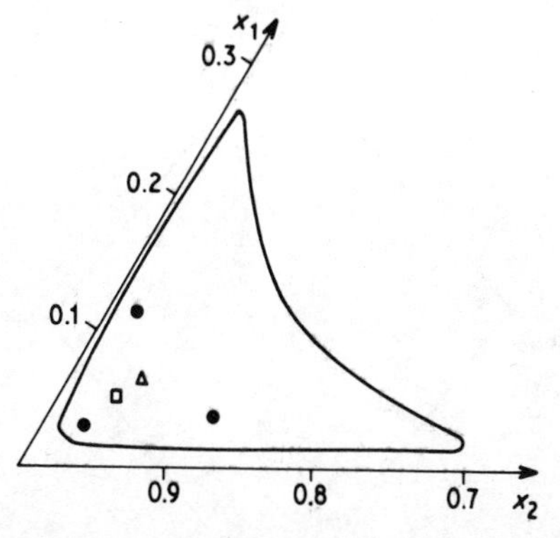

Fig. 11.2. *Three compositions (solid circles) of vegetation types with arithmetic average (open triangle) and estimated composition (open square) formed from the geometric means of components, together with 95 per cent confidence region for the true composition.*

11.3 Analysis of sources of imprecision

The model introduced in Section 11.2 for measurement error is easily adapted to the study of 'observer' error by linear modelling of the mean vector of the logistic normal perturbation error distribution. Expressed in another way, we simply transform compositions to their corresponding logratio vector forms and apply multivariate analysis of variance tests to the transformed vectors, the loglinear modelling and hypothesis-testing of Sections 7.6 and 7.7. We illustrate the methods in an investigation of the variability of the serum protein compositions of Data 33.

We first transform the 4-part composition $\mathbf{x}_{jk}$ associated with the jth analyst and kth specimen to its 3-dimensional logratio form $\mathbf{y}_{jk} = \text{alr}(\mathbf{x}_{jk})$ and then consider the model

$$\mathbf{y}_{jk} = \boldsymbol{\mu} + \boldsymbol{\alpha}_j + \boldsymbol{\beta}_k + \text{error} \qquad (j = 1, 2, 3; k = 1, \ldots, 7).$$

Within this model we can then test the lattice of hypotheses of Fig. 11.3 by the methods of Sections 7.6 and 7.7. The significance probabilities are shown at the appropriate levels of the lattice. All the hypotheses are rejected. Not only are there obvious differences in the specimens, but significant differences between the analysts have been detected.

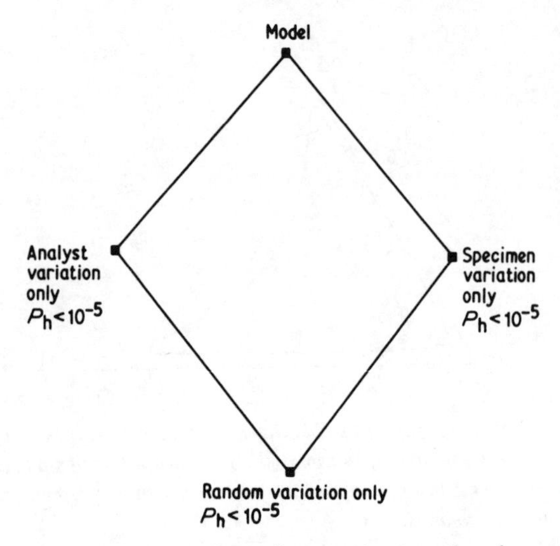

Fig. 11.3. *Lattice of hypotheses for investigation of analyst variation.*

In such studies in the past there has been a tendency to treat the parts of the composition separately. One such study is that of Chayes and Fairbairn (1951), who investigate observer error in determining compositions by asking five geologists to analyse five thin sections cut from the same specimen. For each thin section proportions of seven minerals are reported by each of the analysts. To examine hypotheses concerning analysts and sections, Chayes and Fairbairn perform 2-way analyses of variance *separately* for each component mineral of the composition. Apart from the standard and usually substantial statistical criticism that this employs a number of univariate and therefore partial analyses in an attempt to resolve a single multivariate problem, the interpretation of the univariate analyses is confounded by the unit-sum constraint. For example, Chayes and Fairbairn find that the inter-slide differences are significant at the 0.05 level for quartz, plagioclase and opaque minerals. In the face of the surreptitious effects of the unit-sum constraint, how are we to be sure that the slide effects in quartz and plagioclase are not just consequences of a real slide effect in opaque minerals *together with a unit-sum effect*, and even indeed due to some extent to an absence of effects in the other minerals?

A similar difficulty arises when Krumbein and Tukey (1956,

Table 1) try to analyse their data on pebble counts in glacial drift. Their univariate 3-way analysis of variance, treating components as one of the factors, is carried out by neglecting the fact that the pebble compositions expressed in proportions are correlated. The arcsine square root transformation which Krumbein and Tukey use does not remove the effect of the constraint but replaces it by another constraint.

These difficulties can be avoided by consideration of each composition as a single multivariate entity through the use of multivariate analysis of variance on logratio compositions.

The ineffectiveness of univariate analyses can be well illustrated by the data shown in the ternary diagram of Fig. 11.4. Here 14 aerial photographs, denoted by different letters, were each shown to two subjects, distinguished by capital and lower case letters, who estimated proportions of three types of vegetation. Univariate analyses of variance for each vegetation type find no significant differences between either photographs or subjects, whereas a multivariate analysis in terms of loglinear modelling of the 2-dimensional logratio

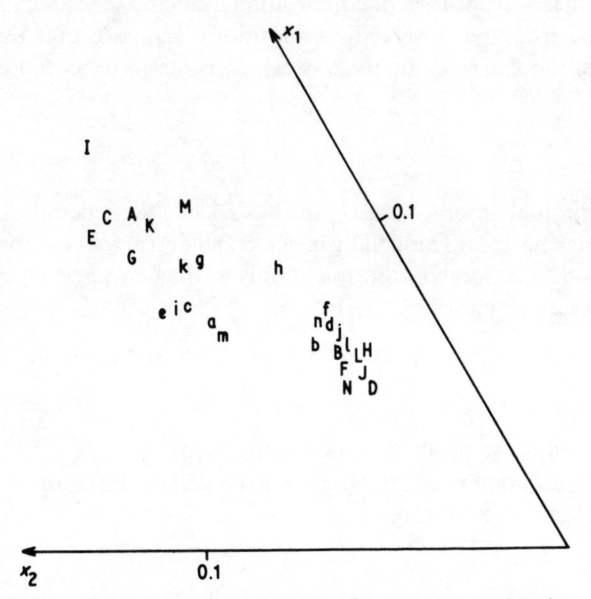

Fig. 11.4. *Twenty-eight compositions of vegetation types on 14 photographs, represented by different letters, estimated by two subjects, distinguished by capital and lower case letters.*

compositions detects a significant difference between photographs at the 0.01 level.

11.4 Imprecision and tests of independence

A question of some importance is whether measurement error in compositions may seriously affect the testing of some hypotheses of independence concerning compositions. We shall here consider two such hypotheses, compositional invariance (Definition 9.3) and complete subcompositional independence (Definition 10.1).

Compositional invariance is a hypothesis concerning a basis $\mathbf{w}$ and its relation to its composition $\mathbf{x} = \mathscr{C}(\mathbf{w})$. There is no doubt that measurement error, particularly of an additive kind, will produce the unhappy situation that compositional invariance of the true composition neither implies, nor is implied by, compositional invariance of the observed or estimated composition. However, when measurement error is multiplicative, the effect is very different. Our sole purpose here is to demonstrate that it is possible to model in such a way that the hypothesis of compositional invariance is identical in both the true and observed compositions. Suppose that the true composition $\boldsymbol{\xi}$ has a true basis $\boldsymbol{\omega}$ which is observed as $\mathbf{w}$. Let

$$\tau = \sum_{i=1}^{D} \omega_i, \qquad t = \sum_{i=1}^{D} w_i$$

be the true and observed sizes of the basis. Then the generalization of the 2-stage process of modelling measurement error for compositions in Section 11.2 requires adjustment only at the first stage, giving the following error model:

1. A multiplicative measurement error in the size determination
$$t = \tau u, \tag{11.5}$$
where u is the positive multiplicative error.
2. A perturbation $\mathbf{v} \in \mathscr{S}^d$ of the composition $\boldsymbol{\xi}$ leading to an observed composition $\mathbf{x}$, where
$$\mathbf{x} = \boldsymbol{\xi} \circ \mathbf{v}. \tag{11.6}$$

It is possible to derive from (11.5) and (11.6) the corresponding error model for the basis $\mathbf{w}$ but we shall not require this explicitly.

Compositional invariance for the true basis corresponds to the hypothesis that $\boldsymbol{\beta}=\mathbf{0}$ in the conditional multivariate regression model

$$\eta_i = \log(\xi_i/\xi_D) = \alpha_i + \beta_i \log \tau + e_i \qquad (i=1,\ldots,d).$$

Now

$$\begin{aligned}\log(x_i/x_D) &= \log(\xi_i/\xi_D) + \log(v_i/v_D)\\ &= \alpha_i + \beta_i \log t + \{e_i - \beta_i \log u + \log(v_i/v_D)\},\end{aligned}$$

so that $\boldsymbol{\beta}=\mathbf{0}$ corresponds also to compositional invariance in the observable basis.

Consider now the hypothesis of complete subcompositional independence, and recall from Property 10.3 that this hypothesis is characterized by a special structure for the logratio covariance matrix

$$\boldsymbol{\Sigma} = \operatorname{diag}(\alpha_1,\ldots,\alpha_d) + \alpha_D \mathbf{J}_d. \tag{11.7}$$

In the modelling of Section 11.2 the estimated composition $\mathbf{x}=\mathscr{C}(\mathbf{w})$ is related to the true composition $\boldsymbol{\xi}$ through the perturbation $\mathbf{x}=\boldsymbol{\xi}\circ\mathbf{v}$ of (11.2) and so

$$\mathrm{V}\{\log(\mathbf{x}_{-D}/x_D)\} = \mathrm{V}\{\log(\boldsymbol{\xi}_{-D}/\xi_D)\} + \mathrm{V}\{\log(\mathbf{v}_{-D}/v_D)\}.$$

It follows that complete subcompositional independence of the true and observed compositions coincide if and only if the covariance matrix $\operatorname{cov}\{\log(\mathbf{v}_{-D}/v_D)\}$ of the error vector $\mathbf{v}$ has the complete subcompositional independence pattern of (11.7).

If there are sufficient replicate determinations of a composition we can estimate this error covariance matrix, as in our analysis of Data 32 in Section 11.2, and use it to test for complete subcompositional independence along the lines of Section 10.2. Application of this test to Data 32 yields an observed value 0.45 of the test statistic of Property 10.4(a) to be compared against upper $\chi^2(2)$ values. Thus there is no evidence of departure from the hypothesis of complete subcompositional independence in the measurement errors.

We have attempted to devise measurement error models which are compatible with the peculiar nature of compositions, namely their confinement to the simplex. The models are flexible in the sense that they allow for the possibility of different precisions in the determination of the magnitudes of the different components and for dependence between the measurement errors in these different component proportions. It would be possible to investigate more fully special patterns of measurement error. For example, for the case where all the components are measured independently and with the same coefficient of variation, the logratio covariance matrix $\boldsymbol{\Sigma}$ takes the isotropic form $\alpha\mathbf{H}$ of Definition 5.5.

As with other forms of data subject to imprecision, it is important to be able to assess the effect that such imprecision may have on statistical inferences. For example, many attempts at discriminating between rock types have been made by standard methods of comparing the major oxides composition of a rock sample of unknown type with the compositions of a 'training' set of rock specimens of known type. The methods applied ignore the possibility of measurement errors and the question 'To what extent may the diagnostic or classification probabilities be affected if we take account of the measurement error present in the determinations?' Such imprecision effects can easily be determined by a simple adaptation of the method of incorporating imprecision into diagnostic models developed in Aitchison and Lauder (1979).

11.5 Rounded or trace zeros

Throughout the monograph so far attention has been confined to the strictly positive simplex. The reason is the obvious one that we cannot take logarithms of zero in the application of logratio transformations. And yet zero components do occur in a number of applications as, for example, a household spending nothing on the commodity group 'tobacco and alcohol' or a rock specimen containing 'no trace' of a particular mineral. In the absence of a one-to-one monotonic transformation between the real line and its non-negative subset, the problem of zeros is unlikely ever to be satisfactorily and generally resolved. It should be emphasized that the problem of zeros is not confined to the logistic normal classes of distributions on the simplex. The Dirichlet class fares no better. For example, in any attempt to fit a $\mathscr{D}^d(\boldsymbol{\alpha})$ distribution to a set of D-part compositions a zero component, say the first, in any composition will cause havoc; for example, any analysis involving the likelihood function is doomed since only two values of the likelihood are possible, 0 for $\alpha_1 \geqslant 1$ and ∞ for $0 < \alpha_1 < 1$. The problem of zeros is familiar in other areas of statistics, for example in lognormal modelling (Aitchison and Brown, 1957, Sections 9.7–9.9), and solutions depend on the frequency and nature of the zeros.

In this section we consider compositional data sets in which only a moderate number of zeros or traces are recorded. We consider a 4-stage approach and illustrate each stage by application to the compositional data set of Data 34, consisting of 4-part foraminiferal compositions at 30 different depths.

Amalgamation feasibility

In most compositional problems we could probably induce zero components by insisting on a fine enough division of the whole into parts. In a household expenditure enquiry, for example, if we successively refine the category 'clothing' we may produce a sequence of increasingly fine parts such as footwear, socks, tartan socks, Stuart tartan socks, and in this way obviously create zero components. A first question which should be asked for compositional data sets with zeros is whether the division of the whole into parts is finer than is strictly necessary to achieve the main aims of the inquiry. If the answer is in the affirmative the follow-up question is obvious: is there a feasible amalgamation which still allows investigation of the main issues and at the same time removes all zero components?

For the foraminiferal compositions of Data 34 suppose that the aim is to investigate whether there is any dependence of composition on depth. For illustrative purposes we restrict attention to a loglinear model with the possibility only of a linear dependence on depth u. Then ideally we would adopt the model

$$\log(x_i/x_4) = \alpha_i + \beta_i u + e_i \qquad (i = 1, 2, 3) \tag{11.8}$$

and test the parametric hypothesis $\beta_i = 0$ $(i = 1, 2, 3)$ within this model. The five zero components in species 3 and 4 prevent us from achieving this. If, however, we know that species 3 and 4 are similar, at any rate more related than any other pair, then perhaps little will be lost if we decide to amalgamate species 3 and 4. Since either species 3 or 4 is present at every level, we rid ourselves of the zero problem by the amalgamation

$$t_1 = x_1, \qquad t_2 = x_2, \qquad t_3 = x_3 + x_4. \tag{11.9}$$

We then investigate the hypothesis $\beta_1 = \beta_2 = 0$ within the model

$$\log(t_i/t_3) = \alpha_i + \beta_i u + e_i \qquad (i = 1, 2) \tag{11.10}$$

by the methods of Sections 7.6 and 7.7. In the notation of Property 7.4 we have

$$N = 30, \quad d = 2, \quad p_{\mathrm{m}} = 2, \quad p_{\mathrm{h}} = 1,$$

so that

$$k = 1, \quad a = \tfrac{1}{2}, \quad b = 1, \quad c = 27.$$

The computed test statistic $q = 0.0348$ and the significance probability is 0.38, so that there is no evidence of a depth trend in the compositions.

Zero and trace replacement

If a zero-removing amalgamation is not feasible then the next question to be posed is the following. Is it possible that the zero denotes not that the part is completely absent but rather that no quantifiable proportion could be recorded to the accuracy or rounding of the measurement process? In other words, can the zeros be regarded as in the same category as traces? If so, then we should investigate the possibility of replacement of zeros and traces by positive values smaller than the smallest recordable value. Such a replacement will certainly allow us to complete any form of logratio analysis on the adjusted compositional data set. Two further questions now arise. How can we choose reasonable replacement values? How sensitive is our analysis to this choice? We address ourselves now to the first of these questions.

We can find a practical answer to the question of replacement values by considering the rounding-off process which leads to any recorded composition. Thus the 3-part composition recorded discretely as $(0.54, 0.19, 0.27)$ could be any composition within the hexagon shown in Fig. 11.5. Since the recorded composition is at the centre of this region there seems no reason for making any adjustment for the rounding-off process. Consider, however, the composition recorded as $(0.00, 0.53, 0.47)$ and its associated half-hexagonal region of possible unrounded compositions. Since the recorded composition is on the boundary of this region there does seem to be a case in practice for replacement of the recorded composition by that corresponding to some interior point of the form $(0.00 + \varepsilon, 0.53 - \frac{1}{2}\varepsilon, 0.47 - \frac{1}{2}\varepsilon)$. For example, if we chose the geometric centre of the half-hexagon we would set ε equal to 4/9 times the maximum possible rounding-off error 0.005. If we further consider the recorded composition $(0.00, 0.00, 1.00)$ and its triangular region of unrounded compositions we see that a reasonable centre for determining the replacement composition would take the form $(\varepsilon, \varepsilon, 1 - 2\varepsilon)$, with ε taken for a geometric centre as 1/3 times the maximum possible rounding error.

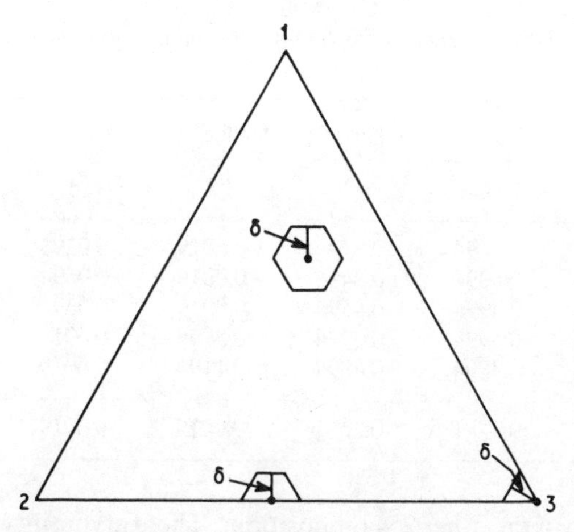

Fig. 11.5. *Regions of possible unrounded compositions corresponding to recorded, rounded compositions, with δ as the maximum rounding error.*

Arguing in this way, we can arrive at a general procedure which replaces any composition with C zero and $D - C$ non-zero components by another composition in which the zeros and traces become $\delta(C+1)(D-C)/D^2$ and the positive components are each reduced by $\delta C(C+1)/D^2$, where δ is the maximum rounding-off error.

Applying this replacement procedure to the five cases with zero components in Data 34 with $C = 1$, $D = 4$ and $\delta = 0.005$, we obtain in Table 11.2 the replacement compositions. We also show in Table 11.2 how this replacement procedure would affect the composition $(0.74, 0.26, 0, 0)$, labelled F31, with two zero components.

Outlier investigation

An important diagnostic tool in assessing the validity of any statistical analysis is the detection of outliers. An obvious next question in our treatment of zero and trace components is therefore the following: even after adjustment for zeros and traces, are the replacement compositions essentially outliers relative to the pattern of variability in the compositions with wholly positive components? We have already considered a means of investigating the possibility of outliers in the computation of atypicality indices as in Property 7.5. For the foraminiferal compositions of Data 34 we can compute the atypicality of each of the five replacement compositions relative to the

Table 11.2. *Replacement compositions for foraminiferal compositions with zero components*

Sample no.	Replacement proportions 1	2	3	4	Atypicality index
F7	0.5894	0.3794	0.0019	0.0294	0.917
F17	0.6894	0.2494	0.0019	0.0594	0.901
F21	0.4994	0.4594	0.0019	0.0394	0.902
F25	0.5994	0.0994	0.2994	0.0019	0.923
F30	0.3894	0.4894	0.1194	0.0019	0.9976
F31	0.7281	0.2681	0.0019	0.0019	0.9995

experience of the other 25 compositions. These atypicality indices are recorded in Table 11.2. It seems that replacement composition F30 must be regarded as an outlier and we would hardly be justified in retaining it in any further analysis without good cause. Since the presence of an apparent outlier can be very informative, we should certainly search for reasons for its existence. In the present case is it simply the result of a less careful sample assessment or greater difficulty in differentiating species found at such depths? Or is this perhaps evidence of an ecological change, with some corroboration from the outlier status of composition F31?

Sensitivity analysis

Even after a compositional data set has been adjusted for zero and trace components, with or without rejection of outliers, it should still be subjected to some form of sensitivity analysis. How sensitive is the statistical analysis under consideration to variation in the replacement procedure? A reasonable way to proceed here seems to be to study the effect on the statistical inference for variation in the choice of δ in the zero replacement procedure. A word of caution is required here. Since $\log \delta \rightarrow -\infty$ as $\delta \rightarrow 0$, we could almost certainly choose δ so small that we could claim that the inference is sensitive to variation of the replacement procedure. In practical terms such a claim would hardly be reasonable. In any specific situation there is a limited range of δ values which we would consider to be justifiable in practice. If δ_r is the maximum rounding-off error we would suggest that

$$\delta_r/5 \leqslant \delta \leqslant 2\delta_r \tag{11.11}$$

Table 11.3. *Sensitivity of significance probability to zero replacement procedure*

δ	Residual determinants $\|\mathbf{R}_m\|$	$\|\mathbf{R}_h\|$	Significance probability
0.001	18517	21403	0.297
0.0025	11550	13231	0.326
0.005	8278	9376	0.366
0.0075	6959	7813	0.399
0.01	6241	6956	0.429

is a reasonable range for a sensitivity investigation.

For the foraminiferal compositions of Data 34, $\delta_r = 0.005$ and so the suggested range for sensitivity analysis is $0.001 \leqslant \delta \leqslant 0.01$. As an illustration Table 11.3 shows the effect on the significance probability associated with testing the hypothesis of no compositional dependence on depth for compositions F1–F29; in other words, we test the hypothesis $\boldsymbol{\beta} = \mathbf{0}$ within the model

$$\mathbf{y} = \boldsymbol{\alpha} + \boldsymbol{\beta} u + \mathbf{e}. \tag{11.12}$$

It is clear that in this case the inference is not very sensitive to the choice of replacement procedure.

11.6 Essential zeros

If there are a fair number of real zeros it may be worth considering the device of three-parameter lognormal modelling (Aitchison and Brown, 1957, p. 14), whereby a constant, either known or to be estimated, is added to every observation. One compositional counterpart would be to apply the logratio transformations, not to $\mathbf{x}$ but to $\mathscr{C}(\mathbf{x} + \boldsymbol{\tau})$ where the $D \times 1$ vector $\boldsymbol{\tau}$ is either chosen or estimated. For example, for 2-part compositions we would be considering $\log\{(x + \tau_1)/(1 + \tau_2 - x)\}$ of $\mathscr{N}^1(\mu, \sigma^2)$ form. This is simply the four-parameter lognormal model of Johnson (1949). Clearly if $\boldsymbol{\tau}$ has to be estimated there are substantial estimation and interpretation problems even for small D. An alternative approach, replacing the logratio transformation by a modified Box and Cox (1964) transformation, is discussed briefly in Section 13.2

If there is a substantial number of zeros mostly in a few components and if amalgamations of components are ruled out, then some form of conditional modelling, separating out the zeros, may be possible.

For example, if the zeros are confined to the first component x_1 of a D-part composition $\mathbf{x}$ we can start our modelling by supposing that $x_1 = 0$ with probability π and $x_1 > 0$ with probability $1 - \pi$. Conditional on $x_1 = 0$ we have then to specify a distribution for $\mathbf{x}_{(1)} = (x_2, \ldots, x_D)$, or equivalently the subcomposition $\mathscr{C}(\mathbf{x}_{(1)})$; suppose that we adopt a $\mathscr{L}^{d-1}(\boldsymbol{\mu}_0, \boldsymbol{\Sigma}_0)$ form for this. Conditional on $x_1 > 0$ we have to specify a distribution for the full composition $\mathbf{x}$, and suppose that we adopt a $\mathscr{L}^d(\boldsymbol{\mu}, \boldsymbol{\Sigma})$ form. We are then in a position to construct the likelihood function corresponding to a compositional data set with such zero components. If $p_d(\cdot|\boldsymbol{\mu},\boldsymbol{\Sigma})$ denotes the density function of the $\mathscr{L}^d(\boldsymbol{\mu}, \boldsymbol{\Sigma})$ distribution, then a composition with $x_1 = 0$ contributes a factor

$$\pi p_{d-1}(\mathbf{x}_{(1)}|\boldsymbol{\mu}_0, \boldsymbol{\Sigma}_0), \tag{11.13}$$

and a composition with $x_1 > 0$ contributes a factor

$$(1 - \pi)p_d(\mathbf{x}|\boldsymbol{\mu}, \boldsymbol{\Sigma}), \tag{11.14}$$

to the likelihood.

With an explicit likelihood function we are in a position to estimate the parameters $\pi, \boldsymbol{\mu}_0, \boldsymbol{\Sigma}_0, \boldsymbol{\mu}, \boldsymbol{\Sigma}$ and to test any parametric hypothesis. Note that there is no assumption in the modelling that the parameters $\boldsymbol{\mu}_0, \boldsymbol{\Sigma}_0$ of the zero component specification are a subset of, or related to, the parameters $\boldsymbol{\mu}, \boldsymbol{\Sigma}$. Indeed, some relationship between the two sets may be the hypothesis of main interest as illustrated in the simple example below. We note that, whatever the distributional forms adopted, the maximum likelihood estimate of π is

$$\hat{\pi} = N_0/(N_0 + N_1), \tag{11.15}$$

where N_0 and N_1 are the numbers of compositions with $x_1 = 0$ and $x_1 > 0$, respectively.

We now consider an ecological application about the relationship of a predator species P to its two prey species Q and R. Of particular interest is the question of whether the relative proportions of the prey species depend on the proportion of the predator species present. The (P, Q, R) compositions at 25 sites are recorded in Data 35, with all three species present at 15 sites, S1–S15, but only the prey species present at the remaining 10 sites, T1–T10. Figure 11.6 shows the ternary diagram for the predator–prey compositions at the 25 sites.

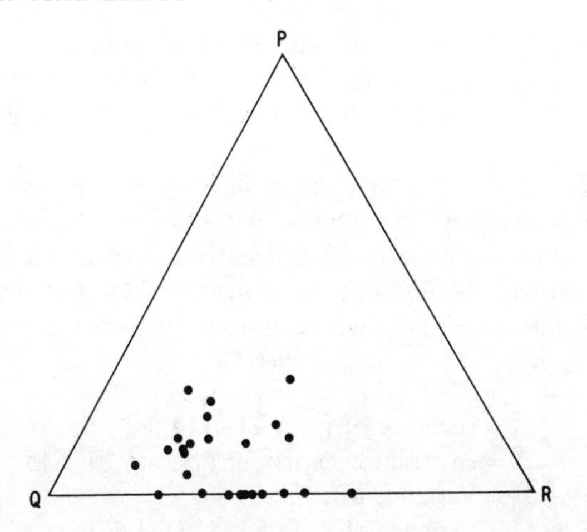

Fig. 11.6. *Ternary diagram for predator–prey compositions at 25 sites.*

In terms of the full composition (x_P, x_Q, x_R) the above question can be formulated in the familiar terms of neutrality on the right of Definition 10.4 for the partition $(x_P | x_Q, x_R)$: is x_Q/x_R independent of x_P? If we carry out the test of this hypothesis using compositions S1–S15 as in either Property 10.7 or Property 10.10, which are identical for this simple partition, we obtain a significance probability of 0.084. We must conclude therefore that there is no evidence of dependence of the relative proportions of Q and R on the proportion of predator, *when the predator is present.*

Suppose, however, that we wish to remove the above conditional clause from any conclusion we make concerning neutrality on the right. In other words, is x_Q/x_R independent of x_P, whether or not x_P is positive? We can formulate this hypothesis within a conditional model as follows. Let π denote the probability that $x_P = 0$ and let the distribution of the logratio $y_2 = \log(x_Q/x_R)$ be $\mathcal{N}(\mu_0, \sigma_0^2)$ conditional on $x_P = 0$. Let the distribution of $y_1 = \log\{x_P/(1 - x_P)\}$, $y_2 = \log(x_Q/x_R)$, conditional on $x_P > 0$, be $\mathcal{N}^2(\boldsymbol{\mu}, \boldsymbol{\Sigma})$, and write

$$\boldsymbol{\mu} = \begin{bmatrix} \mu_1 \\ \mu_2 \end{bmatrix}, \qquad \boldsymbol{\Sigma} = \begin{bmatrix} \sigma_{11} & \sigma_{21} \\ \sigma_{12} & \sigma_{22} \end{bmatrix}.$$

The unconditional form of neutrality on the right takes a simple parametric form:

$$\sigma_{12} = 0, \qquad \mu_2 = \mu_0, \qquad \sigma_{22} = \sigma_0^2. \tag{11.16}$$

In this specification the first constraint ensures neutrality when the predator is present, and the second and third conditions ensure identicality of the distributions of x_Q/x_R for predator present and absent.

Explicit forms of the maximum likelihood estimates are easily obtained under both the model and the hypothesis and so the construction of the Wilks (1938) test statistic (7.16) is straightforward. We can express the test statistic in terms of the following sample variances and covariances, where the divisors used in our application are $N_0, N_1, N_1, N_0 + N_1$, respectively:

$$\begin{aligned}\hat{\sigma}_0^2 &= \text{variance of } y_2 \text{ in T1–T10},\\ \hat{\boldsymbol{\Sigma}} &= \text{covariance matrix of } y_1, y_2 \text{ in S1–S15},\\ \hat{\sigma}_1^2 &= \text{variance of } y_1 \text{ in S1–S15},\\ \hat{\sigma}_2^2 &= \text{variance of } y_2 \text{ in S1–S15, T1–T10}.\end{aligned}$$

The test statistic is then

$$N_0 \log(\hat{\sigma}_2^2/\hat{\sigma}_0^2) + N_1 \log(\hat{\sigma}_1^2\hat{\sigma}_2^2/|\hat{\boldsymbol{\Sigma}}|) \qquad (11.17)$$

to be compared against upper percentage points of the $\chi^2(3)$ distribution, since the hypothesis places three constraints on the parameters.

In our application

$$\hat{\sigma}_0^2 = 0.1987, \quad \hat{\sigma}_1^2 = 0.2355, \quad \hat{\sigma}_2^2 = 0.2175,$$

$$\hat{\boldsymbol{\Sigma}} = \begin{bmatrix} 0.2355 & -0.0989 \\ -0.0989 & 0.2300 \end{bmatrix}$$

and the computed value of the test statistic is 9.59. The significance probability is 0.022, so that we conclude that the relative proportions of the prey species Q and R do depend on the proportion, possibly zero, of the predator P.

11.7 Missing components

If we consider how missing components may arise in a composition, we see that there are two situations:

1. In the determination of the quantities of the parts from which the composition is computed, something has gone wrong in the experimental or observational process for one or more of the parts. From the closure procedure we cannot obtain the components of

any parts of the full composition but only the subcomposition associated with the determined parts.

2. Determination of the full composition has already been completed but, through carelessness, information about one or more components has been lost. The sum of the missing components can therefore be computed as the make-up proportion, and so the information for the composition is essentially an amalgamation with its only summation over the missing parts.

Since the first situation seems likely to be much more frequent than the second, we shall confine our attention to the first. It must be admitted that it is difficult to envisage a fully efficient treatment of (2) within the context of logistic normal modelling. One way out of the difficulty is to ignore the information on the component values in such amalgamations and confine attention to the subcomposition formed from the recorded parts. Such a treatment converts situation (2) into a problem of type (1).

A general approach to this problem is simply to suppose that the full composition follows an $\mathscr{L}^d(\boldsymbol{\mu}, \boldsymbol{\Sigma})$ distribution. Suppose that for the rth composition there is recorded only the subcomposition associated with the selecting matrix $\mathbf{S}_r$ $(r = 1, \ldots, N)$ of order $D_r \times D$. Note that the notation here allows the possibility of the full composition being recorded, with the selecting matrix being the identity matrix $\mathbf{I}_D$. The distribution of the logratio vector $\mathbf{y}_r$ corresponding to the rth subcomposition or composition is then readily expressed as $\mathscr{L}^{d_r}(\mathbf{Q}_{S_r}\boldsymbol{\mu}, \mathbf{Q}_{S_r}\boldsymbol{\Sigma}\mathbf{Q}'_{S_r})$, where $d_r = D_r - 1$ and $\mathbf{Q}_S$ is defined in Property 6.3. The loglikelihood function is then

$$-\frac{1}{2}\sum_{r=1}^{N} d_r \log(2\pi) - \frac{1}{2}\sum_{r=1}^{N} \log|\mathbf{Q}_{S_r}\boldsymbol{\Sigma}\mathbf{Q}'_{S_r}|$$

$$-\frac{1}{2}\sum_{r=1}^{N} (\mathbf{y}_r - \mathbf{Q}_{S_r}\boldsymbol{\mu})'(\mathbf{Q}_{S_r}\boldsymbol{\Sigma}\mathbf{Q}'_{S_r})^{-1}(\mathbf{y}_r - \mathbf{Q}_{S_r}\boldsymbol{\mu}), \qquad (11.18)$$

to which standard estimation and testing procedures may be applied. Clearly, in general, any procedure based on maximum likelihood estimation is going to require numerical methods of solution, and we suggest resort to some readily available package such as MAXLIK (Kaplan and Elston, 1972). There are special cases where simpler methods are available but we shall not here attempt to go into the details of these.

We consider an extremely simple application to the reconciliation of probability assessments assigned by a subject to a finite number D of mutually exclusive and exhaustive hypotheses. A full composition here corresponds to a probability distribution on the D hypotheses and a subcomposition is identified with a conditional probability distribution on a subset of the hypotheses. In this area of application the experimenter may deliberately solicit information on such conditional distributions and hence on subcompositions. Lindley, Tversky and Brown (1979, Example 1) consider the principal cause of death divided into four categories:

1. Unnatural causes.
2. Heart disease.
3. Cancer.
4. Other natural causes.

A subject, when asked to assess the probabilities $(\xi_1, \xi_2, \xi_3, \xi_4)$ of death from each of these causes assigns the vector

$$(w_1, w_2, w_3, w_4) = (0.12, 0.33, 0.27, 0.23). \tag{11.19}$$

The subject is then asked to assess the probabilities of death from the last three causes for a person known to have died from natural causes and assigns the vector

$$(s_2, s_3, s_4) = (0.41, 0.31, 0.28). \tag{11.20}$$

In (11.19) and (11.20) we have expressed the data in the notation of compositional data analysis. In (11.19) the components do not sum to 1 so we would replace the basis $\mathbf{w}$ by its composition

$$\mathbf{x} = \mathscr{C}(\mathbf{w}) = (0.1263, 0.3474, 0.2842, 0.2421),$$

and thus a first estimate by the subject of the full composition $\boldsymbol{\xi} = (\xi_1, \xi_2, \xi_3, \xi_4)$. In (11.20) the subject is essentially supplying an estimate of the subcomposition $\mathscr{C}(\xi_2, \xi_3, \xi_4)$. If we then adopt the perturbation model for imprecision we would set

$$\mathbf{x} = \boldsymbol{\xi} \circ \mathbf{u}, \qquad \mathbf{s} = \mathscr{C}(\xi_2, \xi_3, \xi_4) \circ \mathbf{v},$$

where $\mathbf{u}$ and $\mathbf{v}$ are independent with $\mathscr{L}^3(\mathbf{0}, \boldsymbol{\Sigma})$ and $\mathscr{L}^2(\mathbf{0}, \boldsymbol{\Sigma}_1)$ distributions, where

$$\boldsymbol{\Sigma}_1 = \begin{bmatrix} 0 & 1 & 0 \\ 0 & 0 & 1 \end{bmatrix} \boldsymbol{\Sigma} \begin{bmatrix} 0 & 0 \\ 1 & 0 \\ 0 & 1 \end{bmatrix}$$

With these distributional assumptions it is then easy to show that the maximum likelihood estimate $\hat{\xi}$ of ξ is given by

$$\log(\hat{\xi}_1/\hat{\xi}_4) = \log(x_1/x_4),$$
$$\log(\hat{\xi}_i/\hat{\xi}_4) = \tfrac{1}{2}\{\log(x_i/x_4) + \log(s_i/s_4)\} \qquad (i = 2, 3),$$

from which we obtain

$$\hat{\xi}_1 = 0.127, \quad \hat{\xi}_2 = 0.353, \quad \hat{\xi}_3 = 0.277, \quad \hat{\xi}_4 = 0.243$$

as the reconciliation of the subject's inconsistent assignments of probabilities.

11.8 Bibliographic notes

The central role of the perturbation as a means of modelling imprecision in compositional data was identified in the reply to the discussion of Aitchison (1982), and then extended and applied by Aitchison and Shen (1984).

The long-standing confrontation between zero observations and the logarithmic transformation is well documented; see, for example, the accounts in Aitchison and Brown (1957, Sections 9.7–9.9) for lognormal distributions and in Aitchison (1982, Section 7.4) for logistic normal distributions. The zero-replacement procedure of Section 11.5 is based on the treatment of Aitchison and Lauder (1985) and the conditional modelling in Section 11.6 is built on the approach to essential zeros in the lognormal modelling of Aitchison (1955).

There is an extensive literature on the problem of missing data. The most general technique for coping with missing data is probably the EM algorithm of Dempster, Laird and Rubin (1977); see also Beale and Little (1975). Since the additive logratio transformation, with a never-missing component chosen as common divisor, transforms the problem into the problem of multivariate normal data with missing components, useful references are Murray (1979), Titterington (1984) and Titterington and Jiang (1983).

Problems

11.1 Twelve thin sections of a rock specimen were analysed by the same method and the 5-part mineral compositions are reported in Data 36.

(a) What estimate would suggest if asked what is the true mineral composition of the rock specimen?
(b) Can you provide a visual indication of the reliability of the consequent estimate of the true (microcline, plagioclase, biotite) subcomposition of the specimen?
(c) Test this compositional data set for complete subcompositional independence. Can you draw any conclusions about the nature of the determination process?

11.2 Data 37 records the 7-part mineral compositions as determined on five slides by five analysts. Construct a suitable lattice for investigation of the possibility of analyst differences in reporting mineral compositions. What conclusions do you draw?

11.3 Reconsider the questions you posed in Problem 1.8 about the four different methods of assessing leucocyte compositions in Data 23. Are you now in a position to provide answers to these questions?

11.4 Suppose that two diagnostic assessments had been omitted from Data 17 because zero probabilities had been assigned, in the assessments (0.91, 0, 0.09) by clinician C16 and (0, 0.10, 0.90) by statistician S16. Investigate fully how these may affect the conclusions you have reached in your previous analyses of this data set.

11.5 Suppose that the steroid metabolite excretions (0.71, 0.25, 0) and (1.56, 0.21, 0) of two children C31 and C32 were omitted from Data 9. Review the analysis of Section 9.7 in the light of this additional information.

11.6 One of the new samples C5 with (sand, silt, clay) composition (0.78, 0.22, 0) was omitted from Data 20 because of its zero clay component. Do you consider it reasonable to include this sample to help in discrimination and, if so, what effect does its inclusion have?

11.7 The proportions of two predator species P1, P2 and their common prey species Q, R were estimated at each of 25 sites and are recorded in Data 38. Investigate whether there is any evidence that the ratio of prey species depends on the predator proportions

(a) when both predators are present;
(b) when only predator P1 is present;
(c) whatever predators are present.

11.8 Each of the single male households M1–M20 of Data 8 actually had zero expenditure on alcohol and tobacco. To what extent is it possible to compare the budget patterns of these households with ml–m20 of Data 30?

11.9 In the assay of the serum protein compositions of a blood sample of a Pekin duck it is customary to use four aliquots to obtain replicate determinations. In one such assay there were experimental difficulties and only subcompositions were obtained for two of the aliquots. The compositions and subcompositions are recorded in Table 11.4.

Table 11.4. *Serum protein compositions of a Pekin duck*

Aliquot	*Proportions*		
no.	*Pre-albumin*	*Albumin*	*Globulins*
A1	0.135	0.352	0.513
A2	0.162	0.338	0.500
A3	0.236		0.764
A4		0.412	0.588

On the assumption that the pattern of variability of the full composition over aliquots is $\mathscr{L}^3(\boldsymbol{\mu}, \boldsymbol{\Sigma})$, find the likelihood function and hence obtain maximum likelihood estimates for $\boldsymbol{\mu}$ and $\boldsymbol{\Sigma}$.

What estimate of the serum protein composition would you suggest for this duck?

11.10 A certain syndrome has four possible causes, three located in the adrenal gland or glands,

A: adenoma,
B: bilateral hyperplasia,
C: carcinoma,

and the fourth

D: ectopic carcinoma,

located elsewhere but seriously affecting adrenal function. Of these A and B are benign, and C and D are malignant causes. In a study of subjective probability assessments a clinician, having studied the record of a patient known to be suffering from this syndrome, is asked to assess certain probabilities.

1. Assign probabilities to the four possible causes A, B, C, D. *Answer* (0.22, 0.48, 0.13, 0.25).
2. Given that the cause is benign, assign probabilities to A, B. *Answer* (0.35, 0.65).
3. Given that the cause is malignant, assign probabilities to C, D. *Answer* (0.40, 0.60).
4. Given that the cause is of adrenal location, assign probabilities to A, B, C. *Answer* (0.25, 0.60, 0.15).

Are these assignments consistent? If not, how would you reconcile them?

Suppose that assignment (4) is replaced by the following.

4(a). Assign probabilities to benign (A + B) and malignant (C + D) causes. *Answer* (0.65, 0.35).

How would you reconcile assignments 1, 2, 3, 4(a)?

11.11 Reconsider the questions you posed in Problem 1.3 about the glacial tills of Data 18 and suggest methods of incorporating the compositions with zero components into your analysis.

CHAPTER 12

Compositions in a covariate role

Explanations explanatory of things explained.

Abraham Lincoln: *Lincoln–Douglas Debates*

12.1 Introduction

In our discussion of compositional data analysis so far we have concentrated on the pattern of variability of compositions *per se* and also the manner in which that pattern may depend on other factors or features. In this chapter we examine a number of problems in which we attempt to explain how the variability of some factor or feature depends on a composition. For instance, we may view the differential discrimination between hongite and kongite of Data 1 and 2 in this way by rephrasing Question 1.2.6: to what extent does the variability of type of rock depend on the 5-part mineral composition? Again, in the experiments with firework mixtures in Data 13 the whole purpose of experimentation is, as expressed in Questions 1.12.1–1.12.3, to discover how the variability of brilliance and vorticity depends on the 5-part composition of ingredients. Another problem of this kind is to be found in the calibration of white-cell compositions of Data 11 with Questions 1.10.1 and 1.10.2 asking whether the variability of the microscopically determined composition is sufficiently dependent on the image-analysed composition for the image analysis to prove a good enough substitute for microscopic inspection. In this example we are seeking an explanation of the dependence of one composition on another.

In all of these examples we are involved with conditional distributions in which the conditioning feature, or at least part of it, is a composition. In the varied jargon of regression analysis the composition is playing the role of explanatory variable, 'independent' variable, regressor, or concomitant. We shall settle for the term *covariate* in the remainder of this chapter, and refer to the 'dependent' variable or regressand as the *response* variable or vector.

12.2 Calibration

We can easily model the calibration problem of the white-cell compositions of Data 11 by the now familiar transformation to logratio compositions. Let **z** denote the logratio composition of a typical microscopically determined composition, and **y** the logratio composition of the corresponding image-determined composition. Since the 30 blood samples are assumed to be a random sample from clinical experience we have, in the terminology of Aitchison and Dunsmore (1975), a random calibration experiment. We can then attempt to relate the response logratio vector $\mathbf{z}=(z_1, z_2)$ to the covariate logratio composition $\mathbf{y}=(y_1, y_2)$ by the model:

$$[z_1 \quad z_2] = [1 \quad y_1 \quad y_2] \begin{bmatrix} \alpha_1 & \alpha_2 \\ \beta_1 & \beta_2 \\ \gamma_1 & \gamma_2 \end{bmatrix} + [e_1 \quad e_2], \tag{12.1}$$

where $[e_1 \quad e_2]$ is $\mathcal{N}^2(\mathbf{0}, \boldsymbol{\Sigma})$. This calibration problem thus falls within the scope of logratio linear modelling as described in Sections 7.6 and 7.7. The estimate of the parameter matrix $\boldsymbol{\Theta}$ under the model is

$$\hat{\boldsymbol{\Theta}} = \begin{bmatrix} -0.055 & 0.167 \\ 0.968 & -0.011 \\ 0.054 & 0.998 \end{bmatrix}.$$

The standard test of the hypothesis $\beta_1=\gamma_1=\beta_2=\gamma_2=0$ of no dependence has a significance probability of less than 10^{-5} so that there is little doubt that there is a strong dependence.

Suppose now that we were to switch wholly to the image-processing method of determination of composition. For a new blood sample we could determine the logratio vector **y** of the image determined composition and would then use

$$\hat{\mathbf{z}} = [1 \quad y_1 \quad y_2]\hat{\boldsymbol{\Theta}} \tag{12.2}$$

as a calibrated value, an estimate of the undetermined microscopic value. An important question is how reliable this is.

A simple way of assessing the reliability of calibration is in terms of, say, a 95 per cent prediction region for **z** in the form of Property 7.6.

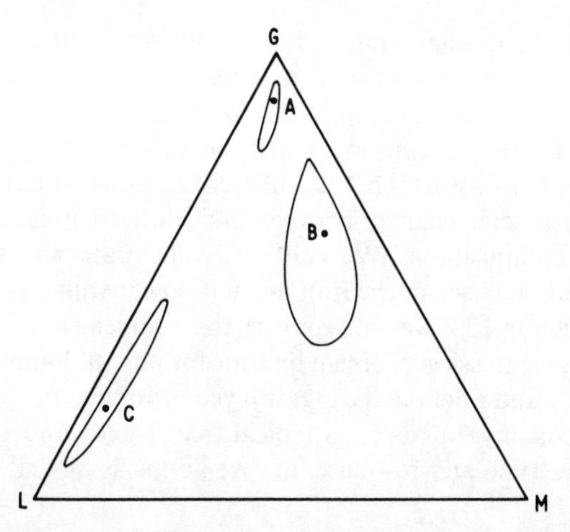

Fig. 12.1. *Predictive regions for microscope-determined calibrates of three image-determined white-cell compositions, A (0.90,0.05,0.05), B (0.6,0.1,0.3), C (0.20,0.75,0.05).*

The predictive distribution for $\mathbf{z}|\mathbf{y}$ with a vague prior on $\boldsymbol{\Theta}$ is, in general notation,

$$\mathrm{St}^d[N-1, \hat{\mathbf{z}}, \{1 + N^{-1} + (\mathbf{y}-\bar{\mathbf{y}})'\mathbf{S}^{-1}(\mathbf{y}-\bar{\mathbf{y}})\}\hat{\boldsymbol{\Sigma}}], \qquad (12.3)$$

where $\bar{\mathbf{y}}$ and $\mathbf{S}$ are the estimates of the logratio mean vector and covariance matrix of the image-analysis compositions. In our particular case $N = 30$, $D = 3$, and we can readily draw the elliptical 95 per cent prediction region in $\mathscr{R}^2$ and translate this by the additive logistic transformation to the ternary diagram. The results for three image-determined compositions are shown in Fig. 12.1. Whether or not these are reliable enough determinations is clearly a matter for the histologist. What we have attempted to illustrate here is an approach which allows him insight into the nature of the calibration and its reliability.

12.3 A before-and-after treatment problem

The state of a specimen, such as a blood sample, is sometimes considered to be its D-part composition, such as the white-cell composition of the blood sample. The purpose of treatment of a patient may be to alter the composition from one that corresponds to 'disease' to one of comparative 'health'. In industrial chemistry the objective of treatment or refining may be to turn a raw input, such as crude oil, of some composition to an output, possibly a gasoline, of some more favourable composition. In such situations a problem that often arises is to assess whether different treatments have different effects in producing changes between the before-treatment and after-treatment compositions. We can easily illustrate an appropriate analysis here in terms of the fruit evaluation experiment of Data 12.

As in Section 12.2 we can convert the problem into a standard multivariate regression problem by transforming to logratio compositions. Let **y** and **z** denote the logratio vectors of the past and present compositions of yatquats for a typical tree. Then **y** and **z** are in the roles of covariate and response and we adopt as model

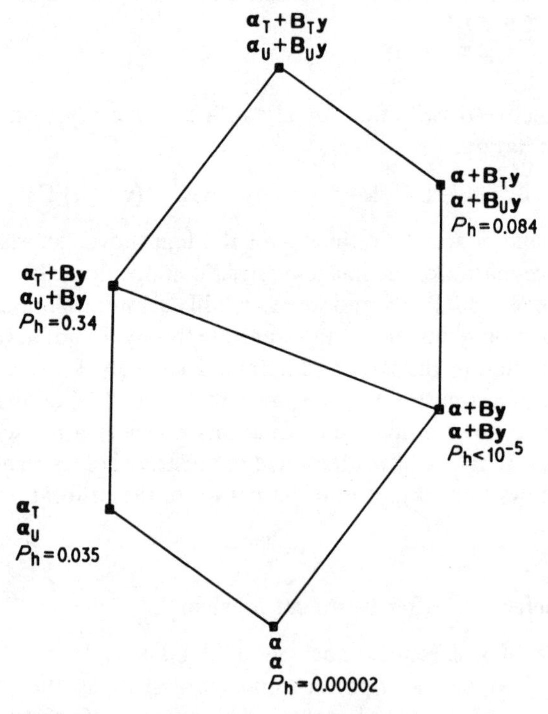

Fig. 12.2 *Lattice of hypotheses for investigation of fruit evaluation experiment.*

$$\begin{aligned} \mathbf{z} &= \boldsymbol{\alpha}_{\mathrm{T}} + \boldsymbol{B}_{\mathrm{T}}\mathbf{y} + \mathbf{e} \qquad \text{for treated trees,} \\ \mathbf{z} &= \boldsymbol{\alpha}_{\mathrm{U}} + \boldsymbol{B}_{\mathrm{U}}\mathbf{y} + \mathbf{e} \qquad \text{for untreated trees,} \end{aligned} \tag{12.4}$$

where the error vector $\mathbf{e}$ is $\mathcal{N}^2(\mathbf{0}, \boldsymbol{\Sigma})$, and investigate the obvious lattice of hypotheses of Fig. 12.2. Estimation and hypothesis testing here are completely standard, as presented in Sections 7.6 and 7.7, and we show only the significance probabilities associated with each hypothesis. It is clear that we can reject the hypotheses until we reach

$$\begin{aligned} \mathbf{z} &= \boldsymbol{\alpha}_{\mathrm{T}} + \boldsymbol{B}\mathbf{y} + \mathbf{e} \qquad \text{for treated trees,} \\ \mathbf{z} &= \boldsymbol{\alpha}_{\mathrm{U}} + \boldsymbol{B}\mathbf{y} + \mathbf{e} \qquad \text{for untreated trees,} \end{aligned}$$

which thus becomes our working model. In reaching this level we have thus detected a significant difference between the compositions of treated and untreated trees. The estimates for this working model are

$$\boldsymbol{\alpha}_{\mathrm{T}} = (0.6649, \quad -0.0763), \qquad \boldsymbol{\alpha}_{\mathrm{U}} = (0.5921, \quad 0.0291),$$

$$\boldsymbol{B} = \begin{bmatrix} 0.1156 & -0.0415 \\ -0.0012 & 0.1121 \end{bmatrix}, \qquad \mathbf{R}_{\mathrm{m}} = \begin{bmatrix} 0.1527 & -0.0148 \\ -0.0148 & 0.1012 \end{bmatrix}.$$

Comparison of $\boldsymbol{\alpha}_{\mathrm{T}}$ and $\boldsymbol{\alpha}_{\mathrm{U}}$ shows the advantageous flesh-to-skin and flesh-to-stone ratios of the treated trees.

12.4 Experiments with mixtures

The statistical theory of experiments with mixtures is concerned with the conditional distribution of a quantitative response z given the composition $\mathbf{x} = (x_1, \ldots, x_D)$ of a mixture of D components. Our fireworks problem of Section 1.12 is of this form, with Questions 1.12.1–1.12.3 enquiring how the two responses, brilliance and vorticity, depend on the covariate composition, the mixture of five ingredients. The standard assumption in such experiments with mixtures is that the expected response $\zeta(\mathbf{x})$ depends only on the composition $\mathbf{x}$, and not on the amount of the mixture. Previous work in this area has adopted a least squares approach, with differences occurring mainly in the choice of the form adopted for $\zeta(\mathbf{x})$. As far as estimation of the composition which optimizes expected response is concerned, there seems to be little to distinguish between forms with comparable goodness of fit. Much discussion has turned on attempts to provide interpretations for the parameters of the model. As indicated in Section 3.5, the difficulties in such interpretations spring from the unit-sum constraint $x_1 + \cdots + x_D = 1$ on the component proportions: it is impossible to alter one proportion without altering at least one of the other proportions. The simplicity of interpretation of additive and interactive effects found in standard factorial experiments is therefore not attainable.

Our success in previous sections in using the logratio vector $\mathbf{y}$ of a composition $\mathbf{x}$ as covariate encourages the exploration of models in which the expected response $\zeta(\mathbf{x})$ is a polynomial in the components of $\mathbf{y}$. Since the components of $\mathbf{y}$ can be varied independently, the polynomial forms adopted can be full in the sense of including all the terms of appropriate degree, in contrast, for example, to Scheffé's (1958) polynomial forms in $(x_1, \ldots, x_D)$ which require the omission of certain terms to ensure identifiability. Thus the linear and quadratic logcontrast models for D-component mixtures take the form

$$\zeta_1(\mathbf{x}) = \beta_0 + \sum_{i=1}^{d} \beta_i y_i, \qquad \zeta_2(\mathbf{x}) = \zeta_1(\mathbf{x}) + \sum_{i=1}^{d} \sum_{j=1}^{d} \gamma_{ij} y_i y_j, \tag{12.5}$$

with obvious extensions to higher-degree polynomials.

The expected responses (12.5) can be more conveniently expressed in symmetric form as follows:

$$\zeta_1(\mathbf{x}) = \beta_0 + \sum_{i=1}^{D} \beta_i \log x_i \qquad (\beta_1 + \cdots + \beta_D = 0), \tag{12.6}$$

$$\zeta_2(\mathbf{x}) = \zeta_1(\mathbf{x}) + \sum_{i=1}^{d} \sum_{j=i+1}^{D} \beta_{ij} (\log x_i - \log x_j)^2. \tag{12.7}$$

The new parameters $\beta_{ij}\,(i = 1, \ldots, d; j = i+1, \ldots, D)$ of (12.6) and (12.7) are simply related to the parameters γ_{ij} $(i = 1, \ldots, d; j = i, \ldots, d)$ of (12.5):

$$\begin{aligned} \beta_{ij} &= -\tfrac{1}{2}\gamma_{ij} \qquad (i = 1, \ldots, d; j = i+1, \ldots, d), \\ \beta_{i,D} &= \gamma_{ii} + \tfrac{1}{2}\sum_{k<i} \gamma_{ki} + \tfrac{1}{2}\sum_{k>i} \gamma_{ik} \qquad (i = 1, \ldots, d). \end{aligned} \tag{12.8}$$

The symmetric forms (12.6) and (12.7) are more convenient for interpretative purposes, while those of (12.5) are simpler for parameter estimation and optimization with the use of (12.8) to interrelate the two versions. Extensions of the symmetric forms (12.6) and (12.7) to higher-degree polynomials lose much of the simplicity of (12.7).

Since our primary concern here is with interpretative differences between the linear and quadratic forms, we shall not require such higher-degree forms.

Since we cannot take logarithms of zero, the model may be applied only to mixtures which are in the strictly positive simplex, that is to mixtures in which all the components are present. While this obviously rules out a number of applications, there remain enough to make the model worthy of fuller investigation. The main interest may be in the relative behaviour of quite complicated mixtures and the region of experimentation may well exclude mixtures with zero components. Moreover, for situations where something goes critically wrong with the response as the proportion of any component approaches zero, the logcontrast model may prove attractive.

The parameters of the linear and quadratic models (12.6) and (12.7) can be simply related to three important forms of hypotheses about the nature of the effect of the mixture on the response. The hypotheses are most simply described in terms of a single division $(x_1,\ldots,x_c|x_C,\ldots,x_D)$ of the composition and the associated partition $(\mathbf{t};\mathbf{s}_1,\mathbf{s}_2)$.

Definition 12.1 A D-part mixture is *inactive in parts* $1,\ldots,c$ if the expected response depends only on the subcomposition $\mathbf{s}_2=\mathscr{C}(\mathbf{x}_{(c)})$. In the extreme case when $c=D$ the mixture has no influence on the expected response and is said to be *completely inactive.*

Hypotheses of inactivity have obvious parametric forms.

Property 12.1 Within the quadratic logcontrast model the hypothesis that a D-part compositional mixture is inactive in parts $1,\ldots,c$ is equivalent to the parametric hypothesis $\beta_i=0\ (i=1,\ldots,c)$, $\beta_{ij}=0\,(i=1,\ldots,c;j=i+1,\ldots,D)$. Moreover, complete inactivity is equivalent to the parametric hypothesis

$$\beta_i=0\,(i=1,\ldots,D),\quad \beta_{ij}=0\,(i=1,\ldots,d;j=i+1,\ldots,D),$$

that is that $\zeta_2(\mathbf{x})=\beta_0$, a constant.

The advantage of the use of logarithms of ratios as in the symmetric logcontrast form (12.7) is that, by making the study of subcompositions easy, it allows consideration of the hypothesis of inactivity of any subvector of components. Moreover, the testing of such hypotheses is standard since, in least squares terms, each is a linear hypothesis within a linear model.

Logcontrast modelling is useful to the experimenter who wishes to investigate the consequences of a change in some components while holding fixed the relative proportions of, or equivalently the subcomposition formed from, the other components. Suppose that for the partition $(x_1,\ldots,x_c|x_C,\ldots,x_D)$ we are interested in varying $(x_1,\ldots;x_c)$ while holding $\mathscr{C}(x_C,\ldots,x_D)$ fixed. A convenient method of studying the feasibility and nature of such variation is through the multiplicative logratio transformation of Definition 6.8 with $\mathbf{y}=\text{mlr}(\mathbf{x})$. We note that since $\mathbf{y}\in\mathscr{R}^d$, its components can be varied independently of each other, unlike those of $\mathbf{x}$. Moreover, by Note 1 of Definition 6.5, the subvector $\mathbf{x}^{(c)}$ and the subcomposition $\mathscr{C}(\mathbf{x}_{(c)})$ are uniquely and separately determined by, and so their independent variation can be studied in terms of, $\mathbf{y}^{(c)}$ and $\mathbf{y}_{(c)}$, respectively.

With this facility for the study of variation in a composition we may, by analogy with standard factorial experiments, consider an expected response $\zeta(\mathbf{x})$ as being additive with respect to the independently variable $\mathbf{x}^{(c)}$ and $\mathscr{C}(\mathbf{x}_{(c)})$ if $\zeta(\mathbf{x})$ is expressible as a sum of separate functions of $\mathbf{y}^{(c)}$ and $\mathbf{y}_{(c)}$. Using the transformations we readily see that the model (12.7) for $\zeta(\mathbf{x})$ has this additivity property if and only if

$$\beta_{ij}=0 \qquad (i=1,\ldots,c;j=C,\ldots,D). \tag{12.9}$$

The symmetry of (12.9) implies that $\zeta(\mathbf{x})$ is also additive with respect to $\mathbf{x}_{(c)}$ and $\mathscr{C}(\mathbf{x}^{(c)})$ and indeed that $\zeta(\mathbf{x})$ is expressible as the sum of separate functions of $\mathscr{C}(\mathbf{x}^{(c)})$ and $\mathscr{C}(\mathbf{x}_{(c)})$. We make this last form the feature of our formal definition.

Definition 12.2 A D-part mixture is said to have $(\mathbf{x}^{(c)}|\mathbf{x}_{(c)})$ or $(\mathbf{t};\mathbf{s}_1,\mathbf{s}_2)$ *partition additivity* if the expected response is the sum of a function of $\mathbf{s}_1$ only and a function of $\mathbf{s}_2$ only.

As noted above, partition additivity implies that the expected response can also be expressed as a sum of separate functions of $\mathbf{x}^{(c)}$ and $\mathscr{C}(\mathbf{x}_{(c)})$ and as a sum of separate functions of $\mathscr{C}(\mathbf{x}^{(c)})$ and $\mathbf{x}_{(c)}$.

We now record the parametric form of this hypothesis.

Property 12.2 Within the quadratic logcontrast model the hypothesis that a D-part compositional mixture has $(\mathbf{x}^{(c)}|\mathbf{x}_{(c)})$ additivity is equivalent to the parametric hypothesis

$$\beta_{ij}=0 \qquad (i=1,\ldots,c; j=C,\ldots,D).$$

Partition additivity is thus an additive property similar to its counterpart in standard factorial analysis where, of course, the simpler rectangular factor space leads to easier expression of additivity. This hypothesis of partition additivity is, in least squares terms, a linear hypothesis within a linear model and so its testing falls within standard linear theory.

The concept of partition additivity can be readily illustrated geometrically for mixtures of three components representable in the ternary diagram of Fig. 12.3. For the partition $(x_1 | x_2, x_3)$ the points A and B, on the same ray through the vertex 1, correspond to the same subcomposition $\mathscr{C}(x_2, x_3)$; similarly for C and D. Points A and C correspond to the same x_1 value; similarly for B and D. Partition additivity can thus be expressed in terms of relationships of the form $\zeta(\mathrm{A})-\zeta(\mathrm{B})=\zeta(\mathrm{C})-\zeta(\mathrm{D})$, or as the zero trapezoidal contrast

$$\zeta(\mathrm{A})-\zeta(\mathrm{B})-\zeta(\mathrm{C})+\zeta(\mathrm{D})=0. \tag{12.10}$$

In the construction of simplex designs for the efficient investigation of partition additivity, clearly one aim should be to choose design points to provide as many such trapezoids as possible.

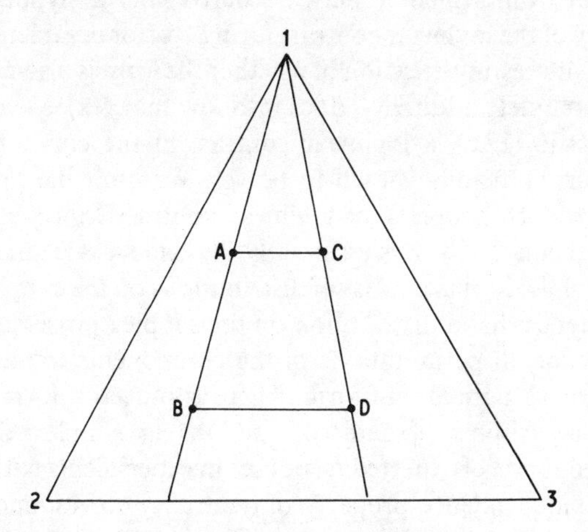

Fig. 12.3. *A trapezoidal contrast associated with the investigation of additivity with respect to the partition* $(x_1 | x_2, x_3)$.

The relationship of hypotheses of inactivity and additivity is contained within the following implication statement.

Property 12.3 If a D-part compositional mixture is inactive in parts $1, \ldots, c$ then it has $(\mathbf{x}^{(c)} | \mathbf{x}_{(c)})$ partition additivity.

The extreme form of partition additivity is readily defined, and its corresponding parametric form ascertained.

Definition 12.3 A D-part compositional mixture is *completely additive* if it is partition additive for every possible partition of order 1 of the composition.

Property 12.4 Within the quadratic logcontrast model the hypothesis that a D-part compositional mixture is completely additive is equivalent to the parametric hypothesis

$$\beta_{ij} = 0 \qquad (i = 1, \ldots, d; j = i + 1, \ldots, D),$$

so that the expected response $\zeta(\mathbf{x})$ takes the linear form

$$\zeta_1(\mathbf{x}) = \beta_0 + \sum_{i=1}^{D} \beta_i \log x_i \qquad (\beta_1 + \cdots + \beta_D = 0).$$

It is easy to verify that $\zeta_1(\mathbf{x})$ possesses complete additivity whatever the model from which it may be derived as a hypothesis. This sufficiency of the loglinear contrast form $\zeta_1(\mathbf{x})$ for complete additivity raises the interesting question of whether the form is also necessary. If $\zeta(\mathbf{x})$ has complete additivity, does it follow that $\zeta(\mathbf{x})$ is necessarily of form $\zeta_1(\mathbf{x})$ in (12.6), a loglinear contrast in the components? The answer can be readily shown to be yes. We omit the proof of this characterization property of loglinear contrast expected responses since, mathematically, it is essentially the same as a characterization property of the Dirichlet class of distributions on the simplex, namely that insistence that a distribution on the simplex possesses complete neutrality for all permutations of the components implies that the distribution is Dirichlet in form. There is indeed a form of duality between logarithms of density functions in simplex distribution theory and forms of expected responses in experiments with mixtures, with the independence property of neutrality corresponding to the response property of additivity; see Chapter 13 for the distribution aspects of this duality.

As for the previously discussed hypotheses, complete additivity is a linear hypothesis within a linear model and clearly implies any hypothesis of partition additivity. In the context of Fig. 12.3, complete additivity requires that the trapezoidal contrast property (12.10) holds for every trapezoid with parallel sides parallel to any side of the triangle.

If the symmetries of model (12.7) and of partition additivity of Definition 12.2 are regarded as undesirable, then an alternative, asymmetric model might take $\zeta_2(\mathbf{x})$ to be a quadratic polynomial in $\mathbf{y} = \text{mlr}(\mathbf{x})$. Such a model would be well suited to partitions based on a single ordering of the components but could prove troublesome if different orderings are involved.

For any experiment with N mixtures and the corresponding quantitative responses, there is clearly a hierarchy of hypotheses which may be considered for testing. For such a lattice of hypotheses we shall again adopt the simplicity postulate of Jeffreys (1961), testing each hypothesis h in turn within the quadratic logcontrast model m of (12.7), starting with the simplest and moving on to more composite hypotheses only on the rejection of all the more simple hypotheses under consideration, and stopping at the first non-rejection. For a quantitative response we use the standard F test with test statistic

$$\frac{(R_{\text{h}} - R_{\text{m}})/(p_{\text{m}} - p_{\text{h}})}{R_{\text{m}}/(N - p_{\text{m}})} \tag{12.11}$$

to be compared against the upper percentage points of the $F(p_{\text{m}} - p_{\text{h}}, N - p_{\text{m}})$ distribution, where $R_{\text{h}}, R_{\text{m}}$ are the residual sums of squares and $p_{\text{h}}, p_{\text{m}}$ the number of parameters under h, m, respectively.

12.5 An application to firework mixtures

To illustrate the hypothesis-testing procedure we analyse the experiment with fireworks of Data 13 and investigate separately the dependence of the quantitative responses of brilliance and vorticity, on the proportions of five ingredients (1, 2, 3, 4, 5). Of these, we recall that 1 and 2 are chemicals included for their light-producing properties, and 3 is primarily a propellent powder for which 4 and 5 act as binders.

Particular interest is in investigating different partitions for the two responses, for brilliance whether there is (1, 2, 3|4, 5) additivity or indeed whether (4, 5) are inactive; for vorticity whether there is

(1, 2|3, 4, 5) additivity and possibly even (1, 2) inactivity. This suggests testing the two lattices of hypotheses in Fig. 12.4 for brilliance and vorticity. Included in each lattice at the first stage is the hypothesis that the responses do not depend at all on the mixture composition. If this hypothesis of purely random variability cannot be rejected by the experimental data, then there is no point in investigating any more composite hypothesis about special effects of the mixture. Such a precautionary first step is commonly overlooked in the literature; see, for example, Cornell (1981, Tables 6.1, 6.6).

Consider the testing of $(x_1, x_2 | x_3, x_4, x_5)$ partition additivity for the vorticity response, a position reached only because the hypotheses at lower levels have been rejected. From Property 12.2 this hypothesis h places the following constraints on the quadratic logcontrast model:

$$\beta_{13} = \beta_{14} = \beta_{15} = \beta_{23} = \beta_{24} = \beta_{25} = 0. \tag{12.12}$$

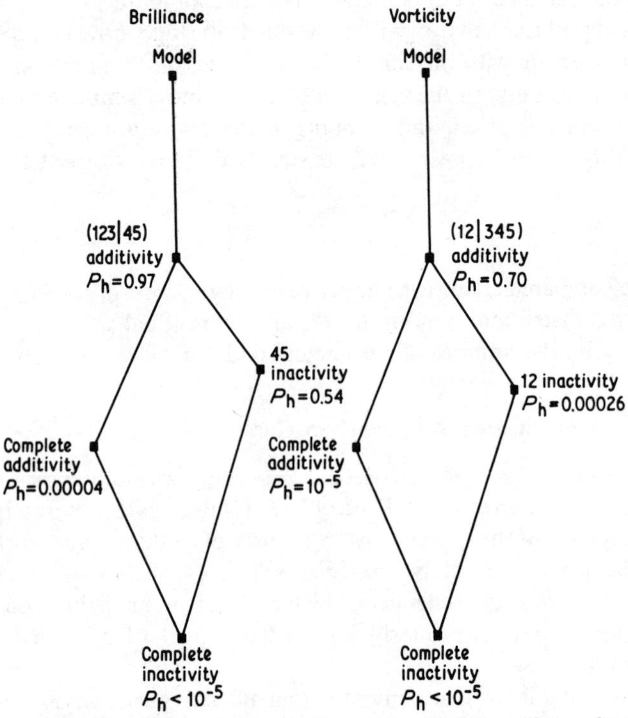

Fig. 12.4. *Lattice of hypotheses for investigation of effects of firework mixture on brilliance and vorticity.*

Here $N = 81$, $p_m = 15$ and $p_h = 9$ since there are six independent constraints in (12.12). Standard linear regression computations lead to

$$R_m = 976.12, \qquad R_h = 1032.89,$$

so that the computed value of the test statistic is 0.64 to be compared against upper percentage points of the $F(6, 66)$ distribution, a clearly non-significant result. As an alternative to this comparison we may use Property 7.2(b) to compute the significance probability in terms of the incomplete beta function, in this instance obtaining a significance probability of 0.70. Thus we would certainly stop at this position on the vorticity lattice.

The complete set of significance probabilities are shown in Fig. 12.4. From these we would conclude that, as a working model for any further investigation of girandole performance, we should incorporate the assumption of (4, 5) inactivity for brilliance and (1, 2 | 3, 4, 5) additivity for vorticity. This is consistent with the role of 4 and 5 as binders for propellent 3, with 1 and 2 the primary light-producing ingredients.

12.6 Classification from compositions

In Section 7.11 we examined the problem of discrimination between hongite and kongite through the assumption that for each type the distribution of logratio composition is normal. Here we re-examine the general problem of classification from compositional information from a different viewpoint. Since much of the activity in this area of compositional data analysis has been towards resolving geological problems we relate the development here within the history and towards the needs of this applied area.

A popular pursuit of numerical geologists is to find ways of classifying rocks on the basis of their geochemical compositions. In such studies there is invariably a 'training set' consisting of the geochemical compositions of rock specimens whose types have been authoritatively identified on the basis of some criterion, such as stratigraphic or potassium–argon age. The objective is then to try to identify characteristics of the geochemical compositions which will hopefully allow classification in the absence of a firm criterion. A recent example is the study by Carr (1981), who compares two main approaches, graphical display of subcompositions and linear discriminant analysis of the crude, untransformed, complete compositions, to the construction of a system for distinguishing between Permian and post-Permian igneous rocks. The training set consists of 65 Permian and 37 post-Permian 10-part compositions, involving the major oxides SiO_2, TiO_2, Al_2O_3, TotFe, MnO, MgO, CaO, Na_2O, K_2O, P_2O_5. In the subcompositional approach scattergrams for selected pairs of oxides, such as SiO_2, TiO_2 in Carr (1981, Fig. 2), or ternary diagrams for triplets of oxides, such as CaO, Na_2O, K_2O in Carr (1981, Fig. 3), are drawn and some simple boundary drawn by eye to separate as far as possible the two types. In the widely employed crude linear discriminant approach, as for example in Le Maître (1976), a routine computational process leads to a linear combination of the crude proportions, whose value, when applied to the geochemical composition of a new rock sample, is used to decide the type. Both approaches are open to a number of serious criticisms.

1. The criterion by which success of a classification method is judged is usually the percentage of correct classifications obtained when the system of classification is applied to the training set. Thus the CNK diagram of Carr (1981) with its 94.1 per cent correct classification appears less good than a crude linear discriminant analysis using the complete compositions with its 95.1 per cent correct classification rate. If, however, we tried to repeat the process with another training set we would be unlikely to obtain the same percentages for correct classifications. Just as in analysis of data in other applications where we encounter variability we have to take account of sampling variability, so in this context we should be asking whether there is a statistically significant difference between the observed percentage correct classification rates. No such significance test has so far been considered for comparisons of this kind.

A related criticism is that such percentages, obtained by applying a system of classification constructed from a training set back to the specimens in the training set, give an invariably optimistic picture of what the correct classification rate is likely to be when applied to new rock samples. This phenomenon of 'resubstitution bias' is well recognized in other disciplines such as medical diagnosis (Lachenbruch and Mickey, 1968), where similar problems involving statistical classification are involved.

2. In some ways a more fundamental issue is whether any statistical use of geochemical data should be aimed at actually assigning a rock sample to a type. In addition to the geochemical composition of the

rock sample there will usually be other information which may be of use in classification. In such circumstances, therefore, it would seem reasonable to use the compositional data to assign probabilities to each of the possible types and so allow the possibility of modifying these probabilities on the basis of the further information. The use of graphical methods such as ternary diagrams does not lend itself easily to the production of such probabilities. Standard linear discriminant analysis does, but only if we can provide some realistic incidence rates at which the various types are likely to be referred to the system.

3. An essential ingredient of crude linear discriminant analysis is the covariance matrix constructed from the crude proportions. We have seen in Section 3.3 that the crude covariance structure has little validity in providing useful information about the nature of dependence between the various components of the composition. Moreover, in Section 8.4 we have seen how inadequate this covariance matrix can be in coping with commonly occurring data sets which are curved or concave in shape. Since crude linear discriminant analysis subsumes that the scattergrams of data points for each type are approximately elliptical, there must be doubts about its appropriateness for classification problems with compositional data, particularly in the assessment of probabilities assigned to the types. One way out of this difficulty is to adopt the approach of Section 7.11 and to apply linear discriminant analysis to logratios of components rather than to the crude components.

4. Dawid (1976) has shown that, in the context of medical diagnosis, linear discriminant analysis can be sensitive to variations of the referral process. Although geological classification is in many ways very different from medical diagnosis, the formal mathematical and statistical structure is similar and the collection of rock specimens is often such a poorly described process that one cannot dismiss this criticism of the linear discriminant approach too lightly. Dawid, in this context, argues in favour of modelling directly the probability of type for a given vector of diagnostic features, and it is on the adaptation of this form of classification for the special features of compositional data that we now concentrate. We shall find that it allows us to assess various aspects of geological classification in a more satisfactory statistical way than the other methods just criticized.

A well-established method of modelling type probabilities for given vector $\mathbf{y}$ of diagnostic features is logistic discriminant analysis (Cox, 1966; Day and Kerridge, 1967; Anderson, 1972; Dawid, 1976). This simply adopts a suitable mathematical form, the additive logistic function of a linear combination of the components of **y**, as the type probabilities. Thus if t denotes type, either type 1 or 2, and **y** denotes a d-dimensional vector of diagnostic features, the logistic form for the probabilities that a case is of type 1, or of type 2, given that its measurement vector is **y**, are

$$p(t=1\,|\,y_1,\ldots,y_d)=\frac{\exp(\beta_0+\beta_1 y_1+\cdots+\beta_d y_d)}{1+\exp(\beta_0+\beta_1 y+\cdots+\beta_d y_d)}, \tag{12.13}$$

$$p(t=2\,|\,y_1,\ldots,y_d)=\frac{1}{1+\exp(\beta_0+\beta_1 y_1+\cdots+\beta_d y_d)}. \tag{12.14}$$

For k types we require a more general logistic function, namely

$$\begin{aligned} p(t=c\,|\,\mathbf{y}) = \exp(\beta_{c0}+\beta_{c1}y_1+\cdots+\beta_{cd}y_d)/\{1 \\ +\exp(\beta_{10}+\beta_{11}y_1+\cdots+\beta_{1d}y_d)+\cdots \\ +\exp(\beta_{k-1,0}+\beta_{k-1,1}y_1+\cdots+\beta_{k-1,d}y_d)\} \\ (c=1,\ldots,k-1), \end{aligned} \tag{12.15}$$

$$\begin{aligned} p(t=k\,|\,\mathbf{y}) = 1/\{1+\exp(\beta_{10}+\beta_{11}y_1+\cdots+\beta_{1d}y_d)+\cdots \\ +\exp(\beta_{k-1,0}+\beta_{k-1,1}y_1+\cdots+\beta_{k-1,d}y_d)\}. \end{aligned} \tag{12.16}$$

When, as in geological classification from geochemistry, the measurements which are to be used for discrimination form a D-part composition, we are again faced with the unit-sum constraint problem. As in previous situations we can use the d-dimensional vector $\mathbf{y}=\mathrm{alr}(\mathbf{x})$ of logratios directly in (12.13)–(12.16) and apply the standard available logistic discriminant analysis procedures. We could express this form of modelling much more concisely by relating the k-vector **p** of type probabilities $p(t\,|\,\mathbf{y})$ $(t=1,\ldots,k)$ to the composition **x** by use of the generalized transformations alg and alr:

$$\mathbf{p}=\mathrm{alg}\{\boldsymbol{\beta}_0+\boldsymbol{B}\,\mathrm{alr}(\mathbf{x})\},$$

where $\boldsymbol{\beta}_0=(\beta_{10},\ldots,\beta_{k-1,0})$ and $\boldsymbol{B}=[\beta_{ci}{:}c=1,\ldots,k-1;i=1,\ldots,d]$. Technically we have here again a compositional covariate with a binomial response in (12.13)–(12.14) and a categorical response in (12.15)–(12.16).

We can rewrite the expression for these probabilities more

symmetrically in terms of the components as

$$p(t=1|\mathbf{x}) = \frac{\exp(\beta_0 + \beta_1 \log x_1 + \cdots + \beta_D \log x_D)}{1 + \exp(\beta_0 + \beta_1 \log x_1 + \cdots + \beta_D \log x_D)}, \quad (12.17)$$

where the parameters are now subject to the constraint

$$\beta_1 + \cdots + \beta_D = 0. \quad (12.18)$$

Note that this constraint is introduced simply as a mathematical device to gain the advantages of symmetry. It is in no way related to the unit-sum constraint and has no interpretative difficulties. The advantage of this symmetric version is that it is nicely poised for the investigation of the adequacy of some subcomposition as a substitute for the complete composition in the process of classification. For example, if we wish to ask whether the subcomposition formed from the subvector $(x_1, \ldots, x_c)$ is an adequate substitute for the complete composition, then in terms of the logistic discriminant model we are asking whether the hypothesis that

$$\beta_\mathrm{C} = \cdots = \beta_\mathrm{D} = 0 \quad (12.19)$$

is tenable.

This expression of the adequacy of a subcomposition as a parametric hypothesis makes available the whole statistical theory of parametric hypothesis testing. In effect, what we have to do to test the adequacy of a subcomposition for classification purposes is to carry out two logistic discriminant analyses, one using the whole composition and the other using the subcomposition as the discriminating vector of measurements, in each case determining the value of the maximized likelihood, say l_C and l_S for the whole composition and for the subcomposition respectively. Then a straightforward test (Wilks, 1938) of whether the restriction to the subcomposition is reasonable is to determine whether $2\log(l_\mathrm{C}/l_\mathrm{S})$ exceeds the upper percentage point of the $\chi^2(d-c)$ distribution.

12.7 An application to geological classification

We illustrate the techniques briefly for two types only, first for discriminating between hongite and kongite. With the transformed logratio composition **y** we may write the logistic discriminant model in form (12.13) with $d=4$, from which, by standard maximum likelihood estimation methods, we obtain maximum likelihood estimates of $\boldsymbol{\beta}$ as

$$\hat{\boldsymbol{\beta}} = (-258.5, 100.8, 113.1, 110.0, -162.1). \qquad (12.20)$$

We then regard classification as consisting of the assignment of type probabilities to new rock samples and for this purpose, for the many reasons advocated in Aitchison and Dunsmore (1975), we use the predictive diagnostic method. For the logistic discriminant case, this can be expressed in the approximate form (Lauder, 1978):

$$\begin{aligned} p(\text{hongite}|\mathbf{x}, \text{data}) &= \Phi\{\hat{\boldsymbol{\beta}}'\mathbf{y}^*/\sqrt{(2.89 + \mathbf{y}^{*\prime}\mathbf{V}\mathbf{y}^*)}\}, \qquad (12.21)\\ p(\text{kongite}|\mathbf{x}, \text{data}) &= 1 - p(\text{hongite}|\mathbf{x}, \text{data}), \end{aligned}$$

where $\Phi(\cdot)$ is the $\mathcal{N}(0, 1)$ distribution function, $\mathbf{y}^*$ is the extended logratio vector $[1 \quad \mathbf{y}']'$ and $\mathbf{V}$ is the estimated covariance matrix of the estimator of $\boldsymbol{\beta}$, here

$$\mathbf{V} = \begin{bmatrix} 7126 & -2785 & -3113 & -3023 & 4487 \\ -2785 & 1095 & 1203 & 1173 & -1747 \\ -3113 & 1203 & 1385 & 1336 & -1974 \\ -3023 & 1173 & 1336 & 1292 & -1912 \\ 4487 & -1747 & -1974 & -1912 & 2835 \end{bmatrix}. \qquad (12.22)$$

When reapplied to the 25 hongite and 25 kongite cases of Data 1 and 2 the probabilities assigned to the type hongite are distributed as shown in Table 12.1. We thus seem to have a reasonable method of classification even making allowances for the well-known fact that assessing a system by resubstitution of the cases from which the system was constructed always gives an overly optimistic view. Application to the new cases of Table 7.11 produces diagnostic assessments similar to those obtained by the predictive linear discriminant method of Section 7.11. The complete failure of the corresponding crude analysis in which $y_1, \ldots, y_d$ in (12.13) are replaced by any d of the raw proportions is demonstrated in Table 12.1 by the very poor discrimination.

For the classification of hongite and kongite we can readily test the effectiveness of each of the five 4-part subcompositions obtained by dropping out just one of the components. Table 12.2 shows the maximized loglikelihoods, and it is clear that we have highly significant test statistics when compared against upper percentage points of the $\chi^2(1)$ distribution. It is thus all too clear that attempting to use any subcomposition for this classification purpose is

Table 12.1. *Distributions of predictive probabilities for classification as type hongite for the 25 hongite and 25 kongite cases of Data 1 and 2 by the crude and logratio methods*

Probability interval	Crude method Hongite	Crude method Kongite	Logratio method Hongite	Logratio method Kongite
0–0.05				
0.05–0.10				4
0.10–0.15				6
0.15–0.20			1	4
0.20–0.25				1
0.25–0.30				2
0.30–0.35				2
0.35–0.40			1	1
0.40–0.45				1
0.45–0.50	13	17	1	2
0.50–0.55	12	8	1	1
0.55–0.60				1
0.60–0.65				
0.65–0.70			2	
0.70–0.75			1	
0.75–0.80			4	
0.80–0.85			3	
0.85–0.90			6	
0.90–0.95			5	
0.95–1.00				

Table 12.2. *Maximized loglikelihoods and values of the test statistics for investigating the discriminatory power of subcompositions in the classification of hongite and kongite*

Composition or subcomposition	*Maximized loglikelihood*	$2(l_C - l_S)$
(A, B, C, D, E)	− 11.42	
Omitting A	− 31.25	39.7
B	− 34.00	45.2
C	− 34.35	45.9
D	− 34.06	45.3
E	− 34.44	46.0

inadvisable. This can easily be verified by assessing the classification probabilities obtained by using a subcomposition.

For this important technique we also report a classification study of the Permian and post-Permian major-oxide compositions used to motivate the methodology of Section 12.6 and contrast our findings with those of Carr (1981).

We first examine the two subcompositions picked out by Carr (1981), namely the (SiO_2, TiO_2) and the (CaO, Na_2O, K_2O) subcompositions, in his study of methods of classification of Permian and post-Permian igneous rocks. First we report that logistic discriminant analysis applied with the full composition yields a maximized loglikelihood value -20.9 and that the number of misclassifications is three Permian misclassified as post-Permian and six post-Permian misclassified as Permian. This overall misclassification rate seems higher than that claimed for graphical methods by Carr. It should be emphasized that visual methods, by adopting a sufficiently complex boundary, could easily devise a 100 per cent correct classification when applied to the training set.

The true test of a system comes when it is subjected to proper statistical evaluation such as an assessment of its performance to prospective samples or when, as we aim to do here, the adequacy of subcompositional analysis is tested by proper statistical tests of parametric hypotheses. The maximized loglikelihoods from logistic discriminant analysis using the (SiO_2, TiO_2) and (CaO, Na_2O, K_2O) subcompositions are -62.6 and -46.0, respectively. As described in the previous section, we test the adequacy of these subcompositions by comparing the values of the two test statistics, here $2 \times (-20.9 + 62.6) = 83.4$ and, similarly, 50.2, against the upper percentage points of the $\chi^2(8)$ and $\chi^2(7)$ distributions, respectively. Thus we would strongly reject (with significance probability $P < 0.001$) the hypotheses associated with the adequacy of each of these two subcompositions. It would therefore seem extremely inadvisable to exclude so many of the oxides from consideration in the construction of a classification system.

With such a simple model-testing procedure we can in fact carry out a systematic investigation to find out just what we can get away with in the process of discarding oxides. As a first step we consider the ten 9-part subcompositions which can be formed from the full 10-part composition by omitting only one of the oxides. The maximized loglikelihoods for these ten subcompositions are shown in Table 12.3.

Table 12.3. *Maximized loglikelihoods for 9-part subcompositions formed by omissions of the specified oxides*

SiO_2	TiO_2	Al_2O_3	TotFe	MnO	MgO	CaO	Na_2O	K_2O	P_2O_5
−25.5	−21.2	−21.0	−33.9	−24.8	−23.3	−21.3	−25.6	−28.5	−21.2

Applying the $\chi^2(1)$ test at the 0.05 per cent significance level to each of these shows immediately that SiO_2, TotFe, MnO, Na_2O, K_2O contribute significantly to the model fit and should not be dropped, whereas any one of TiO_2, Al_2O_3, CaO, P_2O_5 could be dropped without any significant reduction in the quality of the fit. We cannot infer from this that we could drop all four of these oxides, but our next step is clearly to test the hypothesis that the subcomposition (SiO_2, TotFe, MnO, MgO, Na_2O, K_2O) is not significantly poorer than the full composition. Since among those six oxides the omission of MgO from the full composition led to the smallest change in maximized loglikelihood, it seems sensible also to consider (SiO_2, TotFe, MnO, CaO, Na_2O, K_2O), dropping out MgO and bringing in CaO which in Table 12.3 is MgO's nearest rival. The maximized loglikelihoods for these two subcompositions are −22.1 and −24.5, leading to test statistic values of 2.5 and 7.3, neither significant when compared with 9.49, the upper 5 per cent point of the $\chi^2(4)$ distribution. It can also be readily shown that dropping any oxide from either of those two subcompositions results in a significant deterioration of the gooodness of fit.

From such testing we can draw the following conclusions. No 5-part subcomposition gives as adequate an explanation of the relation of type to oxide variation as the full composition. There are, however, two 6-part subcompositions which do not differ significantly in this respect from the whole composition: these are

$$(SiO_2, \text{TotFe}, \text{MnO}, \text{MgO}, Na_2O, K_2O),$$
$$(SiO_2, \text{TotFe}, \text{MnO}, \text{CaO}, Na_2O, K_2O),$$

with the exponents in (12.13)–(12.14) taking the estimated form

$$-30.4 + 13.5 \log SiO_2 - 11.2 \log \text{TotFe} - 2.46 \log \text{MnO} + 3.13 \log \text{MgO} - 8.81 \log Na_2O + 5.83 \log K_2O, \quad (12.23)$$

$$-32.3 + 11.9 \log SiO_2 - 9.11 \log \text{TotFe} - 2.95 \log \text{MnO} + 3.66 \log CaO_2 - 7.68 \log Na_2O + 4.18 \log K_2O, \quad (12.24)$$

respectively. In other words, we could drop either TiO_2, Al_2O_3, CaO, P_2O_5 or TiO_2, Al_2O_3, MgO, P_2O_5 from the composition without deterioration of the quality of the statistical fit. With (12.24) the number of misclassifications, is four Permian misclassified as post-Permian and seven post-Permian as Permian; with (12.24) the corresponding numbers are two and six, respectively. Such conclusions are clearly directly opposed to those of Carr (1981) which seem to indicate that a number of 2- or 3-part subcompositions would be perfectly adequate for the task.

12.8 Bibliographic notes

Compositions in the role of covariates seem to have been considered mainly in relationship to experiments with mixtures, first identified by Quenouille (1953) and later developed by Scheffé (1958, 1961, 1963), Quenouille (1959), Becker (1968, 1978), Cox (1971). Much of this work involved attempts to provide interpretations of the parameters of the model and is well documented by Cornell (1981). The logcontrast form of expected response and the inactivity and additivity hypotheses were introduced by Aitchison and Bacon-Shone (1984). Their hypothesis of inactivity of a subvector is a simple generalization of the idea of a single inactive component introduced by Becker (1968) and developed by Cox (1971). Darroch and Waller (1985) introduced a model for experiments with 3-part mixtures with expected response

$$\zeta(\mathbf{x}) = \alpha(\mathbf{x}_1) + \beta(\mathbf{x}_2) + \gamma(\mathbf{x}_3),$$

a sum of separate functions of the three components, and termed this an additive model. Such modelling seems to be still subject to the mixture variation difficulty described in Section 3.5 and does not allow investigation of a partition additivity hypothesis of Definition 12.2.

Problems with a compositional covariate and categorical response seem to have been first considered by Aitchison (1984a) in relation to the classification of rocks. Situations in which the covariate and response are both compositions appear for the first time in this monograph.

Problems

12.1 Suppose that in the investigation of the white-cell compositions of Data 11 it is now realized that the image analysis is very accurate and free from observer variability. For purposes of investigating the feasibility of constructing a diagnostic system, a clinician wishes to augment his image-determined compositions by calibrating a number of blood samples for which only microscopically determined compositions have been recorded to obtain their image-determined equivalents.

What calibrations would you suggest for the following two microscopically determined (G, L, M) compositions:

(a) (0.784, 0.181, 0.035),
(b) (0.350, 0.225, 0.425)?

What indication of the reliability of these calibrated values can you provide?

12.2 The microhardness (kg/mm^2) of a special glass is believed to depend on the relative proportions of Ge, Sb, Se used in its manufacture. Data 39 shows the recorded microhardness of 18 specimens together with the mixture of Ge, Sb, Se used in their manufacture. Construct as complete a lattice as possible for the investigation of hypotheses of inactivity and additivity. What conclusions do you reach on the dependence of microhardness on the constituents?

12.3 For the bayesite fibreboard experiment of Data 21 attempt to answer the questions you posed in Problem 1.6 through the logcontrast modelling techniques of Section 12.4.

12.4 Investigate the dependence of the porosity of the coxite specimens of Data 4 on mineral composition and depth.

12.5 Reconsider the questions you posed in Problem 1.1 in relation to the possibility of constructing a differential diagnostic system from Data 16 and try to formulate them within the modelling of Section 12.6.

For a system using the full 4-part compositions, what diagnostic probabilities would you assign to the new cases C1–C6?

Investigate whether any subcomposition may be reasonably used in place of the full composition for the purposes of differential diagnosis.

12.6 Investigate the possibility of using Data 3 and 4 to devise a classification system for boxite and coxite, in particular the adequacy of using some subcomposition for this purpose.

12.7 Suppose that the proportions of patients referred to a clinic specializing in two forms 1 and 2 of a particular disease are π_1, π_2, where $\pi_1 + \pi_2 = 1$. Suppose further that the distributions of a D-part serum protein composition $\mathbf{x}$ are known $p(\mathbf{x}|t=1)$, $p(\mathbf{x}|t=2)$ for the given forms $t = 1, 2$, respectively. Then, for a case with serum protein composition $\mathbf{x}$, Bayes's formula provides diagnostic probabilities

$$p(t=1|\mathbf{x}) = \frac{\pi_1 p(\mathbf{x}|t=1)}{\pi_1 p(\mathbf{x}|t=1) + \pi_2 p(\mathbf{x}|t=2)},$$

$$p(t=2|\mathbf{x}) = \frac{\pi_2 p(\mathbf{x}|t=2)}{\pi_1 p(\mathbf{x}|t=1) + \pi_2 p(\mathbf{x}|t=2)}.$$

(a) Show that if $p(\mathbf{x}|t=1)$ and $p(\mathbf{x}|t=2)$ are of $\mathscr{L}^d(\boldsymbol{\mu}_1, \boldsymbol{\Sigma})$ and $\mathscr{L}^d(\boldsymbol{\mu}_2, \boldsymbol{\Sigma})$ forms then $p(t=1|\mathbf{x})$ is of the logistic model form (12.17), and identify $\beta_0, \ldots, \beta_D$ in terms of $\boldsymbol{\mu}_1, \boldsymbol{\mu}_2, \boldsymbol{\Sigma}$.
(b) Show that if $p(\mathbf{x}|t=1)$ and $p(\mathbf{x}|t=2)$ are of $\mathscr{D}^d(\boldsymbol{\alpha}_1)$ and $\mathscr{D}^d(\boldsymbol{\alpha}_2)$ forms then $p(t=1|\mathbf{x})$ is also of form (12.17) but without condition (12.18). Identify $\beta_0, \ldots, \beta_D$ in terms of $\boldsymbol{\alpha}_1, \boldsymbol{\alpha}_2$.

12.8 A radiologist claims that the ratio of heart width w_H to thorax width w_T as measured on a heart X-ray is of considerable diagnostic value. Test this claim in relation to the ratio's ability to differentiate between the normotensive and hypertensive males of Data 29, and in particular examine whether the ratio is as effective as use of the pair of measurements w_H and w_T.

CHAPTER 13

Further distributions on the simplex

No doubt the next chapter
in my book of transformations
is already written.
I am not done with my changes.

Stanley Kunitz: *The Layers*

13.1 Some generalizations of the Dirichlet class

The realization that the Dirichlet class leans so heavily towards independence has prompted a number of authors (Connor and Mosimann, 1969; Darroch and James, 1974; Mosimann, 1975b; James, 1981; James and Mosimann, 1980) to search for generalizations of the Dirichlet class with less independence structure. Their efforts have been met with only limited success and it has remained an open problem to find a useful parametric class of distributions on $\mathscr{S}^d$ which contains the Dirichlet class but also contains distributions which do *not* satisfy any of the simplex independence properties. We describe here, only briefly and for the sake of completeness, two of these generalizations.

If we remove the requirement of equal scale parameters in the relationship of the gamma basis to the $\mathscr{D}^d(\boldsymbol{\alpha})$ composition in Property 3.3 we obtain a first generalization, which we first define in terms of its density function.

Definition 13.1 A D-part composition $\mathbf{x}$ with density function

$$\frac{\Gamma(\alpha_1+\cdots+\alpha_D)}{\Gamma(\alpha_1)\cdots\Gamma(\alpha_D)}\cdot\frac{\prod_{i=1}^{D}\beta_i^{\alpha_i}x_i^{\alpha_i-1}}{(\beta_1 x_1+\cdots+\beta_D x_D)^{\alpha_1+\cdots+\alpha_D}}$$

on $\mathscr{S}^d$ is said to have a *scaled Dirichlet distribution*, denoted by $\mathscr{D}^d(\boldsymbol{\alpha}, \boldsymbol{\beta})$, where $\boldsymbol{\alpha}, \boldsymbol{\beta} \in \mathscr{R}^D_+$.

The nature of the generalization is obvious since $\mathscr{D}^d(\boldsymbol{\alpha}, \boldsymbol{\beta})$ is clearly identical to $\mathscr{D}^d(\boldsymbol{\alpha})$ if and only if $\beta_1 = \cdots = \beta_D$. The $\mathscr{D}^d(\boldsymbol{\alpha}, \boldsymbol{\beta})$ class of distributions on $\mathscr{S}^d$ has the following simple basis property.

Property 13.1 If $\mathbf{w}$ is a D-part basis of independent components with w_i of $\mathrm{Ga}(\alpha_i, \beta_i)$ form $(i = 1, \ldots, D)$, as in Definition 3.3, then $\mathbf{x} = \mathscr{C}(\mathbf{w})$ is $\mathscr{D}^d(\boldsymbol{\alpha}, \boldsymbol{\beta})$ and is independent of $\beta_1 w_1 + \cdots + \beta_D w_D$, which is distributed as $\mathrm{Ga}(\alpha_1 + \cdots + \alpha_D, 1)$.

This scaled class, which involves simply a change of scale from $\mathscr{D}^d(\boldsymbol{\alpha})$, presents the same awkward features of statistical analysis as $\mathscr{D}^d(\boldsymbol{\alpha})$ and still retains much of the overstrong independence structure, such as independence of subcompositions formed from non-overlapping subvectors.

In their attempt to develop classes of distributions in $\mathscr{S}^d$ for the study of neutrality, as defined in Sections 10.3 and 10.4, Connor and Mosimann (1969) introduced a compositional distribution related to a basis of independent beta-distributed components. The beta distribution over $\mathscr{R}^1_+$ is defined in Definition 3.4.

The Connor–Mosimann generalization of the Dirichlet class is first defined in terms of its density function.

Definition 13.2 A D-part composition $\mathbf{x}$ with density function

$$\prod_{i=1}^{d} \frac{x_i^{\alpha_i - 1}}{\mathrm{B}(\alpha_i, \beta_i)} \prod_{i=1}^{d-1} (1 - x_1 - \cdots - x_i)^{\beta_i - (\alpha_{i+1} + \beta_{i+1})}$$
$$\times (1 - x_1 - \cdots - x_d)^{\beta_d - 1} \qquad (\mathbf{x} \in \mathscr{S}^d)$$

where $\boldsymbol{\alpha}, \boldsymbol{\beta} \in \mathscr{R}^d_+$, is said to have a *Connor–Mosimann distribution*, denoted by $\mathscr{E}^d(\boldsymbol{\alpha}, \boldsymbol{\beta})$.

We can identify the form of generalization of the unscaled Dirichlet class $\mathscr{D}^d(\boldsymbol{\alpha})$ by noting that the $d - 1$ constraints,

$$\begin{aligned} \beta_i &= \alpha_{i+1} + \beta_{i+1} \\ &= \alpha_{i+1} + \cdots + \alpha_d + \beta_d \qquad (i = 1, \ldots, d-1), \end{aligned} \tag{13.1}$$

placed on the $2d$ parameters of $\mathscr{E}^d(\boldsymbol{\alpha}, \boldsymbol{\beta})$ produce an unscaled Dirichlet density function with parameters $\alpha_1, \ldots, \alpha_d, \beta_d$. The origin of the $\mathscr{E}^d(\boldsymbol{\alpha}, \boldsymbol{\beta})$ class lies, however, more in its relationship to independent Be^+ distributions through a transformation between $\mathscr{S}^d$ and $\mathscr{R}^d_+$.

Property 13.2 Let $\mathbf{w}^{(d)} \in \mathscr{R}^d_+$ and $\mathbf{x} \in \mathscr{S}^d$ be related through the transformations

$$w_i = x_i/(1 - x_1 - \cdots - x_i) \qquad (i = 1, 2, \ldots, d),$$

$$x_i = w_i \Big/ \prod_{j=1}^{i} (1 + w_j) \qquad (i = 1, \ldots, d), \qquad x_D = 1 \Big/ \prod_{j=1}^{d} (1 + w_j).$$

Then $\mathbf{x}$ is distributed as $\mathscr{E}^d(\boldsymbol{\alpha}, \boldsymbol{\beta})$ if and only if $\mathbf{w}$ has independent components with w_i distributed as $\mathrm{Be}^+(\alpha_i, \beta_i)$ $(i = 1, \ldots, d)$.

This $\mathscr{E}^d(\boldsymbol{\alpha}, \boldsymbol{\beta})$ class is really just as intractable as earlier generalizations, particularly with respect to statistical analysis. Moreover, the class still retains a strong independence structure, so much so that James (1981), in a search for some further generalization, comments that there remains in the literature a lack of tractable, rich distributions for random proportions which are not neutral.

Thus it seems that attempts to obtain a suitably rich class of distributions containing the simple Dirichlet class have so far failed.

13.2 Some generalizations of the logistic normal classes

In our work in this monograph we have concentrated on two transformed normal classes of distributions on the simplex $\mathscr{S}^d$:

1. the additive logistic normal class $\mathscr{L}^d(\boldsymbol{\mu}, \boldsymbol{\Sigma})$,
2. the multiplicative logistic normal class $\mathscr{M}^d(\boldsymbol{\mu}, \boldsymbol{\Sigma})$,

and the substantial technique of Section 6.14 of forming partitioned logistic normal classes based on the classes (1) and (2). Any one-to-one transformation between $\mathscr{R}^d$ and $\mathscr{S}^d$ will of course, when applied to the $\mathscr{N}^d(\boldsymbol{\mu}, \boldsymbol{\Sigma})$ class in $\mathscr{R}^d$, induce a class of distributions in $\mathscr{S}^d$. For example, we could consider what may be termed the *hybrid* logistic transformation from $\mathscr{R}^d$ to $\mathscr{S}^d$:

$$x_1 = \exp(y_1)/\{1 + \exp(y_1)\}$$

$$x_i = \frac{\exp(y_i)}{\{1 + \exp(y_1) + \cdots + \exp(y_{i-1})\}\{1 + \exp(y_1) + \cdots + \exp(y_i)\}} \qquad (i = 2, \ldots, d)$$

$$x_D = 1/\{1 + \exp(y_1) + \cdots + \exp(y_d)\} \qquad (13.2)$$

with inverse transformation from $\mathscr{S}^d$ to $\mathscr{R}^d$:

$$y_1 = \log \frac{x_1}{1 - x_1},$$

$$y_i = \log \frac{x_i}{(1 - x_1 - \cdots - x_{i-1})(1 - x_1 - \cdots - x_i)} \qquad (i = 2, \ldots, d). \tag{13.3}$$

The term 'hybrid' arises from the mixed additive–multiplicative form in which the exponents occur in the denominators of (13.2). An interesting feature is that the sum $x_{c+1} + \cdots + x_D$ can be expressed as $1/\{1 + \exp(y_1) + \cdots + \exp(y_c)\}$ and so is a function of $\mathbf{y}^{(c)}$ only. An immediate implication is that $x_1 + \cdots + x_c$ is also a function of $\mathbf{y}^{(c)}$ only. Since $\mathbf{x}^{(c)}$ is a function of $\mathbf{y}^{(c)}$ only, it follows immediately that $\mathscr{C}(\mathbf{x}^{(c)})$ is a function $\mathbf{y}^{(c)}$ only. This could prove useful in some applications.

Any logistic normal class can be extended by the well-known Box–Cox transformation technique (Box and Cox, 1964). This technique is aimed at finding from within a flexible class of transformations one which in a maximum-likelihood sense gives the nearest-to-normal pattern of variability in the transformed data. For a positive variable w if we make the transformation

$$y = \frac{w^\lambda - 1}{\lambda} \tag{13.4}$$

and assume that y is normally distributed, then we can write down the likelihood associated with a data set. If we then choose the value of λ which maximizes this likelihood we have the best transformation to normality within this class of transformations. Since the limiting case $\lambda \to 0$ leads to the logarithmic transformation, we have certainly a much greater opportunity of obtaining an adequate fit than restricting attention to the logarithmic transformation alone.

The modification of the Box–Cox class of transformations to meet the special requirements of compositional data then takes the following form. First the transformation from $\mathscr{S}^d$ sets

$$y_i = \frac{(x_i/x_D)^\lambda - 1}{\lambda} \qquad (i = 1, \ldots, d), \tag{13.5}$$

where $\lambda > 0$, and $\lambda \to 0$ corresponds to the logratio transformation. Note that there is a theoretical difficulty here, as there is in the

Box–Cox transformations generally, that the range space of each y_i is not the whole of $\mathscr{R}^1$ but only the interval $(-1/\lambda, \infty)$. Nevertheless, although the transformation is not strictly from $\mathscr{S}^d$ to $\mathscr{R}^d$, it may still prove a useful practical tool just as the Box–Cox transformation does in its original form. The inverse transformation to $\mathscr{S}^d$ is then

$$\mathbf{x} = \mathscr{C}\{(1+\lambda y_1)^{1/\lambda}, \ldots, (1+\lambda y_d)^{1/\lambda}, 1\}, \tag{13.6}$$

with the restriction that $-1/\lambda < y_i \; (i = 1, \ldots, d)$.

Just as in the multivariate normal form of the Box–Cox transformation different values of λ may be used for different marginal transformations, so the same generalization may be made here, with (13.5) replaced by

$$y_i = \frac{(x_i/x_D)^{\lambda_i} - 1}{\lambda_i} \qquad (i = 1, \ldots, d), \tag{13.7}$$

and (13.6) replaced by

$$\mathbf{x} = \mathscr{C}\{(1+\lambda_1 y_1)^{1/\lambda_1}, \ldots, (1+\lambda_d y_d)^{1/\lambda_d}, 1\}. \tag{13.8}$$

The transformations (13.5) and (13.7) can be applied to a composition with zero components provided that x_D is positive. Thus for a compositional data set which has an ever-present part, say D, the Box–Cox transformation provides an attractive substitute for the logratio transformation. Unfortunately the resulting transformed normal distributions on the simplex have hardly any of the simple properties of the $\mathscr{L}^d$ class. For example, if $\{(x_1/x_3)^\lambda, (x_2/x_3)^\lambda\}$ has a bivariate normal distribution, then $(x_1/x_2)^\lambda$ is certainly not normally distributed. Thus such transformed normal classes may have only a limited application.

13.3 Recapitulation

In any general study of distributions on the d-dimensional positive simplex $\mathscr{S}^d$ the Dirichlet class $\mathscr{D}^d(\boldsymbol{\alpha})$ of Definition 3.2 must play a central role. This importance arises mainly from its strong independence structure, surely the ultimate achievable by patterns of variability within such a constrained sample space as the simplex: every Dirichlet distribution has, for example, all the independence properties defined in Chapter 10. As pointed out in Section 13.1, attempts so far to extend the Dirichlet class to include members which do not satisfy such simplex independence properties have largely failed.

The recent introduction of the logistic normal classes has provided a framework within which at least some of these independence properties can be expressed as parametric hypotheses and so tested. For example, the additive logistic normal $\mathscr{L}^d(\boldsymbol{\mu}, \boldsymbol{\Sigma})$ of Definition 6.2 possesses complete subcompositional independence under the hypothesis that $\boldsymbol{\Sigma}$ takes the special parametric form of Property 10.3. Unfortunately none of these logistic normal classes contains the Dirichlet class so that some of the especially strong forms of independence, such as complete neutrality for any ordering of the parts, which characterizes the Dirichlet class, cannot be tested within logistic normal classes. This separateness of these classes is therefore a major theoretical obstacle to the analysis of compositional data.

There thus remain two major open problems. Those who wish to extend the Dirichlet class have still to invent a more general class which allows sufficient dependence among components of a composition $\mathbf{x}$. Those who advocate the use of logistic normal classes have still to produce a class of distributions within which extreme forms of independence such as complete neutrality for any ordering of parts can be discussed. If we can define a new and more general class, the $\mathscr{A}^d(\boldsymbol{\alpha}, \boldsymbol{B})$ class, of distributions which contains both the Dirichlet and logistic normal classes as special cases, we shall have solved in a substantial way both these open problems.

13.4 The $\mathscr{A}^d(\alpha, B)$ class

From the $\mathscr{D}^d(\boldsymbol{\alpha})$ class with its D parameters and the $\mathscr{L}^d(\boldsymbol{\mu}, \boldsymbol{\Sigma})$ class with its $\frac{1}{2}d(D+2)$ parameters, it would be possible to form a general class as a $(\lambda, 1-\lambda)$ mixture of the two classes with $\frac{1}{2}D(D+3)$ parameters $(\lambda, \boldsymbol{\alpha}, \boldsymbol{\mu}, \boldsymbol{\Sigma})$. Such a generalization would, however, be at the expense of a substantial increase in the number of the parameters, $D+1$ more than for the $\mathscr{L}^d$ class, with no advantage in their interpretability. Moreover, it would introduce the many statistical difficulties associated with mixture distributions.

In contrast, the generalization which we now introduce is parsimonious in having only one more parameter than the $\mathscr{L}^d$ class, the parameters have simple interpretations in relation to independence hypotheses of interest, and satisfactory forms of statistical analysis can be devised.

For the density function $p_{\mathscr{D}}(\mathbf{x}|\boldsymbol{\alpha})$ of the $\mathscr{D}^d(\boldsymbol{\alpha})$ distribution we have

$$\log p_{\mathscr{D}}(\mathbf{x}|\boldsymbol{\alpha}) = k(\boldsymbol{\alpha}) + \sum_{i=1}^{D} (\alpha_i - 1)\log x_i, \tag{13.9}$$

with $\alpha_i > 0$ $(i = 1, \ldots, D)$. This loglinear form of the logarithm of the density function can indeed be made the target in a proof of the complete neutrality characterization of the Dirichlet class: a composition with complete neutrality for every ordering of its parts must be Dirichlet distributed. The corresponding form for the $\mathscr{L}^d(\boldsymbol{\mu}, \boldsymbol{\Sigma})$ density function can be expressed in a more convenient symmetrical reparametrized version as follows:

$$\log p_{\mathscr{L}}(\mathbf{x} \mid \boldsymbol{\alpha}, \boldsymbol{B}) = k(\boldsymbol{\alpha}, \boldsymbol{B}) + \sum_{i=1}^{D} (\alpha_i - 1) \log x_i - \frac{1}{2} \sum_{i=1}^{d} \sum_{j=i+1}^{D} \beta_{ij} (\log x_i - \log x_j)^2, \qquad (13.10)$$

where $\alpha_1 + \cdots + \alpha_D = 0$. There are simple relationships between the two forms of parametrization $(\boldsymbol{\alpha}\ \boldsymbol{B})$ and $(\boldsymbol{\mu}, \boldsymbol{\Sigma})$ with $\boldsymbol{\Sigma}^{-1} = [\sigma^{ij}]$:

$$\begin{aligned} \beta_{ij} &= -\sigma^{ij} \qquad (i = 1, \ldots, d; j = i+1, \ldots, D), \\ \beta_{i,D} &= \sum_{j=1}^{d} \sigma^{ij} \qquad (i = 1, \ldots, d), \end{aligned} \qquad (13.11)$$

$$\boldsymbol{\alpha}^{(d)} = \boldsymbol{\Sigma}^{-1} \boldsymbol{\mu}, \qquad \alpha_D = -\sum_{i=1}^{d} \alpha_i, \qquad (13.12)$$

where $\boldsymbol{\alpha}^{(d)} = (\alpha_1, \ldots, \alpha_d)$; and

$$\begin{aligned} \sigma^{ij} &= -\beta_{ij} \qquad (i = 1, \ldots, d; j = i+1, \ldots, D), \\ \sigma^{ii} &= \sum_{j=1}^{i-1} \beta_{ji} + \sum_{j=i+1}^{D} \beta_{ij} \qquad (i = 1, \ldots, d), \end{aligned} \qquad (13.13)$$

$$\boldsymbol{\mu} = \boldsymbol{\Sigma} \boldsymbol{\alpha}^{(d)}. \qquad (13.14)$$

The similarity of the loglinear parts of (13.9) and (13.10) suggests the introduction of a new class of distributions on the simplex.

Definition 13.3 The $\mathscr{A}^d(\boldsymbol{\alpha}, \boldsymbol{B})$ *distribution* on the positive simplex $\mathscr{S}^d$ has probability density function $p_{\mathscr{A}}(\mathbf{x} \mid \boldsymbol{\alpha}, \boldsymbol{B})$ defined by

$$\log p_{\mathscr{A}}(\mathbf{x} \mid \boldsymbol{\alpha}, \boldsymbol{B}) = k(\boldsymbol{\alpha}, \boldsymbol{B}) + \sum_{i=1}^{D} (\alpha_i - 1) \log x_i - \frac{1}{2} \sum_{i=1}^{d} \sum_{j=i+1}^{D} \beta_{ij} (\log x_i - \log x_j)^2. \qquad (13.15)$$

To obtain a proper density function the only restrictions on the parameters $(\boldsymbol{\alpha}, \boldsymbol{B})$ are either

(a) the positive definiteness of the quadratic form and $\alpha_1 + \cdots + \alpha_D \geqslant 0$,

or

(b) the non-negative definiteness of the quadratic form and $\alpha_i > 0$ $(i = 1, \ldots, D)$.

The sufficiency of these restrictions can easily be verified from the forms (13.16) and (13.18) below for the normalizing constant. Thus for the simplest case of $\mathscr{A}^1(\alpha_1, \alpha_2, \beta_{12})$, the parameter space is the set

$$\{(\alpha_1, \alpha_2, \beta_{12}) : \beta_{12} > 0, \alpha_1 + \alpha_2 \geqslant 0 \quad \text{or} \quad \beta_{12} = 0, \alpha_1 > 0, \alpha_2 > 0\}.$$

This general class with $\frac{1}{2}D(D+1)$ parameters includes the Dirichlet class as the special case

$$\beta_{ij} = 0 \quad (i = 1, \ldots, d; j = i+1, \ldots, D) \quad \text{and} \quad \alpha_i > 0 \quad (i = 1, \ldots, D),$$

and the logistic normal class as the special case $\alpha_1 + \cdots + \alpha_D = 0$. Moreover, the generality has been bought at the price of the introduction of only one parameter more than the $\mathscr{L}^d$ class.

Apart from the normalizing term $k(\boldsymbol{\alpha}, \boldsymbol{B})$, the logarithm of the density form, and hence the loglikelihood based on any sample, is conveniently linear in the parameters $(\boldsymbol{\alpha}, \boldsymbol{B})$. Closed forms for $k(\boldsymbol{\alpha}, \boldsymbol{B})$ apparently exist only for the special $\mathscr{D}^d$ and $\mathscr{L}^d$ cases. If, however, we write $k(\boldsymbol{\alpha}, \boldsymbol{B}) = -\log c(\boldsymbol{\alpha}, \boldsymbol{B})$, then the normalizing property yields

$$c(\boldsymbol{\alpha}, \boldsymbol{B}) = (2\pi)^{d/2} |\boldsymbol{\Sigma}|^{1/2} \exp(\tfrac{1}{2}\boldsymbol{\alpha}' \boldsymbol{\Sigma} \boldsymbol{\alpha}) \int_{\mathscr{S}^d} x_D^{\delta} p_{\mathscr{L}}(\mathbf{x} | \boldsymbol{\mu}, \boldsymbol{\Sigma}) d\mathbf{x}, \quad (13.16)$$

where $(\boldsymbol{\mu}, \boldsymbol{\Sigma})$ are related to $(\boldsymbol{\alpha}, \boldsymbol{B})$ by (13.11) and (13.12) and

$$\delta = \alpha_1 + \cdots + \alpha_D \quad (13.17)$$

is a measure of departure from logistic normality. Thus $(2\pi)^{d/2} |\boldsymbol{\Sigma}|^{1/2} \exp(\frac{1}{2}\boldsymbol{\alpha}' \boldsymbol{\Sigma} \boldsymbol{\alpha})$ is an upper bound for $c(\boldsymbol{\alpha}, \boldsymbol{B})$. When $\alpha_i > 0$ $(i = 1, \ldots, D)$ we can obtain another integral form

$$c(\boldsymbol{\alpha}, \boldsymbol{B}) = \Delta(\boldsymbol{\alpha}) \int_{\mathscr{S}^d} \exp\left\{ -\frac{1}{2} \sum_{i=1}^{d} \sum_{j=i+1}^{D} \beta_{ij} (\log x_i - \log x_j)^2 \right\} \times p_{\mathscr{D}}(\mathbf{x} | \boldsymbol{\alpha}) d\mathbf{x}, \quad (13.18)$$

where $\Delta(\boldsymbol{\alpha}) = \Gamma(\alpha_1 + \cdots + \alpha_D)/\{\Gamma(\alpha_1)\cdots\Gamma(\alpha_D)\}$. From (13.18) the inequality

$$1 - \frac{1}{2}\sum_{i=1}^{d}\sum_{j=i+1}^{D} \beta_{ij}(\log x_i - \log x_j)^2$$
$$\leqslant \exp\left\{-\frac{1}{2}\sum_{i=1}^{d}\sum_{j=i+1}^{D} \beta_{ij}(\log x_i - \log x_j)^2\right\} \leqslant 1$$

produces not only an obvious upper bound $\Delta(\boldsymbol{\alpha})$ but also a lower bound expressible in terms of gamma, digamma and trigamma functions.

While such bounds may provide useful computational checks, the main problem of statistical analysis is the actual evaluation of $c(\boldsymbol{\alpha}, \boldsymbol{B})$ and its partial derivatives of first and second order. We shall see that these are readily amenable to Hermitian integration.

13.5 Maximum likelihood estimation

Suppose that we wish to fit an $\mathscr{A}^d(\boldsymbol{\alpha}, \boldsymbol{B})$ model to a compositional data set consisting of N, D-part compositions, and that x_{ri} denotes the ith component of the rth composition. Then from (13.15), the loglikelihood l is given by

$$l = -N\log c(\boldsymbol{\alpha}, \boldsymbol{B}) + \sum_{i=1}^{D}(\alpha_i - 1)U_i - \frac{1}{2}\sum_{i=1}^{d}\sum_{j=i+1}^{D}\beta_{ij}V_{ij}, \tag{13.19}$$

where

$$U_i = \sum_{r=1}^{N}\log x_{ri} \qquad (i = 1,\ldots,D), \tag{13.20}$$

$$V_{ij} = \sum_{r=1}^{N}(\log x_{ri} - \log x_{rj})^2 \qquad (i = 1,\ldots,d; j = i+1,\ldots,d), \tag{13.21}$$

clearly form a minimal sufficient statistic for the parameter $(\boldsymbol{\alpha}, \boldsymbol{B})$.

Maximum likelihood estimation requires an iterative procedure and we adopt a mix of Newton–Raphson and steepest ascent methods similar to that used in Section 10.2. Since the special $\mathscr{L}^d(\boldsymbol{\mu}, \boldsymbol{\Sigma})$ case is obtainable from the $\mathscr{A}^d(\boldsymbol{\alpha}, \boldsymbol{B})$ model by placing a single constraint on the parameters, the natural initial values $\boldsymbol{\alpha}_0$, $\boldsymbol{B}_0$ in most situations will be the explicit estimates for the $\mathscr{L}^d$ case given by

$$\mu_{i0} = N^{-1}\sum_{r=1}^{N}(\log x_{ri} - \log x_{rD}) \qquad (i = 1,\ldots,d) \tag{13.22}$$

$$\sigma_{ij0} = N^{-1} \sum_{r=1}^{N} (\log x_{ri} - \log x_{rD})(\log x_{rj} - \log x_{rD}) - \mu_{i0}\mu_{j0} \qquad (i,j = 1,\ldots,d) \quad (13.23)$$

and the relationships (13.11)–(13.12). In what follows we shall regard $(\boldsymbol{\alpha}, \boldsymbol{B})$ as the basic parametrization but shall move freely between $(\boldsymbol{\alpha}, \boldsymbol{B})$ and $(\boldsymbol{\mu}, \boldsymbol{\Sigma})$ and use δ to denote $\alpha_1 + \cdots + \alpha_D$, where convenient. The subscripts 0 and k will denote the initial values and values at the kth iteration, respectively.

The method is routine except for the evaluation of $c(\boldsymbol{\alpha}, \boldsymbol{B})$ and its first- and second-order partial derivatives at each iteration. These can readily be expressed in the form

$$\int_{\mathscr{S}^d} f(\mathbf{x}, \boldsymbol{\alpha}_k, \boldsymbol{B}_k, \boldsymbol{\alpha}_0, \boldsymbol{B}_0)(x_1 \cdots x_D)^{-1} \times \exp[-\tfrac{1}{2}\{\mathbf{y}(\mathbf{x}) - \boldsymbol{\mu}_0\}' \boldsymbol{\Sigma}_0^{-1} \{\mathbf{y}(\mathbf{x}) - \boldsymbol{\mu}_0\} \, d\mathbf{x}, \qquad (13.24)$$

where $\mathbf{y}(\mathbf{x}) = \operatorname{alr}(\mathbf{x})$ and f is a function which differs for different partial derivatives.

The form (13.24) is readily transformable into a multiple integral over $\mathscr{R}^d$ amenable to numerical evaluation by the multivariate version of Hermitian integration (Stroud, 1971, Chapter 3). The successive transformations $\mathbf{y} = \operatorname{alr}(\mathbf{x})$ and $\mathbf{y} = \boldsymbol{\mu}_0 + \sqrt{2}\mathbf{M}_0\mathbf{z}$, where $\mathbf{M}_0\mathbf{M}_0' = \boldsymbol{\Sigma}_0$, reduce (13.24) to a form

$$\int_{\mathscr{R}^d} g(\mathbf{z}) \exp(-\mathbf{z}'\mathbf{z}) d\mathbf{z}, \qquad (13.25)$$

for which Hermitian integration of order k yields an approximation

$$\sum_{i_1=1}^{k} \cdots \sum_{i_d=1}^{k} w_{i_1} \cdots w_{i_d} g(z_{i_1}, \ldots, z_{i_d}), \qquad (13.26)$$

where $w_1, \ldots, w_k$ and $z_1, \ldots, z_k$ are the weights and points for Hermitian integration of order k on $\mathscr{R}^1$; see, for example, Abramowitz and Stegun (1972, Table 25.10). Use of the same transformation at each iteration not only reduces computational effort but also ensures that the Hermitian grid does not drift away from the data cluster in $\mathscr{S}^d$.

The different forms of f in (13.24) can be expressed simply in terms of the function

$$H(\mathbf{x}) = \exp[\tfrac{1}{2}\boldsymbol{\alpha}_k' \boldsymbol{\Sigma}_k \boldsymbol{\alpha}_k \tfrac{1}{2}\{\mathbf{y}(\mathbf{x}) - \boldsymbol{\mu}_k\}' \boldsymbol{\Sigma}_k^{-1} \{\mathbf{y}(\mathbf{x}) - \boldsymbol{\mu}_k\} + \tfrac{1}{2}\{\mathbf{y}(\mathbf{x}) - \boldsymbol{\mu}_0\}' \boldsymbol{\Sigma}_0^{-1} \{\mathbf{y}(\mathbf{x}) - \boldsymbol{\mu}_0\}] \qquad (13.27)$$

and the $\frac{1}{2}D(D+1)$-vector function $\mathbf{h}(\mathbf{x})$ with components

$$\log x_i \quad (i=1,\ldots,D), \qquad -\tfrac{1}{2}(\log x_i - \log x_j)^2 \quad (i=1,\ldots,d; \; j=i+1,\ldots,D). \qquad (13.28)$$

The forms of f for $c(\boldsymbol{\alpha}, \boldsymbol{B})$, its vector of first derivatives and its matrix of second derivatives, are then respectively $H(\mathbf{x})$, $H(\mathbf{x})\mathbf{h}(\mathbf{x})$ and $H(\mathbf{x})\mathbf{h}(\mathbf{x})\mathbf{h}'(\mathbf{x})$.

The summation (13.26) is over a grid of k^d points so that it is clear that the dimensionality d which can feasibly be tackled will depend on the speed and capacity of the computing facilities available. Experience suggests that compositions of up to five parts can be adequately fitted by an $\mathscr{A}^4(\boldsymbol{\alpha}, \boldsymbol{B})$ model on a standard microcomputer.

Table 13.1 shows the application of the method to the aphyric Skye lava AFM compositions of Data 6. Hermitian integration of order 10 was used with convergence in eight iterations. Comparison with the use of Hermitian integration of order 20 verified the accuracy of the order 10 computations; indeed in this case order 5 would have provided sufficient accuracy.

For comparison purposes Table 13.1 also shows the maximum likelihood estimates and maximized loglikelihoods for the special forms $\mathscr{L}^d$ and $\mathscr{D}^d$ of $\mathscr{A}^d$. Even without any formal testing procedures it is clear that, for this data set, $\mathscr{L}^d$ is an adequate substitute for $\mathscr{A}^d$ whereas $\mathscr{D}^d$ is incapable of describing the observed pattern of compositional variability.

Table 13.1. *Maximum likelihood estimate and maximized loglikelihood l for models $\mathscr{L}^2$, $\mathscr{A}^2$ and $\mathscr{D}^2$ for the Skye lava data*

	Estimates under		
Parameter	$\mathscr{L}^2$	$\mathscr{A}^2$	$\mathscr{D}^2$
α_1	−51.5	−50.1	9.28
α_2	−42.9	−41.4	4.48
α_3	94.4	99.9	3.31
β_{12}	23.9	25.0	0
β_{13}	21.2	20.6	0
β_{23}	31.0	30.3	0
l	74.06	74.23	44.57

13.6 Neutrality and partition independence

We first investigate under what conditions $\mathscr{A}^d(\boldsymbol{\alpha}, \boldsymbol{B})$ has neutrality on the right for the partition $(x_1,\ldots,x_c|x_C,\ldots,x_D)$. A useful transformation for the study of the independence of $(x_1,\ldots,x_c)$ and $(x_C,\ldots,x_D)$ is the following:

$$w_i = x_i/(1 - x_1 - \cdots - x_i) \qquad (i = 1,\ldots,d) \qquad (13.29)$$

with inverse

$$x_i = w_i \Big/ \prod_{j=1}^{i} (1 + w_j) \ (i = 1,\ldots,d), \qquad x_D = 1 \Big/ \prod_{j=1}^{d} (1 + w_j). \ (13.30)$$

Variation of $\mathbf{w}$ is over d-dimensional positive space $\mathscr{R}^d_+$, with $\mathbf{w}_1 = (w_1,\ldots, w_c)$ and $\mathbf{w}_2 = (w_C,\ldots, w_d)$ in one-to-one correspondence with $(x_1,\ldots,x_c)$ and $\mathscr{C}(x_C,\ldots,x_D)$. Let $p(\mathbf{w}_1, \mathbf{w}_2)$ denote the joint density function of $\mathbf{w}_1, \mathbf{w}_2$. It is then trivial to show that $\log p(\mathbf{w}_1, \mathbf{w}_2)$ can be expressed as a sum of separate functions of $\mathbf{w}_1$ and $\mathbf{w}_2$ if and only if

$$\beta_{ij} = 0 \qquad (i = 1,\ldots,c; j = C,\ldots,D), \qquad (13.31)$$

which is thus the parametric hypothesis associated with neutrality on the right. The symmetry of (13.31) means that it is also the condition for neutrality on the left, and hence for neutrality, for the given partition.

Indeed, the condition (13.31) has further implications for the partition. We can investigate the joint distribution of $\mathbf{s}_1 = \mathscr{C}(x_1,\ldots,x_c)$, $\mathbf{s}_2 = \mathscr{C}(x_C,\ldots,x_D)$ and $\mathbf{t} = (x_1 + \cdots + x_c,\ x_C + \cdots + x_D)$ by an obvious transformation and readily find that $\mathbf{s}_1, \mathbf{s}_2, \mathbf{t}$ are independent, so that the parametric hypothesis (13.31) also corresponds to partition independence. Moreover, the distributions of $\mathbf{s}_1, \mathbf{s}_2$ and $\mathbf{t}$ are easily determined as

$$\mathscr{A}^{c-1}(\boldsymbol{\alpha}_1, \boldsymbol{B}_1), \mathscr{A}^{d-c}(\boldsymbol{\alpha}_2, \boldsymbol{B}_2), \mathscr{D}^1(\alpha_1 + \cdots + \alpha_c, \alpha_C + \cdots + \alpha_D),$$

where

$$\boldsymbol{\alpha}_1 = (\alpha_1,\ldots,\alpha_c), \boldsymbol{\alpha}_2 = (\alpha_C,\ldots,\alpha_D),$$
$$\boldsymbol{B}_1 = \{\beta_{ij}: i = 1,\ldots,c-1; j = i+1,\ldots,c\},$$
$$\boldsymbol{B}_2 = \{\beta_{ij}: i = C,\ldots,d; j = i+1,\ldots,D\}$$

and the conditions

$$\alpha_1 + \cdots + \alpha_c > 0, \alpha_C + \cdots + \alpha_D > 0$$

are required to ensure a proper Dirichlet form for $\mathbf{t}$.

This equivalence of several forms of independence within the $\mathscr{A}^d$ model is obviously a hindrance when investigation of different forms of directional independence, such as neutrality on the left and on the right, is required. However, where interest is in symmetric forms of independence such as partition independence, the $\mathscr{A}^d$ model provides great flexibility since hypotheses are easily specified for any partition of the composition and for any permutation of its components. We note that complete neutrality, or equivalently partition independence for every partition, corresponds to the hypothesis that $\boldsymbol{B} = \mathbf{0}$ and thus, in accordance with the well-known characterization, to the Dirichlet hypothesis.

Approximate tests of parametric hypotheses within the $\mathscr{A}^d(\boldsymbol{\alpha}, \boldsymbol{B})$ model can be constructed in terms of the Wilks (1938) asymptotic version of the generalized likelihood ratio test as described in Section 7.4.

For a test procedure for the hypothesis $\mathscr{P}$ of partition independence as given by (13.31) we first require to obtain the maximized loglikelihoods $l_{\mathscr{A}}$ and $l_{\mathscr{P}}$ under the $\mathscr{A}^d(\boldsymbol{\alpha}, \boldsymbol{B})$ model and partition independence, respectively. Since (13.31) imposes $c(D-c)$ constraints on the parameters, we then test for partition independence by comparing $2(l_{\mathscr{A}} - l_{\mathscr{P}})$ against the upper percentage points of the $\chi^2\{c(D-c)\}$ distribution. We can illustrate the procedure simply by considering the partition $(x_1 | x_2, x_3) = (\mathrm{F} | \mathrm{A}, \mathrm{M})$ for the aphyric Skye lava compositions. For this partition the subcomposition $\mathbf{s}_1$ is trivial, being a constant, and so the hypothesis concerns only the independence of $t = x_1$ and the 2-part subcomposition $\mathbf{s}_2 = (x_2, x_3)/(x_2 + x_3)$.

The loglikelihood $l_{\mathscr{P}}$ under partition independence can be expressed as the sum

$$l_{\mathscr{P}}(\alpha_1, \alpha_2, \alpha_3 | \mathbf{x}) = l_{\mathscr{D}}(\alpha_1, \alpha_2 + \alpha_3 | t) + l_{\mathscr{A}}/(\alpha_2, \alpha_3, \beta | \mathbf{s}_2)$$
$$- \sum_{r=1}^{N} \log(1 - x_{r1}),$$

where $l_{\mathscr{D}}$ and $l_{\mathscr{A}}$ are loglikelihoods associated with $\mathscr{D}^1$ and $\mathscr{A}^1$ problems. Note that the computational problem of maximizing under $\mathscr{P}$ involves an $\mathscr{A}^1$ distribution instead of the $\mathscr{A}^2$ of the model specification and so requires only one-dimensional Hermitian integration. Indeed, we can simplify the computational problem further by noting that if separate fitting of $\mathscr{D}^1(\beta_1, \gamma)$ to t and $\mathscr{A}'(\alpha_2, \alpha_3, \beta)$ to

$\mathbf{s}_2$ yield maximized loglikelihoods $\hat{l}_{\mathscr{D}^1}$ and $\hat{l}_{\mathscr{A}^1}$ then

$$\hat{l}_{\mathscr{P}} \leqslant \hat{l}_{\mathscr{D}^1} + \hat{l}_{\mathscr{A}^1} - \sum_{r=1}^{n} \log(1 - x_{r1})$$

$$= 56.45$$

for the Skye lava data. Thus we know that the value of the test statistic $2(l_{\mathscr{A}} - l_{\mathscr{P}})$ exceeds $2(74.23 - 56.45) = 35.6$ and comparison of this with upper percentage points of $\chi^2(2)$ shows very strong rejection of the partition independence hypothesis. This is not surprising from examination of a ternary diagram of the 3-part compositions in Fig. 1.9, since it is clear that the pattern of variability of x_1 or F differs along different rays through the vertex F, and so depends on the subcomposition $\mathbf{s}_2$.

13.7 Subcompositional independence

For the $\mathscr{L}^d(\boldsymbol{\mu}, \boldsymbol{\Sigma})$ distribution the hypothesis of complete subcompositional independence corresponds, by Property 10.3, to a logratio covariance matrix $\boldsymbol{\Sigma}$ of special form

$$\boldsymbol{\Sigma} = \operatorname{diag}(\gamma_1, \ldots, \gamma_d) + \gamma_D \mathbf{J}_d. \tag{13.32}$$

By Property 5.13 this special form can be inverted explicitly and then use of (13.11)–(13.12) readily shows that for $\mathscr{A}^d(\boldsymbol{\alpha}, \boldsymbol{B})$ we have complete subcompositional independence when

$$\sum_{i=1}^{D} \alpha_i = 0, \quad \beta_{ij} = \delta_i \delta_j \qquad (i = 1, \ldots, d; j = i+1, \ldots, D). \tag{13.33}$$

The relationships between the D parameters $\delta_1, \ldots, \delta_D$ of (13.33) and the D parameters $\gamma_1, \ldots, \gamma_D$ of (13.32) are

$$\gamma_i \delta_i = (\gamma_1^{-1} + \cdots + \gamma_D^{-1})^{-1/2} = (\delta_1 + \cdots + \delta_D)^{1/2}.$$

Also we know that every Dirichlet distribution has complete subcompositional independence so that we can also say that $\mathscr{A}^d(\boldsymbol{\alpha}, \boldsymbol{B})$ has complete subcompositional independence when

$$\alpha_i > 0 \quad (i = 1, \ldots, D), \quad \beta_{ij} = 0 \quad (i = 1, \ldots, d; j = i+1, \ldots, D). \tag{13.34}$$

Conditions (13.33) and (13.34) are quite different, reflecting the separateness of the logistic normal and Dirichlet classes, and do not necessarily capture all the parametric constraints under which

$\mathscr{A}^d(\boldsymbol{\alpha}, \boldsymbol{B})$ has complete subcompositional independence. Unfortunately, any attempt to find more general conditions seems bound to involve unwieldly multiple integrals. For any investigation of complete subcompositional independence it would appear that the following sequence is the best that can be recommended. First test for Dirichlet form along the lines of Shen (1982) or even by informal comparison of the $\mathscr{D}^d$, $\mathscr{L}^d$ and $\mathscr{A}^d$ fits, as in Table 13.1. In the likely event of rejection, since compositional data seldom appear to have such strong independence properties as possessed by the Dirichlet class, test for logistic normality as in Section 7.3. If there is reasonable conformity, use the test of complete subcompositional independence within logistic normality as described in Section 10.2. If logistic normality is rejected then there seems no alternative to making some kind of judgement on the basis of conformity of the sample covariance matrix with the special structure of (13.32), a necessary condition for complete subcompositional independence whatever the distributional form.

Use of the form (13.33) instead of (13.32) has some computational advantages but naturally leads to exactly the same test statistic within an $\mathscr{L}^d$ model. Also some forms of partial subcompositional independence are readily expressed as modifications of (13.33). Thus for the partition $(x_1, \ldots, x_c | x_C, \ldots, x_D)$ the hypothesis

$$\begin{aligned} \beta_{ij} &= \delta_i \delta_j \qquad (i = 1, \ldots, c-1; j = i+1, \ldots, c), \\ \beta_{ij} &= 0 \qquad (i = 1, \ldots, c; j = C, \ldots, D), \end{aligned}$$

combines partition independence with complete subcompositional independence within the first subcomposition only.

13.8 A generalized lognormal gamma distribution with compositional invariance

When a composition $\mathbf{x} = (x_1, \ldots, x_D)$ is formed as the relative proportions of a basis $\mathbf{w} = (w_1, \ldots, w_D)$ of D positive quantities, a question of interest, as we have seen in Section 9.5, is whether the basis has compositional invariance, in the sense that shape or composition $\mathbf{x}$ is independent of the additive size $t = w_1 + \cdots + w_D$ of the basis.

Dirichlet and logistic normal compositions are, by Property 3.3 and Property 6.1, related to bases of independent equally scaled gamma and multivariate lognormal forms, respectively, in interestingly contrasting ways. If w_i $(i=1,\ldots,D)$ are independently distributed as $\mathrm{Ga}(\alpha_i,\theta)$ then the basis $(w_1,\ldots,w_D)$ has compositional invariance with composition $\mathbf{x}=\mathscr{C}(\mathbf{w})$ distributed as $\mathscr{D}^d(\boldsymbol{\alpha})$, independently of the basis size $w_1+\cdots+w_D$ which has a $\mathrm{Ga}(\alpha_1+\cdots+\alpha_D,\theta)$ distribution. The composition formed from a basis $\mathbf{w}=(w_1,\ldots,w_D)$ with a multivariate lognormal distribution $\boldsymbol{\Lambda}^D(\boldsymbol{\zeta},\boldsymbol{\Omega})$ has $\mathscr{L}^d$ form but, by Property 9.5, the basis cannot have compositional invariance without being degenerate in the sense of $\boldsymbol{\Omega}$ having less than full rank.

Although we have seen in Section 9.5 how to model and test non-degenerate bases which can possess compositional invariance and have compositions of $\mathscr{L}^d$ form, it is clearly of some interest to ask whether it may now be possible to discover a more general class of non-degenerate distributions for D-dimensional bases which can support compositional invariance and has compositions of either $\mathscr{D}^d$ or $\mathscr{L}^d$ form or indeed of $\mathscr{A}^d$ form. An answer is to be found in a generalization of the independent $\mathrm{Ga}(\alpha_i,\theta)$ $(i=1,\ldots,D)$ and the multivariate lognormal $\Lambda^d(\boldsymbol{\zeta},\boldsymbol{\Omega})$ distributions along lines similar to the $\mathscr{A}^d$ generalization of $\mathscr{D}^d$ and $\mathscr{L}^d$ in Section 13.4. Consider the class $\mathscr{H}^D(\boldsymbol{\alpha},\theta,\boldsymbol{\Delta})$ of distributions on $\mathscr{R}^D_+$, D-dimensional positive space, with density function $p_{\mathscr{H}}(\mathbf{w}|\boldsymbol{\alpha},\theta,\boldsymbol{\Delta})$ given by

$$\begin{aligned}\log p_{\mathscr{H}}(\mathbf{w}|\boldsymbol{\alpha},\theta,\boldsymbol{\Delta}) &= k(\boldsymbol{\alpha},\theta,\boldsymbol{\Delta})\\ &+\sum_{i=1}^{D}(\alpha_i-1)\log w_i\\ &-\theta(w_1+\cdots+w_D)\\ &-\frac{1}{2}\sum_{i=1}^{D}\sum_{j=1}^{D}\delta_{ij}\log w_i\log w_j,\end{aligned} \tag{13.35}$$

where $\boldsymbol{\Delta}=[\delta_{ij}]$ is symmetric and non-negative definite. When $\delta_{ij}=0$ $(i,j=1,\ldots,D)$ and $\alpha_i>0\,(i=1,\ldots,D)$ then $\mathscr{H}^D(\boldsymbol{\alpha},\theta,\boldsymbol{\Delta})$ takes an independent gamma form, and when $\theta=0$, a multivariate lognormal form. Thus we certainly have a generalized form for basis distributions which can obviously yield both $\mathscr{D}^d$ and $\mathscr{L}^d$ compositions.

For the investigation of compositional invariance consider the transformation

$$x_i=w_i/(w_1+\cdots+w_D)\qquad (i=1,\ldots,D),\qquad t=w_1+\cdots+w_D,$$

with inverse

$$w_i = tx_i \qquad (i = 1, \ldots, D)$$

and jacobian $\text{jac}(\mathbf{w}|\mathbf{x}, t) = t^d$. For $\mathscr{H}^D(\boldsymbol{\alpha}, \theta, \boldsymbol{\Delta})$, the joint density function $p(\mathbf{x}, t)$ of composition $\mathbf{x}$ and size t then takes the form

$$\begin{aligned}\log p(\mathbf{x}, t) = k(\boldsymbol{\alpha}, \theta, \boldsymbol{\Delta}) &+ \sum_{i=1}^{D} (\alpha_i - 1) \log t \\ &- \theta t + \sum_{i=1}^{D} (\alpha_i - 1) \log x_i \\ &- \frac{1}{2} \sum_{i=1}^{D} \sum_{j=1}^{D} \delta_{ij} (\log x_i + \log t)(\log x_j + \log t).\end{aligned}$$

Clearly a necessary and sufficient condition for compositional invariance is that the coefficients of $\log x_i \log t (i = 1, \ldots, D)$ should each be zero, or equivalently, $\boldsymbol{\Delta} \mathbf{j}_D = \mathbf{0}$. With this condition the number of parameters in $\boldsymbol{\Delta}$ is reduced from $\frac{1}{2}D(D+1)$ by the D constraints to $\frac{1}{2}dD$. Thus we could take $\gamma_{ij} = -\delta_{ij}$ $(i \neq j)$, with symmetry, as the $\frac{1}{2}dD$ parameters, with

$$\delta_{ii} = \sum_{\substack{j=1 \\ j \neq i}}^{D} \gamma_{ij}.$$

Then

$$\begin{aligned}&\sum_{i=1}^{D} \sum_{j=1}^{D} \delta_{ij} (\log x_i + \log t)(\log x_j + \log t) \\ &\quad = \sum_{i=1}^{D} \sum_{j=i+1}^{D} \gamma_{ij} (\log x_i - \log x_j)^2\end{aligned}$$

the terms in $\log t$ and $(\log t)^2$ vanishing because of the constraints $\boldsymbol{\Delta} \mathbf{j}_D = \mathbf{0}$. Thus an $\mathscr{H}^D(\boldsymbol{\alpha}, \theta, \boldsymbol{\Delta})$ basis with $\boldsymbol{\Delta} \mathbf{j}_D = \mathbf{0}$ has compositional invariance with composition of $\mathscr{A}^d(\boldsymbol{\alpha}, \boldsymbol{B})$ form and size of $\text{Ga}(\sum \alpha_i, \theta)$ form, and is, moreover, non-degenerate. The $\mathscr{H}^D$ class of models therefore provides a framework within which the question of compositional invariance can be discussed more generally than within the logistic normal regression model of Section 9.5.

13.9 Discussion

While the main aim of this chapter has been to extend the Dirichlet and the additive logistic normal classes to allow a broader investigation of simplex independence hypotheses, the forms of independence treated have been symmetric. For the study of hypotheses of directional independence such as neutrality on the right for the ordering $(x_1,\ldots,x_D)$ of the compositional components, we have seen in Section 10.4 that the multiplicative logistic normal model $\mathcal{M}^d(\boldsymbol{\mu},\boldsymbol{\Sigma})$ with $\log\{x_i/(1-x_1-\cdots-x_i)\}$ $(i=1,\ldots,D)$ distributed as $\mathcal{N}^d(\boldsymbol{\mu},\boldsymbol{\Sigma})$ is admirably suited. For the given ordering neutrality on the right for any given partition can be expressed as a parametric hypothesis. The Connor–Mosimann generalized Dirichlet class $\mathscr{E}^d$ of Definition 13.2 already has the property of neutrality on the right for each partition of the ordered composition $(x_1,\ldots,x_D)$ and so within this class only stronger forms of simplex independence can be expressed as parametric hypotheses.

In the same way as we introduced $\mathscr{A}^d$ as a generalization of $\mathscr{D}^d$ and $\mathscr{L}^d$, so we can construct a class $\mathscr{J}^d(\boldsymbol{\gamma},\boldsymbol{\delta},\boldsymbol{B})$ which generalizes the $\mathscr{E}^d$ and $\mathcal{M}^d$ classes, with the logarithm of its density function $p_{\mathscr{J}}$ of the following form:

$$\log p_{\mathscr{J}}(\mathbf{x}) = k(\boldsymbol{\gamma},\boldsymbol{\delta},\boldsymbol{B}) + \sum_{i=1}^{d}(\gamma_i - 1)\log x_i$$
$$+ \sum_{i=1}^{d}\delta_i\log(1-x_1-\cdots-x_i)$$
$$-\frac{1}{2}\sum_{i=1}^{d}\sum_{j=i+1}^{D}\beta_{ij}\log\frac{x_i}{1-x_1-\cdots-x_i}\log\frac{x_j}{1-x_1-\cdots-x_j}. \tag{13.36}$$

For the special cases we have the multiplicative logistic normal $\mathcal{M}^d$ with $\frac{1}{2}d(d+3)$ parameters when $\alpha_i+\beta_i=0$ $(i=1,\ldots,d)$; the $\mathscr{E}^d$ Dirichlet generalization with $2d$ parameters when $\beta_{ij}=0$ $(i,j=1,\ldots,d)$ in a slightly different parametrization from that of Definition 13.3; and the Dirichlet $\mathscr{D}^d$ with D parameters when $\delta_i=0$ $(i=1,\ldots,d-1)$, $\beta_{ij}=0$ $(i=1,\ldots,d;\ j=i+1,\ldots,D)$. We may note, however, that this generalization has been achieved less parsimoniously than the $\mathscr{A}^d$ generalization with a need to introduce d parameters more than the multiplicative logistic normal model $\mathcal{M}^d$.

Another point of interest is that there is a perhaps surprising duality between concepts and results for the $\mathscr{A}^d$ class and experiments with mixtures, as discussed in Section 12.4, where we studied the nature of the expected response to a set of factors in the form of a composition. The duality is between neutrality in compositional distribution and additivity in expected response, between logarithm of density function and expected response. The dual of the complete neutrality characterization of the Dirichlet class is then that complete additivity implies that the expected response is a logcontrast of the components of the mixture, with form

$$\beta_0 + \beta_1 \log x_1 + \cdots + \beta_D \log x, \qquad \text{where} \quad \beta_1 + \cdots + \beta_D = 0.$$

The fact that the $\mathscr{D}^d$ and $\mathscr{L}^d$ classes are identified with parametric hypotheses within the $\mathscr{A}^d$ model encourages the view that it should now be possible to construct simple tests of Dirichlet and logistic normal forms based on the generalized loglikelihood statistics $l_{\mathscr{A}} - l_{\mathscr{D}}$ and $l_{\mathscr{A}} - l_{\mathscr{L}}$, respectively. In particular, we would then be able to avoid the complications of separate family testing in comparing Dirichlet with logistic normal form, as in Shen (1982). Unfortunately these parametric hypotheses correspond to sets on the boundary of the parameter space of the $\mathscr{A}^d$ model so that the Wilks (1938) approximation is not directly applicable. How exactly to adapt this asymptotic result, possibly along the lines of the Chernoff (1954) and Feder (1968) extension, remains an open question.

13.10 Bibliographic notes

Reference has already been made to the main contributions to the generalizations of the Dirichlet class. The first account of a test between the separate Dirichlet and additive logistic normal classes appears in Shen (1982). The general classes $\mathscr{A}^d, \mathscr{H}^d, \mathscr{J}^d$ are defined and developed in Aitchison (1985a).

Problems

13.1 Apply the method of maximum likelihood to estimate the parameters of an $\mathscr{A}^3(\boldsymbol{\alpha}, \boldsymbol{B})$ model for the 4-part shift compositions of Data 22. Compare the maximized loglikelihoods for the $\mathscr{A}^3, \mathscr{D}^3$ and $\mathscr{L}^3$ classes.

13.2 For the partition of order 1 dividing the shifts of Data 22 into productive and non-productive parts, test the hypothesis of partition independence within the $\mathscr{A}^3(\boldsymbol{\alpha}, \boldsymbol{B})$ model of Problem 13.1.

CHAPTER 14

Miscellaneous problems

A wise man has something better to do than boast of his cures, namely to be always self-critical.

Philippe Pinel: *Traité Medico-philosophique sur l'Alienation Mentale*

14.1 Introduction

In this final chapter we examine a miscellany of problems, studying some at length, merely indicating for others a possible line of approach.

14.2 Multi-way compositions

For some compositions the parts are most readily classified by two or more factors and so the components may be set out most naturally in a 2-way or a multi-way array. For example, an ocean sediment may be separated by size, coarseness or fineness, of grain and also by chemical nature, carbonate or non-carbonate, of grain. Thus a typical ocean sediment composition (in proportions by weight) would be as follows.

	Carbonate	Non-carbonate
Coarse	0.14	0.36
Fine	0.29	0.21

An early example of interest in 2-way compositions was the pigmentation study by Tocher (1908) of almost half a million Scottish schoolchildren. The intention was to study for each region and each sex the 2-way 5×4 composition consisting of the proportions of children in each of five hair-colour and four eye-colour categories. Unfortunately Tocher's funding allowed him only to study the two 'marginal' compositions separately, the 5-part hair-colour compositions and the 4-part eye-colour compositions.

We have already met a simple example in the colour–size compositions of Data 14 and 15 and posed Questions 1.13.1 and 1.13.2 about association between the two factors, colour and size. Suppose that the probability that a clam in an East Bay colony (Data 14) will be in the colour–size category (j, k) is π_{jk} with

$$\pi_{11} + \pi_{12} + \pi_{13} + \pi_{21} + \pi_{22} + \pi_{23} = 1.$$

For a 2×3 contingency table with cell probabilities π_{jk} the hypothesis of no association can be represented by parametric constraints in the form of cross-ratios:

$$\frac{\pi_{11}\pi_{23}}{\pi_{21}\pi_{13}} = \frac{\pi_{12}\pi_{23}}{\pi_{22}\pi_{13}} = 1. \tag{14.1}$$

The difference between a contingency table analysis and our present problem is that our assessment of whether (14.1) holds is to be based not on actual numbers in the cells of a single 2-way table but on sets of estimates of the π_{jk} in the form of a composition, one set for each colony.

The cross-ratio nature of the hypothesis constraints in (14.1) suggests that some form of logratio analysis of the compositions may be appropriate. Instead of the additive logratio transformation we may introduce another transformation from $\mathscr{S}^5$ to $\mathscr{R}^5$ which involves the cross-ratios of (14.1) more directly, namely,

$$\begin{aligned} y_1 &= \log(x_{13}/x_{23}), \quad y_2 = \log(x_{21}/x_{23}), \quad y_3 = \log(x_{22}/x_{23}), \\ y_4 &= \log\{x_{11}x_{23}/(x_{21}x_{13})\}, \quad y_5 = \log\{x_{12}x_{23}/(x_{22}x_{13})\}. \end{aligned} \tag{14.2}$$

It is easy to establish that there is a unique inverse transformation from $\mathscr{R}^5$ to $\mathscr{S}^5$; its particular form is of no direct interest to us here.

The test of no association between colour and size is now obvious. We can concentrate on the set of twenty 2-dimensional vectors of the form $\mathbf{y} = (y_4, y_5)$ and ask whether these could reasonably have arisen from an $\mathscr{N}^2(\boldsymbol{\mu}, \boldsymbol{\Sigma})$ distribution with mean $(0, 0)$. The problem is thus identified with testing a multivariate normal mean: with an obvious notation we compare the scaled Hotelling T^2-statistic

$$\frac{(N-2)N}{2(N-1)}\bar{\mathbf{y}}'\hat{\boldsymbol{\Sigma}}^{-1}\bar{\mathbf{y}} \tag{14.3}$$

against the upper percentage points of the $F(2, N-2)$ distribution. From Data 14 we have

$$N = 20, \qquad \bar{\mathbf{y}} = \begin{bmatrix} 0.522 \\ 0.250 \end{bmatrix}, \qquad \hat{\boldsymbol{\Sigma}} = \begin{bmatrix} 0.3056 & 0.0519 \\ 0.0519 & 0.1278 \end{bmatrix};$$

the computed value of the test statistic (14.3) is then 7.62 and the significance probability is 0.005. Thus there is evidence of association between colour and size in East Bay clams. The positive values of $\bar{y}_4$ and $\bar{y}_5$ indicate that there is a tendency for darkening of shell colour as size increases.

From the similar set of 2-way compositions for West Bay colonies in Data 15 we obtain

$$N = 20, \qquad \bar{\mathbf{y}} = \begin{bmatrix} -0.230 \\ 0.111 \end{bmatrix}, \qquad \hat{\boldsymbol{\Sigma}} = \begin{bmatrix} 0.2299 & -0.0102 \\ -0.0102 & 0.0963 \end{bmatrix},$$

and the computed value 2.51 of the test statistic (14.3) leads to a significance probability of 0.15. Thus there is no evidence in this data set of any association between colour and size for West Bay clams.

Extension of this form of analysis from 2-way compositions to compositions whose parts are defined by three or more factors leads in an obvious way to hypotheses involving higher-order interactions and their testing through their parametric cross-ratio equivalents. It is also possible to define forms of independence associated with row and column subcompositions. For example, consider the following question about the clam data: do the size patterns within the colour groups depend on the relative proportions in these colour groups? This question can be rephrased in terms of the partition $(\mathbf{t}; \mathbf{s}_1, \mathbf{s}_2)$ formed from the single division

$$(x_{11}, x_{12}, x_{13} | x_{21}, x_{22}, x_{23})$$

of the composition: are $\mathbf{s}_1$ and $\mathbf{s}_2$ independent of $\mathbf{t}$? The problem then reduces to that of testing for subcompositional invariance as described in Section 10.3.

More complicated independence properties involving row and column subcompositions may have their parametric expression in terms of the pattern of the logratio covariance matrix. Discussion of such covariance patterns would take us beyond the scope of this monograph. The interested reader will find an account in Li (1985).

Problems such as the above clam ecology problem, in which the compositions are formed by application of the constraining operator $\mathscr{C}$ to multinomial or contingency tables based on a counting process, clearly bear a close relationship to categorical data analysis. One view of such compositional data problems is that they are a form of random-effects categorical data analysis with the varying category probabilities being modelled by logistic normal distributions.

14.3 Multi-stage compositions

In a number of applications the formation of a composition may take place in stages or, for modelling purposes, we may find it convenient to envisage such a staged process. For example, each yatquat fruit of Data 12 goes through a two-stage process before being marketed. At stage 1 the fruit is separated into its flesh, skin and stone components, say (t_1, t_2, t_3). At stage 2 the flesh component is separated into its ring and pulp components, say (x_{11}, x_{12}); the skin component into its fibrous and non-fibrous components (x_{21}, x_{22}); the stone component into its enzyme and residue components (x_{31}, x_{32}). In this example the six second-stage components refer to different parts so that we can formally represent stages 1 and 2 as the amalgamation and sub-compositions of the partition of order 2 formed from the division

$$(x_{11}, x_{12} | x_{21}, x_{22} | x_{31}, x_{32})$$

of the 6-part final composition.

How we model such a partition $(\mathbf{t}; \mathbf{s}_1, \mathbf{s}_2, \mathbf{s}_3)$ will depend on the purpose of the investigation. As a simple illustrative example, suppose that our main interest is in whether the ring-to-pulp ratio depends on the flesh-to-nonflesh ratio and even on the skin-to-stone ratio. In terms of the amalgamation we may then consider the multiplicative logratio transformation

$$z_1 = \log\{t_1/(1-t_1)\}, \qquad z_2 = \log\{t_2/(1-t_1-t_2)\} = \log(t_2/t_3), \tag{14.4}$$

and adopt the model

$$y_1 = \log(x_{11}/x_{12}) = \alpha_1 + \beta_{11} z_1 + \beta_{12} z_2 + \text{error}, \tag{14.5}$$

with $\beta_{11} = 0$ and $\beta_{12} = 0$ having obvious interpretations. If there is similar interest in the other subcompositions then the univariate regression model of (14.5) can be expanded into a multivariate form by appending

$$y_2 = \log(x_{21}/x_{22}) = \alpha_2 + \beta_{21}z_1 + \beta_{22}z_2 + \text{error},$$
$$y_3 = \log(x_{31}/x_{32}) = \alpha_3 + \beta_{31}z_1 + \beta_{32}z_2 + \text{error}. \quad (14.6)$$

In either case hypothesis testing is standard and we will not pursue this aspect of multi-stage modelling any further.

In other applications the parts at the final stage may be the same irrespective of the particular part from which they arise at the penultimate stage. A simple illustrative example arises in physiological studies of energy metabolism in animals. At the first stage the dry carcase is divided into fat and non-fat components, and at the second stage, the fat and non-fat components are each divided into protein and residue components. We can here set out the details of the 2-stage process in 2-way compositional form:

	Protein	Residue	Total
Fat	x_{11}	x_{12}	x_{1+}
Non-fat	x_{21}	x_{22}	x_{2+}
Total	x_{+1}	x_{2+}	1

showing also the marginal compositions as row and column totals.

How we model multi-stage compositions with common final parts again depends on the aim of the investigation. In the present example, if interest lies in how the proportions of protein within fat and non-fat parts may vary with the ratio of fat to non-fat within the dry carcase, then a simple partition approach with bivariate regression model

$$y_1 = \log(x_{11}/x_{12}) = \alpha_1 + \beta_1 \log(x_{1+}/x_{2+}) + \text{error},$$
$$y_2 = \log(x_{21}/x_{22}) = \alpha_2 + \beta_2 \log(x_{1+}/x_{2+}) + \text{error}, \quad (14.7)$$

should be adequate. If, however, the intention is in future to attempt to estimate or predict the fat to non-fat ratio from information on only the protein to residue ratio within the total dry carcase, then a univariate regression model

$$\log(x_{1+}/x_{2+}) = \alpha + \beta \log(x_{+1}/x_{+2}) + \text{error}, \quad (14.8)$$

involving only the marginal compositions, is indicated.

Note that here and in Section 14.2 the approach has been to model directly the conditional distribution of interest. It is not possible to start with an additive logistic normal distribution for the 2-way or 2-stage composition and then hope to study problems involving the marginal compositions. It is, therefore, necessary to formulate the objective of the investigation clearly and then to select a form of modelling directed towards attaining that objective.

14.4 Multiple compositions

We have already met problems involving two different compositions. In Section 12.2 we were interested in calibrating image-determined compositions to microscopically determined compositions. In Section 12.3 we used a compositional regression model to investigate whether the relation of present fruit composition to past fruit composition was affected by hormone treatment of the fruit tree. There are other types of problem in which more than one composition may be involved. For example, in ecological studies we may have, for each of a number of quadrats within a geographical region, two compositions in the form of the proportions of the different species of animals present and the proportions of the different species of plants present. Within a similar context, the compositions may be the proportions of different predators and of different prey. In such circumstances we may wish to define some general measure of association between the two compositions and then study how this measure may vary from one region to another.

The general problem here is to define some measure of association or correlation between two compositions

$$\mathbf{x}_1 = (x_{11}, \ldots, x_{1C}), \qquad \mathbf{x}_2 = (x_{21}, \ldots, x_{2D})$$

of C and D parts respectively. The corresponding problem for variation of $\mathbf{y}_1$ and $\mathbf{y}_2$ in $\mathscr{R}^c$ and $\mathscr{R}^d$ is well known under the terminology of canonical component analysis. The measure used is the canonical correlation defined as the maximum product-moment correlation attainable between two linear combinations $\mathbf{b}_1'\mathbf{y}_1$ and $\mathbf{b}_2'\mathbf{y}_2$ of the vectors $\mathbf{y}_1$ and $\mathbf{y}_2$. Mathematically this problem is equivalent to maximization of the covariance

$$\operatorname{cov}(\mathbf{b}_1'\mathbf{y}_1, \mathbf{b}_2'\mathbf{y}_2) \tag{14.9}$$

subject to standardizing constraints

$$\operatorname{var}(\mathbf{b}_1'\mathbf{y}_1) = \operatorname{var}(\mathbf{b}_2'\mathbf{y}_2) = 1. \tag{14.10}$$

Computationally the canonical correlation λ is the maximum eigenvalue of the eigenproblem

$$\begin{bmatrix} -\lambda\boldsymbol{\Sigma}_{11} & \boldsymbol{\Sigma}_{12} \\ \boldsymbol{\Sigma}_{21} & -\lambda\boldsymbol{\Sigma}_{22} \end{bmatrix} \begin{bmatrix} \mathbf{b}_1 \\ \mathbf{b}_2 \end{bmatrix} = \mathbf{0}, \tag{14.11}$$

where, in an obvious matrix partitioning,

$$\begin{bmatrix} \boldsymbol{\Sigma}_{11} & \boldsymbol{\Sigma}_{12} \\ \boldsymbol{\Sigma}_{21} & \boldsymbol{\Sigma}_{22} \end{bmatrix} = \mathrm{V}(\mathbf{y}_1, \mathbf{y}_2). \tag{14.12}$$

The linear combinations $\mathbf{b}_1'\mathbf{y}_1$ and $\mathbf{b}_2'\mathbf{y}_2$, standardized by the equivalents of (14.10),

$$\mathbf{b}_1'\boldsymbol{\Sigma}_{11}\mathbf{b}_1 = \mathbf{b}_2'\boldsymbol{\Sigma}_{22}\mathbf{b}_2 = 1, \tag{14.13}$$

are the canonical components. Canonical component analysis can be extended to further roots of (14.11) in a way analogous to principal component analysis, but we shall confine our attention to the first or maximal canonical components.

We have seen in earlier analyses, particularly in Chapter 8, that the compositional counterpart of a linear combination of a d-dimensional vector $\mathbf{y}$ is a logcontrast $\mathbf{a}'\log\mathbf{x}$, with $\mathbf{a}'\mathbf{j} = 0$. With this in mind we can easily arrive at a suitable definition of canonical correlation for compositions, together with a computational procedure for estimating it.

Definition 14.1 The *canonical correlation* between a C-part composition $\mathbf{x}_1$ and a D-part composition $\mathbf{x}_2$ is the maximum correlation between two logcontrasts $\mathbf{a}_1'\log\mathbf{x}_1$ and $\mathbf{a}_2'\log\mathbf{x}_2$. When the two maximizing logcontrasts are standardized by the unit-variance constraints

$$\mathrm{var}(\mathbf{a}_1'\log\mathbf{x}_1) = \mathrm{var}(\mathbf{a}_2'\log\mathbf{x}_2) = 1$$

then $\mathbf{a}_1'\log\mathbf{x}_1$ and $\mathbf{a}_2'\log\mathbf{x}_2$ are termed *logcontrast canonical components.*

Property 14.1

(a) The canonical correlation between a C-part composition $\mathbf{x}_1$ and a D-part composition $\mathbf{x}_2$ is the maximum eigenvalue λ of the eigenproblem

$$\begin{bmatrix} -\lambda\boldsymbol{\Sigma}_{11} & \boldsymbol{\Sigma}_{12} \\ \boldsymbol{\Sigma}_{21} & -\lambda\boldsymbol{\Sigma}_{22} \end{bmatrix} \begin{bmatrix} \mathbf{b}_1 \\ \mathbf{b}_2 \end{bmatrix} = \mathbf{0},$$

where

$$\begin{bmatrix} \boldsymbol{\Sigma}_{11} & \boldsymbol{\Sigma}_{12} \\ \boldsymbol{\Sigma}_{21} & \boldsymbol{\Sigma}_{22} \end{bmatrix} = \mathrm{V}(\mathbf{y}_1, \mathbf{y}_2)$$

and $\mathbf{y}_1, \mathbf{y}_2$ are the additive logratio vectors of $\mathbf{x}_1, \mathbf{x}_2$.

(b) The logcontrast canonical components are $\mathbf{b}_1'\mathbf{y}_1$ and $\mathbf{b}_2'\mathbf{y}_2$ where $\mathbf{b}_1$ and $\mathbf{b}_2$ are subject to the constraints

$$\mathbf{b}_1'\boldsymbol{\Sigma}_{11}\mathbf{b}_1 = \mathbf{b}_2'\boldsymbol{\Sigma}_{22}\mathbf{b}_2 = 1.$$

Notes

1. In Section 8.3 we saw that it was necessary to make an isotropic covariance adjustment to the logratio covariance matrix version of the eigenproblem of principal component analysis in order to ensure permutation invariance. We may wonder why a similar adjustment is not required in Property 14.1(a). The answer is that here we are involved in comparison of covariance structures, and it is easy to verify, along the lines of Section 5.5, that the canonical correlation and canonical components are invariant under permutations of the parts of both compositions.

2. For computational purposes the eigenproblem of Property 14.1(a) can be reduced to a more standard form; for example, by elimination of $\mathbf{b}_1$ we have the eigenproblem

$$(\boldsymbol{\Sigma}_{21}\boldsymbol{\Sigma}_{11}^{-1}\boldsymbol{\Sigma}_{12} - \lambda^2\boldsymbol{\Sigma}_{22})\mathbf{b}_2 = \mathbf{0}$$

for the determination of λ and $\mathbf{b}_2$. Then $\mathbf{b}_1$ is readily determined by

$$\mathbf{b}_1 = \lambda^{-1}\boldsymbol{\Sigma}_{11}^{-1}\boldsymbol{\Sigma}_{12}\mathbf{b}_2.$$

3. In applications the covariance matrix in Property 14.1(a) is simply replaced by its sample estimate computed from the joint logratio vectors of form $(\mathbf{y}_1, \mathbf{y}_2)$.

There are other problems where we may be naturally led to consider two compositions. Suppose that household income t is derived from C different sources so that

$$t = u_1 + \cdots + u_C,$$

and that income is disposed of into D activities so that

$$t = w_1 + \cdots + w_D.$$

It may then be of interest to study the interrelations of t and the two compositions, source pattern $\mathbf{v} = \mathscr{C}(u_1, \ldots, u_C)$ and expenditure pattern $\mathbf{x} = \mathscr{C}(w_1, \ldots, w_D)$. To model this kind of input–output situation we may consider the logratio vectors

$$\mathbf{y} = \mathrm{alr}(\mathbf{v}), \quad \mathbf{z} = \mathrm{alr}(\mathbf{x})$$

and the conditional model:

$$p(\mathbf{y}, \mathbf{z} \mid t) = N\left\{\begin{bmatrix} \boldsymbol{\alpha} + \boldsymbol{\beta} \log t \\ \boldsymbol{\gamma} + \boldsymbol{\delta} \log t \end{bmatrix}, \begin{bmatrix} \boldsymbol{\Sigma}_{11} & \boldsymbol{\Sigma}_{12} \\ \boldsymbol{\Sigma}_{21} & \boldsymbol{\Sigma}_{22} \end{bmatrix}\right\}.$$

The now familiar range of parametric hypotheses is again available, providing interpretations of such questions as 'Does disposal pattern depend on source pattern?' and 'Does disposal pattern depend on income?'

14.5 Kernel density estimation for compositional data

Some compositional data sets will fail the battery of tests of logistic normality so substantially that even the more general $\mathscr{A}^d$ class of Definition 13.3 is unlikely to provide a successful substitute. As an example of such a data set we may cite the 30 microscopically determined white-cell compositions of Data 11. Inspection of the ternary diagram of Fig. 14.1 suggests that the density function may be

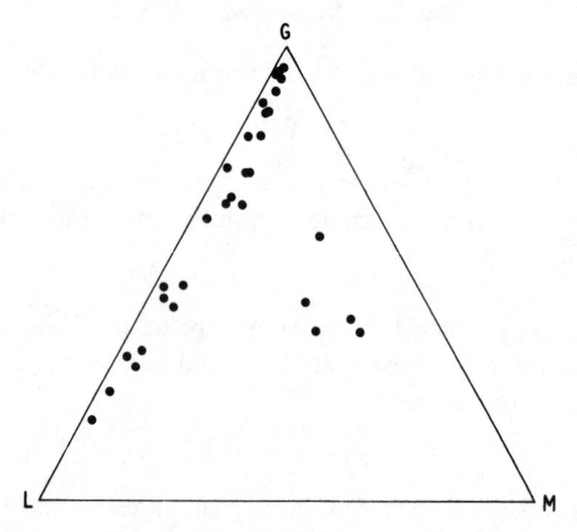

Fig. 14.1. *Ternary diagram for 30 microscopically determined white-cell compositions.*

multi-modal. It may still be necessary to obtain some reasonable assessment of such a density function, for example as a step towards assessment of diagnostic or classification probabilities. In such circumstances kernel density estimation can prove an attractive alternative.

Let $\mathbf{x}_1, \ldots, \mathbf{x}_N$ be the D-part compositions of the compositional data set and let $\boldsymbol{\xi} \in \mathscr{S}^d$ denote a typical D-part composition. To apply a kernel method of density estimation we must first define a kernel $K(\boldsymbol{\xi} \,|\, \mathbf{x}, \lambda)$, a density function for $\boldsymbol{\xi} \in \mathscr{S}^d$, centred, in some sense, on data point $\mathbf{x}$ and with λ a 'smoothing factor' to be determined. The kernel density estimate at any composition $\boldsymbol{\xi}$ is then taken to be

$$N^{-1} \sum_{r=1}^{N} K(\boldsymbol{\xi} \,|\, \mathbf{x}_r, \lambda). \tag{14.14}$$

The smoothing factor is determined by the method of Habbema, Hermans and van den Broek (1974) as the λ which maximizes the pseudo-likelihood

$$\prod_{r=1}^{N} n^{-1} \sum_{s \neq r} K(\mathbf{x}_r | \mathbf{x}_s, \lambda), \tag{14.15}$$

where $n = N - 1$. Note that recent studies by Bowman (1984) and Stone (1984) indicate that for 'long-tailed' data this pseudo-likelihood method can yield inconsistent kernel densities.

As a step towards a suitable kernel for compositional data let us first recall a standard method of kernel density estimation for a data set of vectors $\mathbf{y}_1, \ldots, \mathbf{y}_N$ in $\mathscr{R}^d$, with $\boldsymbol{\eta}$ denoting a typical vector in $\mathscr{R}^d$. For such variation a popular kernel is

$$\Phi(\boldsymbol{\eta} \,|\, \mathbf{y}, \lambda \mathbf{T}), \tag{14.16}$$

where $\Phi(\cdot | \boldsymbol{\mu}, \boldsymbol{\Sigma})$ denotes the density function of the $\mathscr{N}^d(\boldsymbol{\mu}, \boldsymbol{\Sigma})$ distribution. Popular choices for $\mathbf{T}$ are $\mathbf{T} = \mathbf{I}_d$, giving a spherical normal kernel, and $\mathbf{T} = \mathrm{diag}(t_1^2, \ldots, t_d^2)$, where t_i is the estimated standard deviation of the ith component of the d-dimensional vector, giving a multinormal kernel with independent components.

The success of such kernel methods for variation in $\mathscr{R}^d$ may encourage us to exploit the logratio transformation between $\mathscr{R}^d$ and $\mathscr{S}^d$ with $\mathbf{y}_r = \mathrm{alr}(\mathbf{x}_r)$ $(r = 1, \ldots, N)$. Under the logratio transformation $\boldsymbol{\eta} = \mathrm{alr}(\boldsymbol{\xi})$ and by (6.3), the kernel (14.16) becomes

$$K(\boldsymbol{\xi} \,|\, \mathbf{x}, \lambda) = (\xi_1 \cdots \xi_D)^{-1} \Phi(\boldsymbol{\eta} \,|\, \mathbf{y}, \lambda \mathbf{T}). \tag{14.17}$$

To be effective the kernel must be invariant under the group of permutations of the parts of the composition and unfortunately this is not so for the simple identity and diagonal forms of $\mathbf{T}$ used in the common multinormal kernels. It is, however, easy to ensure this invariance by a simple alternative form for $\mathbf{T}$, namely the estimate $\hat{\boldsymbol{\Sigma}}$ of the logratio covariance matrix. The invariance of the kernel $K(\boldsymbol{\xi}\,|\,\mathbf{x},\lambda)$ defined by

$$\begin{aligned}(2\pi)^{d/2}|\hat{\boldsymbol{\Sigma}}|^{1/2}(\xi_1\cdots\xi_D)K(\boldsymbol{\xi}\,|\,\mathbf{x},\lambda)\\ =\lambda^{-d/2}\exp\{-\tfrac{1}{2}\lambda^{-1}(\boldsymbol{\eta}-\mathbf{y})'\hat{\boldsymbol{\Sigma}}^{-1}(\boldsymbol{\eta}-\mathbf{y})\}\end{aligned} \quad (14.18)$$

is assured by Property 5.3(a), (b). The logarithm of the pseudo-likelihood (14.15), in so far as it involves λ, can then be expressed as

$$-\tfrac{1}{2}Nd\log\lambda+\sum_{r=1}^{N}\log\left\{\sum_{s\neq r}\exp(-\tfrac{1}{2}\lambda^{-1}q_{rs})\right\}, \quad (14.19)$$

where

$$q_{rs}=(\mathbf{y}_r-\mathbf{y}_s)'\hat{\boldsymbol{\Sigma}}^{-1}(\mathbf{y}_r-\mathbf{y}_s). \quad (14.20)$$

The maximization of (14.19) with respect to λ is computationally straightforward with its dependence on the compositional data only through the quadratic forms q_{rs}. With the maximizing value $\hat{\lambda}$ of the smoothing factor, the density at any composition $\boldsymbol{\xi}$ can then be assessed by the kernel density estimate (14.14).

To illustrate the procedure we contrast the results for two compositional data sets, the 30 microscopically determined white-cell compositions of Data 11 and the 23 aphyric Skye lava compositions of Data 6. Both data sets consist of 3-part compositions, but as we have indicated above the pattern of variability of the white-cell compositions is far from logistic normal, whereas the lava compositions are reasonably logistic normal. For the white-cell compositions we might therefore expect the kernel density function to differ from a fitted logistic normal density function, whereas these two density functions should be close for the lava compositions.

For the white-cell and lava compositions the maximizing $\hat{\lambda}$ are 0.151 and 0.427, respectively. Figure 14.2 shows for each data set the plots of the estimated densities by the two methods at the compositions of the data sets, confirming our expectation of discrepancy for the first, and rough agreement for the second, data set.

Other forms of kernel are possible. For example, since the $\mathscr{D}^d(\boldsymbol{\alpha})$

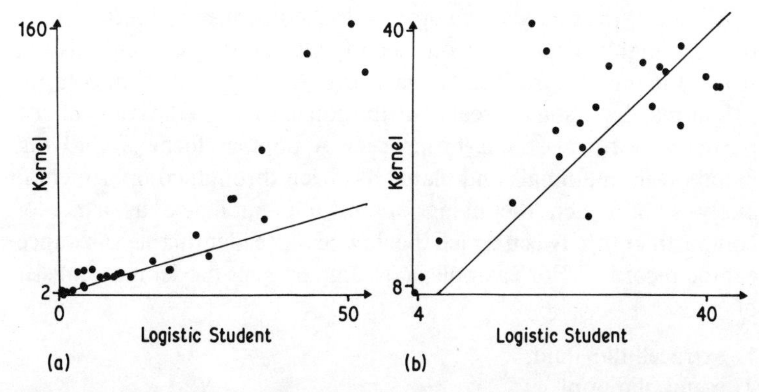

Fig. 14.2. *Comparison of densities assigned by the kernel method and the logistic Student fit for (a) the 30 microscopically determined white-cell compositions of Data 11 and (b) the 23 aphyric Skye lava compositions of Data 6.*

distribution has its mode at $\mathscr{C}(\boldsymbol{\alpha} - \mathbf{j})$, the density function of $\mathscr{D}^d(\mathbf{j} + \lambda^{-1}\mathbf{x})$ with its mode at $\mathbf{x}$ could be considered as an alternative to (14.18). It is, however, less computationally attractive than the logistic normal, though it has certain advantages for sparse data: see Aitchison and Lauder (1985) for further details.

14.6 Compositional stochastic processes

The chemical composition of a product or specimen may vary over time, the proportions of different colours of strands in the cross-section of a thread may vary along its length, foraminiferal and pollen compositions may vary with depth, and so with time, in a core sample. These are examples of stochastic processes in which the states are compositions and the state space is a simplex. A first step in any attempt to describe and analyse such processes is probably to transform to logratio compositions and ask whether the question of interest can be posed and answered within the context of the corresponding process with $\mathscr{R}^d$ state space.

As a simple example, consider a sequential sampling scheme for monitoring inspection of phials containing a 2-part biochemical mixture (x_1, x_2), with acceptable quality defined in terms of a limited range of compositions. Conversion to logratio composition $y = \log(x_1/x_2)$ will then allow access, for example, to the standard $\mathscr{R}^1$ random walk approach to sequential sampling inspection.

Although the analysis of compositional stochastic processes may thus be highly dependent on identifying counterparts in existing stochastic theory for $\mathscr{R}^d$ state space, the statistical theory of compositions may have some direct contribution to make in the study of one particular form of stochastic process. A popular form of studying metabolism in animals and plants has been through compartmental analysis, in which the living organism is modelled as a set of compartments between which the flow of some identifiable substance can be recorded. For example, a 3-compartment model may consist of:

1. extracellular fluid,
2. metabolic pool,
3. urine,

in the study of radioactive sulphate location subsequent to intramuscular injection. Occupancy proportions of the three compartments by the radioactive sulphate at successive times $t_1, t_2, \ldots$ after injection form a sequence of compositions $\mathbf{x}_1, \mathbf{x}_2, \ldots$. A common form of modelling of such a process has been by a Markov chain

$$\mathbf{x}_n = \mathbf{P}_n \mathbf{x}_{n-1} \qquad (n = 1, 2, \ldots), \tag{14.21}$$

where the transition matrix $\mathbf{P}_n$ depends on the time interval (t_{n-1}, t_n). The perturbation operation of Definition 2.13 offers an alternative modelling opportunity with a perturbation step

$$\mathbf{x}_n = \mathbf{x}_{n-1} \circ \mathbf{u}_n, \tag{14.22}$$

in which $\mathbf{u}_n$ is of $\mathscr{L}^d$ form with parameters depending on the time interval (t_{n-1}, t_n). For example, we might consider $\mathscr{L}^d(\boldsymbol{\mu}, \boldsymbol{\Sigma})$ as a unit-time perturbation and consequently take $\mathbf{u}_n$ to be of $\mathscr{L}^d\{(t_n - t_{n-1})\boldsymbol{\mu}, (t_n - t_{n-1})\boldsymbol{\Sigma}\}$ form.

One substantial difference between (14.21) and (14.22) is that the latter cannot accommodate stationary states but this can be overcome by introducing an autoregressive factor, most simply expressed in logratio terms as

$$\mathbf{y}_n = \boldsymbol{\alpha} + \boldsymbol{B}\mathbf{y}_{n-1} + \mathbf{v}_n, \tag{14.23}$$

where $\mathbf{y}_n = \operatorname{alr}(\mathbf{x}_n)$, $\mathbf{v}_n = \operatorname{alr}(\mathbf{u}_n)$. One advantage of the perturbation or logratio approach to compartmental analysis compared with the Markov chain approach is the greater availability of estimation and hypothesis-testing techniques.

14.7 Relation to Bayesian statistical analysis

The natural parameter space for a multinomial experiment with D categories is the simplex

$$\mathcal{S}^d = \{(\theta_1,\ldots,\theta_D):\theta_i > 0 \quad (i=1,\ldots,D), \theta_1 + \cdots + \theta_D = 1\}$$

where $\theta_1,\ldots,\theta_D$ denote the category probabilities. There has thus been an intensive study of distributions on the simplex in this area of Bayesian statistical analysis. The sequence of ideas has been roughly as follows.

1. Meaningful hypotheses about categorical probabilities are associated with zero loglinear contrasts of $(\theta_1,\ldots,\theta_D)$.
2. With conjugate Dirichlet priors, the posterior distributions of loglinear contrasts of $(\theta_1,\ldots,\theta_D)$ are approximately normal (Lindley, 1964; Swe, 1964; Bloch and Watson, 1967).
3. Because of its strong independence structure the Dirichlet class is not rich enough to provide an adequate range of prior distributions.
4. A rich class of prior distributions can be specified directly in terms of normal loglinear contrasts (Leonard, 1973).

Bayesian analysts have thus been led to consideration of a reparametrization ψ of θ through the transformation

$$\theta_i = \exp(\psi_i) \Big/ \sum_{i=1}^{D} \exp(\psi_i) \qquad (i=1,\ldots,D)$$

and to specification of the prior distribution by way of an $\mathcal{N}^D(\boldsymbol{\mu},\boldsymbol{\Sigma})$ distribution on $\boldsymbol{\psi}$. There is, of course, a small price to be paid for the loss of conjugacy but the transformation provides a flexible class of additive logistic normal priors within which forms of exchangeability, quasi-independence and smoothness can be readily introduced by further modelling of $\boldsymbol{\mu}$ and $\boldsymbol{\Sigma}$, for example as in Leonard (1973, 1975, 1977).

Aitchison (1985b) has speculated on what advantages may accrue to compositional data analysts and to Bayesian categorical data analysts through a cross-fertilization of ideas from their separate interests in distributions in the simplex. In this a main contribution from compositional data analysis may be the extension of the prior distributions available from the $\mathcal{L}^d$ class to other transformed normal class. We confine ourselves to two comments here.

1. The importance of the $\mathcal{M}^d$ class of Definition 6.6 in compositional data analysis is that it allows simple specification and testing of hypotheses of neutrality on the right at any level c. Because of its relation to a specific ordering of the components, may the $\mathcal{M}^d$ class have some relevance to the specification of prior distributions on category probability parameters when the categories are ordered? Are there any circumstances in prior specification where we might envisage the assignments to the category probabilities as taking place sequentially, and if so is neutrality a useful concept of assignment? Similar questions could also be asked about the possible use of the partitioned logistic normal classes of Definition 6.7.
2. The Dirichlet class is a conjugate class for multinomial experiments, whereas the logistic normal classes do not have this conjugate property. The general class $\mathcal{A}^d(\boldsymbol{\alpha}, \boldsymbol{B})$ of Definition 13.3 does, however, have this conjugate property: if $\mathcal{A}^d(\boldsymbol{\alpha}, \boldsymbol{B})$ is a prior distribution on the multinomial probability vector $\boldsymbol{\theta} \in \mathcal{S}^d$ and the multinomial experiment results in $\mathbf{n} = (n_1, \ldots, n_D)$ in the D categories, then the posterior distribution is $\mathcal{A}^d(\boldsymbol{\alpha} + \mathbf{n}, \boldsymbol{B})$.

For Bayesians, another area of application of particular interest may be the use of logistic normal distributions to describe and analyse intra-person and inter-person variability in inferential trials in which subjects (such as clinicians) assign probabilities to a finite number of hypotheses (disease types) on the basis of information (diagnostic test results) presented for a sequence of cases (patients); Problem 1.2 was a simple example of such a study. For an account of various useful measures of performance and the simplex distributional properties associated with their analysis, see Aitchison (1981c). A related problem is in the reconciliation of subjective probability assessments by Lindley, Tversky and Brown (1979), who in their use of normal log-odds models to describe subjective assessments implicitly use logistic normal modelling. Our analysis of a simple example in Section 11.7 suggests that a fuller exploitation of the properties of the $\mathcal{L}^d$ class, such as the simple distributional properties of subcompositions, has still much to offer in this area of Bayesian investigation.

14.8 Compositional and directional data

A D-part composition $(x_1, \ldots, x_D)$ is characterized by the non-negativity of its components and by the unit-sum constraint

$$x_1 + \cdots + x_D = 1. \tag{14.24}$$

There is a substantial methodology for the statistical analysis of vectors subject to another constraint, namely directional data; see, for example, the texts by Mardia (1972) and Watson (1983). A direction in d dimensions can be represented by a $D \times 1$ vector $\mathbf{u} = (u_1, \ldots, u_D)$, where $D = d + 1$, with the unit-length constraint

$$u_1^2 + \cdots + u_D^2 = 1. \tag{14.25}$$

Directional data applications are almost entirely confined to $d = 1$ and 2 with representation of directions as points on the unit circle and sphere and usually in terms of the useful polar coordinate system; in contrast, many compositional data applications have d as large as 10.

At first sight the transformation

$$u_i = \sqrt{x_i} \qquad (i = 1, \ldots, D) \tag{14.26}$$

appears to convert a compositional data problem into a directional data problem since constraint (14.25) is then satisfied. Unfortunately this is not the case since (14.26) takes the simplex $\mathscr{S}^d$ into only part of the d-dimensional unit sphere

$$\mathscr{F}^d = \{\mathbf{u}{:}\mathbf{u} \in \mathscr{R}^D, \mathbf{u}'\mathbf{u} = 1\},$$

namely that part in the positive orthant: Fig. 14.3(a) shows the transformation for $d = 1$. Indeed the simplex $\mathscr{S}^d$ and the unit sphere are topologically different and there is no way of transforming one into the other.

Even with this fundamental topological difficulty it could be argued that the transformation provides a practical tool of exploiting

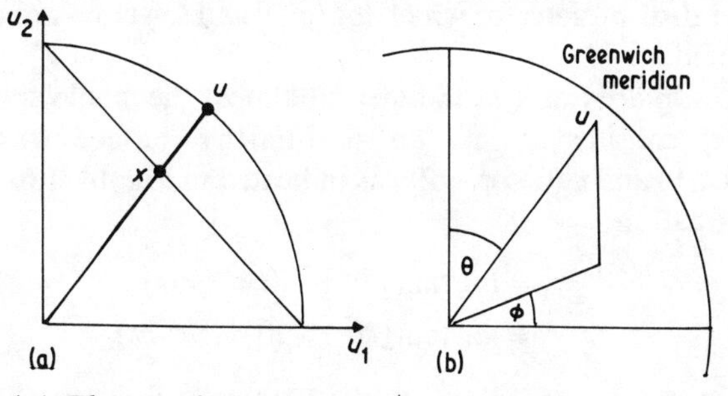

Fig. 14.3. *(a) The transformation* $u_i = \sqrt{x_i}$ *(i = 1, 2). (b) Spherical polar coordinates.*

existing directional data methodology for compositional data analysis. For example, Stephens (1982) uses the transformation (14.26) and the von Mises spherical distribution to describe the patterns of variability of the activities of different groups of students. While such a transformation technique may be adequate for descriptive and discriminatory purposes, the theory of spherical distributions seems to have little to offer compositional data analysis, which is now adequately served by its own simplex distributions. Moreover, such fundamental simplex operations as formation of subcompositions, amalgamations and partitions do not have meaningful counterparts in directional terms, nor is there a satisfactory definition of covariance structure for directional vectors.

We may even conjecture that any flow of ideas between these topologically different sample spaces could be from compositional to directional analysis. For example, in terms of the familiar spherical polar coordinates

$$u_1 = \sin\theta\cos\phi, \qquad u_2 = \sin\theta\sin\phi, \qquad u_3 = \cos\theta,$$

for $d = 2$, the transformation (14.26), followed by the multiplicative transformation

$$y_1 = \log\{x_3/(1 - x_3)\}, \qquad y_2 = \log\{(x_2/(1 - x_3 - x_2)\},$$

takes the positive orthant of the sphere onto $\mathscr{R}^2$. Since

$$\begin{aligned} y_1 &= -2\log\tan\theta, \\ y_2 &= 2\log\tan\phi \qquad (0 \leqslant \theta \leqslant \tfrac{1}{2}\pi, 0 \leqslant \phi \leqslant \tfrac{1}{2}\pi), \end{aligned}$$

our knowledge of the usefulness of transformed normal distributions on the simplex suggests that a useful strategy on the sphere may be to consider distributions for which $\log\tan\theta$ and $\log\tan\phi$ are normally distributed.

We can indeed extend the idea to almost the whole sphere. For example, consider points on the Earth's surface with angles $\theta\,(0 \leqslant \theta \leqslant \pi)$ and $\phi\,(0 \leqslant \phi < 2\pi)$ as indicated in Fig. 14.4(b). Then the transformation

$$\begin{aligned} y_1 &= \log\tan(\tfrac{1}{2}\theta) \qquad (0 < \theta < \pi), \\ y_2 &= \log\tan(\tfrac{1}{4}\phi) \qquad (0 < \phi < 2\pi), \end{aligned}$$

and the assumption of an $\mathscr{N}^2(\boldsymbol{\mu}, \boldsymbol{\Sigma})$ distribution for (y_1, y_2) will induce a distribution on the sphere, except for a slit along 'Greenwich meridian'. Such a class of distributions, with its ease of specifying correlation between θ and ϕ, could prove useful, for example in the study of the longitude–latitude variation of species, provided the Greenwich meridian is a region of low density.

Problems

14.1 Reconsider the questions you posed in Problem 1.10 about the areal compositions of animals and vegetation and try to formulate them as hypotheses about 2-way compositions. What answers to your questions can you now provide?

14.2 Data 40 shows for each of 33 'counties' of Scotland the percentages of boys in each of five hair-colour categories and also in each of four eye-colour categories, from the study by Tocher (1908). Can you provide a measure of the correlation of patterns of hair and eye colour distribution within county? How would your form of analysis differ if Tocher had been able to afford to compute the percentages in each hair-colour category?

APPENDIX A

Algebraic properties of elementary matrices

Table A1 provides definitions and terminology of elementary matrices used throughout this monograph.

Table A1

Notation	*Definition*	*Order*	*Rank*
$\mathbf{I}_d$	identity matrix	$d \times d$	d
$\mathbf{J}_d$	matrix of units	$d \times d$	1
$\mathbf{j}_d$	column vector of units	$d \times 1$	1
$\mathbf{F}_{d,D}$	$[\mathbf{I}_d : -\mathbf{j}_d]$	$d \times D$	d
$\mathbf{G}_D$	$\mathbf{I}_D - D^{-1}\mathbf{J}_D$	$D \times D$	d
$\mathbf{H}_d$	$\mathbf{I}_d + \mathbf{J}_d$	$d \times d$	d

To simplify notation in the recording of properties of these elementary matrices, we drop the suffix notation on the understanding that the orders of the matrices are as in the table above.

$\mathbf{J}_d$ *properties*

J1. $\mathbf{J} = \mathbf{j}\mathbf{j}'$

J2. $\mathbf{J}\mathbf{j} = d\mathbf{j}$

J3. $\mathbf{J}^2 = d\mathbf{J}$

J4. $\mathbf{j}'\mathbf{j} = d$

$\mathbf{F}_{d,D}$ *properties*

F1. $\mathbf{F} = \mathbf{F}\mathbf{G}$

F2. $\mathbf{F}\mathbf{j}_D = \mathbf{0}$

$\mathbf{G}_D$ properties

G1. $\mathbf{G}' = \mathbf{G}$ (symmetric)

G2. $\mathbf{G} = \mathbf{G}^2$

G3. $\mathbf{G} = \mathbf{F}'(\mathbf{F}\mathbf{F}')^{-1}\mathbf{F} = \mathbf{F}'\mathbf{H}^{-1}\mathbf{F}$

G4. $\mathbf{G} = \mathbf{I} - \mathbf{j}(\mathbf{j}'\mathbf{j})^{-1}\mathbf{j}'$

G5. $\mathbf{G} = \mathbf{G}\mathbf{H}$

G6. $\mathbf{G} = \mathbf{G}\mathbf{H}\mathbf{G}$

G7. $\mathbf{G}\mathbf{j} = \mathbf{0}$

G8. $\mathbf{G}\mathbf{J} = \mathbf{0}$

G9. $|\mathbf{G}| = 0$

G10. $\mathbf{G}$ has one zero eigenvalue with eigenvector $\mathbf{j}_D$; the other d eigenvalues are all 1.

G11. A D-vector $\mathbf{a} \in \mathscr{R}^D$ satisfies $\mathbf{a}'\mathbf{j} = 0$ if and only if $\mathbf{G}\mathbf{a} = \mathbf{a}$.

$\mathbf{H}_d$ properties

H1. $\mathbf{H}' = \mathbf{H}$ (symmetry)

H2. $\mathbf{H} = \mathbf{F}\mathbf{F}'$

H3. $\mathbf{H}\mathbf{j} = D\mathbf{j}$

H4. $\mathbf{H}\mathbf{J} = D\mathbf{J}$

H5. $|\mathbf{H}| = D$

H6. $\mathbf{H}^{-1} = \mathbf{I}_d - D^{-1}\mathbf{J}_d$

H7. $\mathbf{H}^{1/2} = \mathbf{I}_d - d^{-1}(1 \pm D^{1/2})\mathbf{J}_d$

H8. $\mathbf{H}^{-1/2} = \mathbf{I}_d - d^{-1}(1 \pm D^{-1/2})\mathbf{J}_d$

APPENDIX B

Bibliography

Abramowitz, M. and Stegun, I.A. (1972). *Handbook of Mathematical Functions.* New York: Dover.

Aitchison, J. (1955). On the distribution of a positive random variable having a discrete probability mass at the origin. *J. Amer. Statist. Assoc.*, **50**, 901–908.

Aitchison, J. (1974). Hippocratic inference. *Bull. Institute of Mathematics and its Applications*, **10**, 48–53.

Aitchison, J. (1976). Goodness of prediction fit. *Biometrika*, **62**, 547–554.

Aitchison, J. (1981a). A new approach to null correlations of proportions. *J. Math. Geol.*, **13**, 175–189.

Aitchison, J. (1981b). Distributions on the simplex for the analysis of neutrality. In *Statistical Distributions in Scientific Work* (C. Taillie, G.P. Patil and B. Baldessari, eds) Vol. 4, pp. 147–156. Dordrecht, Holland: D. Reidel Publishing Company.

Aitchison, J. (1981c). Some distribution theory related to the analysis of subjective performance in inferential tasks. In *Statistical Distributions in Scientific Work* (C. Taillie, G.P. Patil and B. Baldessari, eds) Vol. 5, pp. 363–385. Dordrecht, Holland: D. Reidel Publishing Company.

Aitchison, J. (1982). The statistical analysis of compositional data (with discussion). *J.R. Statist. Soc.* B, **44**, 139–177.

Aitchison, J. (1983). Principal component analysis of compositional data. *Biometrika*, **70**, 57–65.

Aitchison, J. (1984a). The statistical analysis of geochemical compositions. *J. Math. Geol.*, **16**, 531–564.

Aitchison, J. (1984b). Reducing the dimensionality of compositional data sets. *J. Math. Geol.*, **16**, 617–636.

Aitchison, J. (1985a). A general class of distributions on the simplex. *J.R. Statist. Soc.* B, **47**, 136–146.

Aitchison, J. (1985b). Practical Bayesian problems in simplex sample spaces. In *Proceedings of the Second International Meeting on Bayesian Statistics* (J.M. Bernardo, M.H. DeGroot, D.V. Lindley, and A.F.M. Smith, eds). pp. 15–32. Amsterdam: North Holland.

Aitchison, J. and Bacon-Shone, J.H. (1984). Logcontrast models for experiments with mixtures. *Biometrika*, **71**, 323–330.

Aitchison, J. and Begg, C.B. (1976). Statistical diagnosis when the cases are not classified with certainty. *Biometrika*, **63**, 1–12.

Aitchison, J. and Brown, J.A.C. (1957). *The Lognormal Distribution*. Cambridge University Press.

Aitchison, J. and Dunsmore, I.R. (1975). *Statistical Prediction Analysis*. Cambridge University Press.

Aitchison, J., Habbema, J.D.F. and Kay, J.W. (1977). A critical comparison of two methods of statistical discrimination. *Appl. Statist.*, **26**, 15–25.

Aitchison, J. and Lauder, I.J. (1979). Statistical diagnosis from imprecise data. *Biometrika*, **66**, 475–83.

Aitchison, J. and Lauder, I.J. (1985). Kernel density estimation for compositional data. *Appl. Statist.*, **34**, 129–137.

Aitchison, J. and Shen, S.M. (1980). Logistic-normal distributions: some properties and uses. *Biometrika*, **67**, 261–72.

Aitchison, J. and Shen, S.M. (1984). Measurement error in compositional data. *J. Math. Geol.*, **16**, 637–650.

Aitchison, J. and Silvey, S.D. (1960). Maximum-likelihood estimation procedures and associated tests of significance. *J.R. Statist. Soc.* B, **22**, 154–171.

Anderson, J.A. (1972). Separate sample logistic discrimination. *Biometrika*, **59**, 19–35.

Andrews, D.F., Gnanadesikan, R. and Warner, J.L. (1973). Methods for assessing multivariate normality. In *Multivariate Analysis* III (P.R. Krishnaiah, ed.), pp. 95–116. New York: Academic Press.

Barker, D.S. (1978). Magmatic trends on alkali-iron-magnesium diagrams. *Amer. Mineral.*, **63**, 531–534.

Beale, E.M.L. and Little, R.J.A. (1975). Missing values in multivariate analysis. *J.R. Statist. Soc.* B, **37**, 129–145.

Becke, F. (1897). Gesteine der Columbretes, Tschermak's Mineral. *Petrogr. Mitt.*, **16**, 308–336.

Becker, N.G. (1968). Models for the response of a mixture. *J.R. Statist. Soc.* **B**, **30**, 349–358.

Becker, N.G. (1978). Models and designs for experiments with mixtures. *Aust. J. Statist.*, **20**, 195–208.

Birks, H.J.B. and West, R.G. (1973). *Quaternary Plant Ecology*. The 14th Symposium of the British Ecological Society, University of Cambridge, 28–30 March 1972. Oxford: Blackwell Scientific Publications.

Bloch, D.A. and Watson, G.S. (1967). A Bayesian study of the multinomial distribution. *Ann. Math. Statist.*, **38**, 1423–1435.

Bowman, A.W. (1984). An alternative method of cross-validation for the smoothing of density estimates. *Biometrika*, **71**, 353–360.

Box, G.E.P. and Cox, D.R. (1964). The analysis of transformations. *J.R. Statist. Soc.* B, **26**, 211–252.

Brown, J.A.C. and Deaton, A.S. (1972). Surveys in applied economics: models of consumer behaviour. *Econ. J.*, **82**, 1145–1236.

Butler, J.C. (1975). Occurrence of negative open variances in ternary systems. *J. Math. Geol.*, **7**, 31–45.

Butler, J.C. (1976). Principal component analysis using the hypothetical closed array. *J. Math. Geol.*, **8**, 25–36.

Butler, J.C. (1978). Visual bias in R-mode dendograms due to the effect of closure. *J. Math. Geol.*, **10**, 243–252.

Butler, J.C. (1979a). The effects of closure on the moments of a distribution. *J. Math. Geol.*, **11**, 75–84.

Butler, J.C. (1979b). Effects of closure on the measure of similarity between samples. *J. Math. Geol.*, **11**, 431–440.

Butler, J.C. (1979c). Trends in ternary petrologic variation diagrams – fact or fantasy? *Amer. Mineral.*, **64**, 1115–1121.

Butler, J.C. (1979d). Numerical consequences of changing the units in which chemical analyses of igneous rocks are analysed. *Lithos*, **12**, 33–39.

Butler, J.C. (1981). Effect of various transformations on the analysis of percentage data. *J. Math. Geol.*, **13**, 53–68.

Carr, P.F. (1981). Distinction between Permian and post-Permian igneous rocks in the Southern Sydney Basin, New South Wales, on the basis of major-element geochemistry. *J. Math. Geol.*, **13**, 193–199.

Chayes, F. (1956). *Petrographic Modal Analysis.* New York: Wiley.

Chayes, F. (1960). On correlation between variables of constant sum. *J. Geophys. Res.*, **65**, 4185–4193.

Chayes, F. (1962). Numerical correlation and petrographic variation. *J. Math. Geol.*, **70**, 440–452.

Chayes, F. (1964). Variance-covariance relations in some published Harker diagrams of volcanic suites. *J. Petrology*, **5**, 219–237.

Chayes, F. (1971). *Ratio Correlation.* University of Chicago Press.

Chayes, F. (1972). Effect of the proportion transformation on central tendency. *J. Math. Geol.*, **4**, 269–270.

Chayes, F. (1983). Detecting nonrandom associations between proportions by tests of remaining-space variables. *J. Math. Geol.*, **15**, 197–206.

Chayes, F. and Fairbairn, H.W. (1951). A test of the precision of thin-section analysis by point counter. *Amer. Mineral.*, **36**, 707–712.

Chayes, F. and Kruskal, W. (1966). An approximate statistical test for correlations between proportions. *J. Geol.*, **74**, 692–702.

Chayes, F. and Trochimczyk, J. (1978). An effect of closure on the structure of principal components. *J. Math. Geol.*, **10**, 323–333.

Chernoff, H. (1954). On the distribution of the likelihood ratio. *Ann. Math. Statist.*, **25**, 573–578.

Coakley, J.P. and Rust, B.R. (1968). Sedimentation in an Arctic lake. *J. Sed. Petrology*, **38**, 1290–1300.

Connor, J.R. and Mosimann, J.E. (1969). Concepts of independence for proportions with a generalization of the Dirichlet distribution. *J. Amer. Statist. Assoc.*, **64**, 194–206.

Cornell, J.A. (1981). *Experiments with Mixtures.* New York: Wiley.

Cox, D.R. (1966). Some procedures associated with the logistic response curve. In *Research Papers in Statistics: Festschrift for J. Neyman* (F.N. David, ed.), pp. 57–71. New York: Wiley.

Cox, D.R. (1971). A note on polynomial response functions for mixtures. *Biometrika*, **50**, 155–159.

Cramer, H. (1946). *Mathematical Methods of Statistics.* Princeton University Press.

Darroch, J.N. (1969). Null correlations for proportions. *J. Math. Geol.*, **1**, 221–227.

Darroch, J.N. and James, I.R. (1974). *F*-independence and null correlations of bounded-sum, positive variables. *J.R. Statist. Soc.* B, **36**, 467–483.

Darroch, J.N. and Mosimann, J.E. (1985). Canonical and principal components of shape. *Biometrika*, **72**, 247–252.

Darroch, J.N. and Ratcliff, D. (1970). Null correlations for proportions II. *J. Math. Geol.*, **2**, 307–312.

Darroch, J.N. and Ratcliff, D. (1971). A characterization of the Dirichlet distribution. *J. Amer. Statist. Assoc.*, **66**, 641–643.

Darroch, J.N. and Ratcliff, D. (1978). No association of proportions. *J. Math. Geol.*, **10**, 361–368.

Darroch, J.N. and Waller, J. (1985). Additivity and interaction in three-component experiments with mixtures. *Biometrika*, **72**, 153–164.

Dawid, A.P. (1976). Properties of diagnostic data distributions. *Biometrics*, **32**, 647–658.

Day, N.E. and Kerridge, D.F. (1967). A general maximum likelihood discriminant. *Biometrics*, **23**, 313–323.

Deaton, A.S. (1978). Specification and testing in applied demand analysis. *Econ. J.*, **88**, 524–536.

Deaton, A.S. and Muellbauer, J. (1980). *Economics and Consumer Behavior.* Cambridge University Press.

Dempster, A.P., Laird, N.M. and Rubin, D.B. (1977). Maximum likelihood from incomplete data via the EM algorithm (with discussion). *J.R. Statist. Soc.* B, **39**, 1–38.

Edwards, A.W.F. (1977). *Foundations of Mathematical Genetics.* Cambridge University Press.

Effroymson, M.A. (1960). Multiple regression analysis. In *Mathematical Methods for Digital Computers* (A. Ralston and H.S. Will, eds) Vol. 1, pp. 191–203. New York: Wiley.

Fawcett, C.D. (1901). A second study of the variation and correlation of the human skull, with special reference to the Naqada crania. *Biometrika*, **1**, 408–467.

Feder, P.I. (1968). On the distribution of the loglikelihood ratio test statistic when the true parameter is 'near' the boundaries of the hypothesis regions. *Ann. Math. Statist.*, **39**, 2044–2055.

Ferrers, N.M. (1866). *An Elementary Treatise on Trilinear Coordinates.* London: Macmillan.

Fisher, R.A. (1947). The analysis of covariance method for the relation between a part and the whole. *Biometrics*, **3**, 65–66.

Gantmacher, F.R. (1959). *The Theory of Matrices.* New York: Chelsea.

Giridhar, A., Narasimham, P.S.L. and Mahadevan, S. (1981). Density and microhardness of Ge-Sb-Se glasses. *J. Non-Crystalline Solids*, **43**, 29–35.

Gnanadesikan, R. (1977). *Methods for Statistical Data Analysis of Multivariate Observations.* New York: Wiley.

Goodhardt, G.J., Ehrenberg, A.S.C. and Chatfield, C. (1984). The Dirichlet: a comprehensive model of buying behaviour (with discussion). *J.R. Statist. Soc.* A, **147**, 621–655.

Gower, J.C. (1966). Some distance properties of latent root and vector methods used in multivariate analysis. *Biometrika*, **53**, 325–338.

Habbema, J.D.F., Hermans, J. and Van Den Broek, K. (1974). A stepwise discriminant analysis program, using density estimation. In *Compstat 1974, Proceedings in Computational Statistics* (G. Bruckmann, ed.), pp. 101–110. Vienna Physics Verlag.

Hohn, M.E. and Nuhfer, E.B. (1980). Asymmetric measures of association, closed data, and multivariate analysis. *J. Math. Geol.*, **12**, 235–246.

Hotelling, H. (1933). Analysis of a complex of statistical variables into principal components. *J. Educ. Psych.*, **24**, 417–441.

Houthakker, H.S. (1960). Additive preferences. *Econometrica*, **28**, 244–256.

Iddings, J.P. (1903). Chemical composition of igneous rocks. *U.S. Geol. Survey Professional Paper 18.*

James, I.R. (1972). Products of independent beta variables with application to Connor and Mosimann's generalized Dirichlet distribution. *J. Amer. Statist. Assoc.*, **67**, 910–912.

James, I.R. (1975). Multivariate distributions which have beta conditional distributions. *J. Amer. Statist. Assoc.*, **70**, 681–684.

James, I.R. (1981). Distributions associated with neutrality properties for random proportions. In *Statistical Distributions in Scientific Work* (C. Taillie, G.P. Patil and B. Baldessari, eds), pp. 125–136. Dordrecht, Holland: D. Reidel Publishing Company.

James, I.R. and Mosimann, J.E. (1980). A new characterization of the Dirichlet distribution through neutrality. *Ann. Statist.*, **8**, 183–189.

Jeffreys, H. (1961). *Theory of Probability* (3rd edition). Oxford University Press.

Johnson, N.L. (1949). Systems of frequency curves generated by methods of translation. *Biometrika*, **36**, 149–176.

Johnson, N.L. and Kotz, S. (1972). *Distributions in Statistics: Continuous Multivariate Distributions.* Boston: Houghton Mifflin.

Jöreskog, K.J., Klovan, J.E. and Reyment, R.A. (1976). *Geological Factor Analysis.* Amsterdam: Elsevier.

Kaiser, R.F. (1962). Composition and origin of glacial till, Mexico and Kasoag quadrangles, New York. *J. Sed. Petrology*, **32**, 502–513.

Kaplan, E.B. and Elston, R.C. (1972). A subroutine package for maximum likelihood estimation (MAXLIK). *Institute of Statistics Micro Series No. 823.*

Kork, J.O. (1977). Examination of the Chayes–Kruskal procedure for testing correlations between proportions. *J. Math. Geol.*, **9**, 543–562.

Krumbein, W.C. (1962). Open and closed number systems stratigraphic mapping. *Bull. Amer. Assoc. Petrol. Geologists*, **46**, 2229–2245.

Krumbein, W.C. and Tukey, J.W. (1956). Multivariate analysis of mineralogic, lithologic, and chemical composition of rock bodies. *J. Sed. Petrology*, **26**, 322–337.

Krumbein, W.C. and Watson, G.S. (1972). Effect of trends on correlation in open and closed three-component systems. *J. Math. Geol.*, **4**, 317–330.

Kullback, S. and Leibler, R.A. (1951). On information and sufficiency. *Ann. Math. Statist.*, **22**, 525–540.

Lachenbruch, P.A. and Mickey, M.R. (1968). Estimation of error rates in discriminant analysis. *Technometrics*, **10**, 1–10.

Lauder, I.J. (1978). Computational problems in predictive diagnosis. *Compstat 1978*, 186–192.

Le Maître, R.W. (1962). Petrology of volcanic rock, Gough Island South Atlantic. *Geol. Soc. Amer. Bull.*, **73**, 1309–1340.

Le Maître, R.W. (1968). Chemical variation within and between volcanic rock series – a statistical approach. *J. Petrol.*, **9**, 220–252.

Le Maître, R.W. (1976). A new approach to the classification of igneous rocks using the basalt-andesite-dacite-rhyolite suite as an example. *Contrib. Mineral Petrology*, **56**, 191–203.

Le Maître, R.W. (1982). *Numerical Petrography*. Amsterdam: Elsevier.

Leonard, T. (1973). A Bayesian method for histograms. *Biometrika*, **60**, 297–308.

Leonard, T. (1975). Bayesian estimation methods for two-way contingency tables. *J.R. Statist. Soc.* **B**, **37**, 23–37.

Leonard, T. (1977). Bayesian simultaneous estimation for several multinomial distributions. *Comm. Statist.*, **A6**.

Leser, C.E.V. (1963). Forms of Engel functions. *Econometrica*, **31**, 694–703.

Li, C.K. (1985). *The Statistical Analysis of Multi-way and Multiple Compositional Data.* Ph.D. dissertation: University of Hong Kong.

Lindley, D.V. (1964). The Bayesian analysis of contingency tables. *Ann. Math. Statist.*, **35**, 1622–1643.

Lindley, D.V., Tversky, A. and Brown, R.V. (1979). On the reconciliation of probability assessments (with discussion). *J.R. Statist. Soc.* A, **142**, 146–180.

McAlister, D. (1879). The law of the geometric mean. *Proc. R. Soc.*, **29**, 367.

McCammon, R.B. (1975). *Concepts in Geostatistics.* New York: Wiley.

MacDonell, W.R. (1904). A study of the variation and correlation of the human skull, with special reference to English crania. *Biometrika*, **3**, 191–244.

Mardia, K.V. (1972). *Statistics of Directional Data.* London: Academic Press.

Mardia, K.V., Kent. J.T. and Bibby, J.M. (1979). *Multivariate Analysis.* New York: Academic Press.

Marquardt, D.W. (1963). An algorithm for least-squares estimation of nonlinear parameters. *J. SIAM*, **11**, 431–441.

Miesch, A.T. (1969). The constant sum problem in geochemistry. In *Computer Applications in the Earth Sciences*, (D.F. Merriam, ed.), pp. 161–177. New York: Plenum.

Miesch, A.T. (1980). Scaling variables and interpretation of eigenvalues in principal component analysis of geologic data. *J. Math. Geol.*, **12**, 523–538.

Mosimann, J.E. (1962). On the compound multinomial distribution, the multivariate β-distribution and correlations among proportions. *Biometrika*, **49**, 65–82.

Mosimann, J.E. (1963). On the compound negative multinomial distribution and correlations among inversely sampled pollen counts. *Biometrika*, **50**, 47–54.

Mosimann, J.E. (1970). Size allometry: size and shape variables with characterizations of the lognormal and generalized gamma distributions. *J. Amer. Statist. Assoc.*, **65**, 630–645.

Mosimann, J.E. (1975a). Statistical problems of size and shape. I. Biological applications and basic theorems. In *Statistical Distributions in Scientific Work* (G.P. Patil, S. Kotz and J.K. Ord, eds) Vol. 2, pp. 187–217. Dordrecht, Holland: D. Reidel Publishing Company.

Mosimann, J.E. (1975b). Statistical problems of size and shape. II. Characterizations of the lognormal, gamma, and Dirichlet distributions. In *Statistical Distributions in Scientific Work* (G.P. Patil, S. Kotz and J.K. Ord, eds) Vol. 2, pp. 219–239. Dordrecht, Holland: D. Reidel Publishing Company.

Murray, G.D. (1979). The estimation of multivariate normal density functions using incomplete data. *Biometrika*, **66**, 375–380.

Napier, L.E. (1922). The differential blood count. *Ind. Med. Gazette*, **57**, 176–179.

Nash, J.C. (1979). *Compact Numerical Methods for Computers: Linear Algebra and Function Minimisation.* New York: Wiley.

Nisbet, E.G., Bickle, M.J. and Martin, A. (1977). The mafic and ultramafic lavas of the Belingwe Greenstone Belt, Rhodesia. *J. Petrology*, **18**, 521–566.

Obenchain, R.L. (1970). Simplex distributions generated by transformations. *Bell Laboratories Technical Report.*

Pearson, K. (1897). Mathematical contributions to the theory of evolution. On a form of spurious correlation which may arise when indices are used in the measurement of organs. *Proc. R. Soc.*, **60**, 489–498.

Pearson, K. (1901). On lines and planes of closest fit to systems of points in space. *Phil. Mag.*, **2** (6th series), 559–572.

Poore, R.Z. and Berggren, W.A. (1975). Late Cenozoic planktonic foraminiferal biostratigraphy and paleoclimatology of Hatton–Rockall Basin: DSDP Site 116. *J. Foraminiferal Research*, **5**, 270–293.

Quenouille, M.H. (1953). *The Design and Analysis of Experiments.* London: Griffin.

Quenouille, M.H. (1959). Experiments with mixtures. *J.R. Statist. Soc.* **B**, **21**, 201–202.

Rao, C.R. (1965). *Linear Statistical Inference and its Applications.* New York: Wiley.

Reyment, R.A. (1977). Presidential address to the International Association for Mathematical Geology. *J. Math. Geol.*, **9**, 451–454.

Roth, H.D., Pierce, J.W. and Huang, T.C. (1972). Multivariate discriminant analysis of bioclastic turbidites. *J. Math. Geol.*, **4**, 249–261.

Saha, A.K., Bhattacharyya, C. and Lakshmipathy, S. (1974). Some problems of interpreting the correlations between the modal variables in granitic rocks. *J. Math. Geol.*, **6**, 245–258.

Sarmanov, O.V. (1961). False correlations between random variables. *Trudy MIAN SSR*, **64**, 173–184.

Sarmanov, O.V. and Vistelius, A.B. (1959). On the correlation of percentage values. *Dokl. Akad. Nauk. SSSR*, **126**, 22–25.

Scheffé, H. (1958). Experiments with mixtures. *J.R. Statist. Soc.* B, **20**, 344–360.

Scheffé, H. (1961). Reply to Mr Quenouille's comments about my paper on mixtures. *J.R. Statist. Soc.* B, **23**, 171–172.

Scheffé, H. (1963). A simplex centroid design for experiments with mixtures (with discussion). *J.R. Statist. Soc.* B, **25**, 235–263.

Shen, S.M. (1982). A method for discriminating between models describing compositional data. *Biometrika*, **69**, 587–596.

Skala, W. (1977). A mathematical model to investigate distortions of correlation coefficients in closed arrays. *J. Math. Geol.*, **9**, 519–528.

Snow, J.W. (1975). Association of proportions. *J. Math. Geol.*, **7**, 63–73.

Sprent, P. (1972). The mathematics of size and shape. *Biometrics*, **28**, 23–37.

Steiner, A. (1958). Petrographic implications of the 1954 Ngauruhoe lava and its xenoliths. *New Zealand J. Geol. Geophys.*, **1**, 3254–3263.

Stephens, M.A. (1974). EDF statistics for goodness of fit and some comparisons. *J. Amer. Statist. Assoc.*, **69**, 730–737.

Stephens, M.A. (1982). Use of the von Mises distribution to analyse continuous proportions. *Biometrika*, **69**, 197–203.

Stone, C.J. (1984). An asymptotically optimal window selection rule for kernel density estimates. *Ann. Statist.*, **12**, 1285–1297.

Stroud, A.H. (1971). *Approximate Calculation of Multiple Integrals*. New Jersey: Prentice-Hall.

Swe, C. (1964). *The Bayesian analysis of contingency tables*. Ph.D. dissertation, University of Liverpool.

Thompson, R.N., Esson, J. and Duncan, A.C. (1972). Major element chemical variation in the Eocene lavas of the Isle of Skye, Scotland. *J. Petrology*, **13**, 219–253.

Till, R. and Colley, H. (1973). Thoughts on use of principal component analysis in petrogenetic problems. *J. Math. Geol.*, **4**, 341–350.

Titterington, D.M. (1984). Recursive parameter estimation using incomplete data. *J.R. Statist. Soc.* B, **46**, 257–267.

Titterington, D.M. and Jiang, J-M (1983). Recursive estimation procedures for missing-data problems. *Biometrika*, **70**, 613–624.

Tocher, J.F. (1908). Pigmentation survey of school children in Scotland. *Biometrika*, **6**, 129–235.

Trochimczyk, J. and Chayes, F. (1977). Sampling variation of principal components. *J. Math. Geol.*, **9**, 497–506.

Trochimczyk, J. and Chayes, F. (1978). Some properties of principal component scores. *J. Math. Geol.*, **10**, 43–52.

Vistelius, A.B., Ivanov, D.N., Kuroda, Y. and Fuller, C.R. (1970). Variation of modal composition of granitic rocks in some regions around the Pacific. *J. Math. Geol.*, **2**, 63–80.

Wald, A. (1943). Tests of statistical hypotheses concerning several parameters when the number of observations is large. *Trans. Amer. Math. Soc.*, **54**, 426–482.

Watson, G.S. (1983). *Distributions on Spheres.* New York: Wiley.
Webb, W.M. and Briggs, L.I. (1966). The use of principal component analysis to screen mineralogical data. *J. Geol.*, **74**, 716–720.
Wilks, S.S. (1938). The large-sample distribution of the likelihood ratio for testing composite hypotheses. *Ann. Math. Statist.*, **9**, 60–62.
Wilks, S.S. (1941). On the determination of sample sizes for setting tolerance limits. *Ann. Math. Statist.*, **12**, 91–96.
Wilks, S.S. (1942). Statistical prediction with special references to the problem of tolerance limits. *Ann. Math. Statist.*, **13**, 400–409.
Working, H. (1943). Statistical laws of family expenditure. *J. Amer. Statist. Assoc.*, **38**, 43–56.
Zukav, G. (1979). *The Dancing Wu-Li Masters.* New York: Bantam.

APPENDIX C

Computer software for compositional data analysis

In principle the various logratio transformations of compositions applied in this monograph produce vectors which are amenable to standard multivariate statistical analyses. Unfortunately most of the recurring operations on compositions, such as subcompositions, amalgamations, partitions and perturbations, are awkward to handle on available computer packages for multivariate statistics. Moreover, many of the compositional concepts such as complete subcompositional independence, the relation of compositions to bases, logcontrast models in experiments with mixtures, and graphical methods such as the use of the ternary diagram, do not arise in such packages.

In order to encourage the use and development of the statistical methodology for compositional data analysis, a microcomputer package CODA has been developed. CODA is a flexible system by which all the applications of this monograph were analysed and which, through its simple menu and single key response system and graphical facilities, will allow users to explore their own compositional data sets quickly and simply.

CODA is supplied on an IBM PC compatible diskette together with a manual providing full details of the menus and using examples from this monograph as illustrations. For details, apply to Chapman and Hall Ltd.

APPENDIX D

Data sets

Data 1. *Compositions of 25 specimens of hongite*

Specimen no.	*Percentages by weight*				
	A	B	C	D	E
H1	48.8	31.7	3.8	6.4	9.3
H2	48.2	23.8	9.0	9.2	9.8
H3	37.0	9.1	34.2	9.5	10.2
H4	50.9	23.8	7.2	10.1	8.0
H5	44.2	38.3	2.9	7.7	6.9
H6	52.3	26.2	4.2	12.5	4.8
H7	44.6	33.0	4.6	12.2	5.6
H8	34.6	5.2	42.9	9.6	7.7
H9	41.2	11.7	26.7	9.6	10.8
H10	42.6	46.6	0.7	5.6	4.5
H11	49.9	19.5	11.4	9.5	9.7
H12	45.2	37.3	2.7	5.5	9.3
H13	32.7	8.5	38.9	8.0	11.9
H14	41.4	12.9	23.4	15.8	6.5
H15	46.2	17.5	15.8	8.3	12.2
H16	32.3	7.3	40.9	12.9	6.6
H17	43.2	44.3	1.0	7.8	3.7
H18	49.5	32.3	3.1	8.7	6.3
H19	42.3	15.8	20.4	8.3	13.2
H20	44.6	11.5	23.8	11.6	8.5
H21	45.8	16.6	16.8	12.0	8.8
H22	49.9	25.0	6.8	10.9	7.4

Data 1. (*Contd.*)

Specimen no.	*Percentages by weight*				
	A	B	C	D	E
H23	48.6	34.0	2.5	9.4	5.5
H24	45.5	16.6	17.6	9.6	10.7
H25	45.9	24.9	9.7	9.8	9.7

A: albite
B: blandite
C: cornite
D: daubite
E: endite

Data 2. *Compositions of 25 specimens of kongite*

Specimen no.	*Percentages by weight*				
	A	B	C	D	E
K1	33.5	6.1	41.3	7.1	12.0
K2	47.6	14.9	16.1	14.8	6.6
K3	52.7	23.9	6.0	8.7	8.7
K4	44.5	24.2	10.7	11.9	8.7
K5	42.3	47.6	0.6	4.1	5.4
K6	51.8	33.2	1.9	7.0	6.1
K7	47.9	21.5	10.7	9.5	10.4
K8	51.2	23.6	6.2	13.3	5.7
K9	19.3	2.3	65.8	5.8	6.8
K10	46.1	23.4	10.4	11.5	8.6
K11	30.6	6.7	43.0	6.3	13.4
K12	49.7	28.1	5.1	8.0	9.1
K13	49.4	24.3	7.6	8.5	10.2
K14	38.4	9.5	30.6	14.8	6.7
K15	41.6	19.0	17.3	13.8	8.3
K16	42.3	43.3	1.6	5.9	6.9
K17	45.7	23.9	10.3	11.6	8.5
K18	45.5	20.3	13.6	10.9	9.7
K19	52.1	17.9	10.7	7.9	11.4
K20	46.2	14.3	18.5	12.2	8.8
K21	47.2	30.9	4.6	6.3	11.0
K22	45.4	33.3	4.0	11.9	5.4
K23	48.6	23.4	8.7	10.7	8.6
K24	31.2	4.5	47.0	10.2	7.1
K25	44.3	15.0	19.4	10.5	10.8

A: albite
B: blandite
C: cornite
D: daubite
E: endite

Data 3. *Compositions and depths of 25 specimens of boxite*

Specimen no.	*Percentages by weight*					*Depth (m)*
	A	B	C	D	E	
B1	43.5	25.1	14.7	10.0	6.7	1
B2	41.1	27.5	13.9	9.5	8.0	2
B3	41.5	20.1	20.6	11.1	6.7	3
B4	33.9	37.8	11.1	11.5	5.7	4
B5	46.5	16.0	15.6	14.3	7.6	5
B6	45.3	19.4	14.8	13.5	9.3	6
B7	33.2	25.2	15.2	17.1	9.3	7
B8	40.8	15.1	21.7	14.6	7.8	8
B9	33.0	30.8	15.1	12.9	8.2	9
B10	28.2	38.6	12.1	14.1	6.9	10
B11	33.9	31.5	15.4	12.0	7.2	11
B12	48.7	19.3	13.4	10.7	7.9	12
B13	37.8	37.1	10.4	8.6	6.1	13
B14	42.0	26.6	13.7	10.5	7.2	14
B15	44.2	26.5	12.9	9.6	6.8	15
B16	39.7	23.2	20.6	10.2	6.3	16
B17	39.3	28.1	13.0	13.6	6.0	17
B18	34.1	26.7	13.6	17.0	8.6	18
B19	36.2	35.3	11.2	11.9	5.4	19
B20	39.5	36.0	9.4	8.4	6.7	20
B21	39.5	22.5	18.7	11.4	7.9	21
B22	33.0	33.5	17.7	9.8	6.0	22
B23	42.3	16.6	16.9	17.0	7.2	23
B24	39.9	19.0	13.4	21.3	6.4	24
B25	37.8	30.9	11.9	12.9	6.5	25

A: albite
B: blandite
C: cornite
D: daubite
E: endite

Data 4. *Compositions, depths and porosities of 25 specimens of coxite*

Specimen no.	*Percentages by weight* A	B	C	D	E	*Depth (m)*	*Porosity*
C1	44.2	31.9	5.4	10.5	8.0	1	21.8
C2	49.0	25.4	5.8	11.3	8.5	2	25.2
C3	50.2	24.8	5.7	11.1	8.2	3	26.1
C4	49.9	24.7	5.4	11.4	8.6	4	26.3
C5	48.5	27.8	5.9	10.2	7.6	5	22.6
C6	45.9	27.1	6.9	11.5	8.6	6	21.4
C7	44.1	31.9	6.0	10.2	7.8	7	22.0
C8	46.4	29.9	5.5	10.3	7.9	8	22.0
C9	45.7	27.0	6.2	12.0	9.1	9	23.1
C10	46.4	30.0	5.1	10.4	8.1	10	24.0
C11	41.7	30.2	7.7	11.6	8.8	11	18.4
C12	44.9	25.7	7.7	12.4	9.3	12	20.5
C13	48.6	27.7	5.8	10.2	7.7	13	22.9
C14	49.7	26.7	4.9	10.6	8.1	14	27.2
C15	49.6	24.4	6.4	11.2	8.4	15	23.4
C16	46.5	28.6	5.9	10.7	8.3	16	22.5
C17	47.3	24.2	7.9	11.8	8.8	17	21.6
C18	44.7	30.0	6.8	10.5	8.0	18	20.5
C19	48.0	25.6	7.0	11.1	8.3	19	22.7
C20	50.0	23.8	6.6	11.2	8.4	20	23.7
C21	51.4	24.2	5.7	10.7	8.0	21	26.2
C22	53.3	25.1	5.2	9.4	7.0	22	26.4
C23	47.9	25.4	6.7	11.4	8.6	23	22.2
C24	43.5	29.8	6.7	11.2	8.8	24	19.5
C25	44.5	29.2	6.5	11.2	8.6	25	21.3

A: albite
B: blandite
C: cornite
D: daubite
E: endite

Porosity is the percentage of void space that the specimen contains.

Data 5. *Sand, silt, clay compositions of 39 sediment samples at different water depths in an Arctic lake*

Sediment no.	Percentages Sand	Silt	Clay	Water depth (m)	Sediment no.	Percentages Sand	Silt	Clay	Water depth (m)
S1	77.5	19.5	3.0	10.4	S21	9.5	53.5	37.0	47.1
S2	71.9	24.9	3.2	11.7	S22	17.1	48.0	34.9	48.4
S3	50.7	36.1	13.2	12.8	S23	10.5	55.4	34.1	49.4
S4	52.2	40.9	6.6	13.0	S24	4.8	54.7	41.0	49.5
S5	70.0	26.5	3.5	15.7	S25	2.6	45.2	52.2	59.2
S6	66.5	32.2	1.3	16.3	S26	11.4	52.7	35.9	60.1
S7	43.1	55.3	1.6	18.0	S27	6.7	46.9	46.4	61.7
S8	53.4	36.8	9.8	18.7	S28	6.9	49.7	43.4	62.4
S9	15.5	54.4	30.1	20.7	S29	4.0	44.9	51.1	69.3
S10	31.7	41.5	26.8	22.1	S30	7.4	51.6	40.9	73.6
S11	65.7	27.8	6.5	22.4	S31	4.8	49.5	45.7	74.4
S12	70.4	29.0	0.6	24.4	S32	4.5	48.5	47.0	78.5
S13	17.4	53.6	29.0	25.8	S33	6.6	52.1	41.3	82.9
S14	10.6	69.8	19.6	32.5	S34	6.7	47.3	45.9	87.7
S15	38.2	43.1	18.7	33.6	S35	7.4	45.6	46.9	88.1
S16	10.8	52.7	36.5	36.8	S36	6.0	48.9	45.1	90.4
S17	18.4	50.7	30.9	37.8	S37	6.3	53.8	39.9	90.6
S18	4.6	47.4	48.0	36.9	S38	2.5	48.0	49.5	97.7
S19	15.6	50.4	34.0	42.2	S39	2.0	47.8	50.2	103.7
S20	31.9	45.1	23.0	47.0					

Adapted from Coakley and Rust (1968, Table 1).

Data 6. *AFM compositions of 23 aphyric Skye lavas*

Specimen no.	*Percentages* A	F	M	*Specimen no.*	*Percentages* A	F	M
S1	52	42	6	S11	21	60	19
S2	52	44	4	S12	25	53	22
S3	47	48	5	S13	24	54	22
S4	45	49	6	S14	22	55	23
S5	40	50	10	S15	22	56	22
S6	37	54	9	S16	20	58	22
S7	27	58	15	S17	16	62	22
S8	27	54	19	S18	17	57	26
S9	23	59	18	S19	14	54	32
S10	22	59	19	S20	13	55	32
				S21	13	52	35
				S22	14	47	39
				S23	24	56	20

A: $Na_2O + K_2O$
F: Fe_2O_3
M: MgO

Adapted from Thompson, Esson and Duncan (1972, Fig. 7).

Data 7. *Proportions of supervisor's statements assigned to different categories*

Fortnight	*Supervisee* 1				2				3			
	C	D	E	F	C	D	E	F	C	D	E	F
1	0.093	0.043	0.773	0.091	0.081	0.015	0.712	0.077	0.313	0.092	0.538	0.057
2	0.192	0.411	0.282	0.115	0.052	0.209	0.585	0.154	0.375	0.364	0.141	0.119
3	0.072	0.772	0.073	0.084	0.170	0.284	0.397	0.148	0.125	0.262	0.487	0.125
4	0.256	0.360	0.221	0.163	0.183	0.056	0.673	0.088	0.273	0.147	0.450	0.130
5	0.244	0.204	0.387	0.165	0.268	0.546	0.083	0.103	0.318	0.290	0.331	0.062
6	0.195	0.015	0.712	0.077	0.609	0.026	0.272	0.094	0.254	0.027	0.625	0.093

C: commanding
D: demanding
E: expository
F: faulting

Data 8. *Household expenditures (HK$) on four commodity groups of 20 single men (M) and* 20 single women (*W*)

House-hold no.	*Commodity group* 1	2	3	4	*House-hold no.*	*Commodity group* 1	2	3	4
M1	497	591	153	291	W1	820	114	183	154
M2	839	942	302	365	W2	184	74	6	20
M3	798	1308	668	584	W3	921	66	1686	455
M4	892	842	287	395	W4	488	80	103	115
M5	1585	781	2476	1740	W5	721	83	176	104
M6	755	764	428	438	W6	614	55	441	193
M7	388	655	153	233	W7	801	56	357	214
M8	617	879	757	719	W8	396	59	61	80
M9	248	438	22	65	W9	864	65	1618	352
M10	1641	440	6471	2063	W10	845	64	1935	414
M11	1180	1243	768	813	W11	404	97	33	47
M12	619	684	99	204	W12	781	47	1906	452
M13	253	422	15	48	W13	457	103	136	108
M14	661	739	71	188	W14	1029	71	244	189
M15	1981	869	1489	1032	W15	1047	90	653	298
M16	1746	746	2662	1594	W16	552	91	185	158
M17	1865	915	5184	1767	W17	718	104	583	304
M18	238	522	29	75	W18	495	114	65	74
M19	1199	1095	261	344	W19	382	77	230	147
M20	1524	964	1739	1410	W20	1090	59	313	177

1. Housing, including fuel and light
2. Foodstuffs, including alcohol and tobacco
3. Other goods, including clothing, footwear and durable goods
4. Services, including transport and vehicles

Data 9. *Urinary excretions (mg/24hr) of steroid metabolites for 37 adults and 30 normal children*

Case no.	*Adults*			*Case no.*	*Children*		
	1	*2*	*3*		*1*	*2*	*3*
A1	2.47	0.29	0.40	C1	1.78	0.29	0.075
A2	2.96	0.39	1.10	C2	1.77	0.21	0.065
A3	4.09	0.26	0.90	C3	2.01	0.37	0.045
A4	3.27	0.24	1.80	C4	1.17	0.25	0.025
A5	2.30	0.51	0.50	C5	1.29	0.17	0.055
A6	5.06	0.50	1.30	C6	2.80	0.26	0.305
A7	2.86	0.42	1.50	C7	1.36	0.30	0.205
A8	3.38	0.51	0.60	C8	3.31	0.28	0.205
A9	3.18	0.12	0.50	C9	0.52	0.07	0.005
A10	3.49	0.31	1.30	C10	1.97	0.14	0.005
A11	3.56	0.57	1.20	C11	2.10	0.23	0.125
A12	5.24	0.51	0.48	C12	2.41	0.23	0.055
A13	5.62	0.29	0.50	C13	1.60	0.26	0.065
A14	2.99	0.38	0.50	C14	1.14	0.071	0.015
A15	2.18	0.34	0.60	C15	1.44	0.161	0.075
A16	3.86	0.37	0.60	C16	1.96	0.21	0.145
A17	3.04	0.35	0.40	C17	2.01	0.27	0.105
A18	2.82	0.29	0.60	C18	0.83	0.12	0.045
A19	2.40	0.38	0.90	C19	1.58	0.12	0.105
A20	4.73	0.35	1.40	C20	1.82	0.20	0.105
A21	3.49	0.40	0.80	C21	1.49	0.25	0.045
A22	6.32	0.86	2.30	C22	1.96	0.23	0.115
A23	3.88	0.37	0.90	C23	1.97	0.29	0.105
A24	3.79	0.42	1.20	C24	0.83	0.10	0.105
A25	9.95	1.00	0.80	C25	1.58	0.08	0.215
A26	7.03	0.56	1.10	C26	2.84	0.09	0.105
A27	4.23	0.48	1.20	C27	1.77	0.14	0.085
A28	5.60	0.48	1.80	C28	3.02	0.34	0.505

(*Contd.*)

Data 9. (*Contd.*)

Case no.	Adults 1	2	3	Case no.	Children 1	2	3
A29	4.30	0.36	0.60	C29	1.17	0.17	0.205
A30	9.74	0.76	1.10	C30	1.69	0.27	0.085
A31	4.54	0.29	0.60				
A32	6.33	0.92	0.90				
A33	6.65	0.66	2.00				
A34	5.96	0.50	1.90				
A35	1.86	0.50	0.50				
A36	1.33	0.13	0.20				
A37	5.42	0.46	0.60				

1. Total cortisol metabolites
2. Total corticosterone metabolites
3. Pregnanetriol + Δ-5-pregnentriol

Adapted from data on Cushing's syndrome supplied by Dr Meta Damkjaer-Nielsen, Glostrup Hospital, Copenhagen.

Data 10. *Activity patterns of a statistician for 20 days*

Day no.	*Proportion of day in activity*					
	1	*2*	*3*	*4*	*5*	*6*
D1	0.144	0.091	0.179	0.107	0.263	0.217
D2	0.162	0.079	0.107	0.132	0.265	0.254
D3	0.153	0.101	0.131	0.138	0.209	0.267
D4	0.177	0.087	0.140	0.132	0.155	0.310
D5	0.158	0.110	0.139	0.116	0.258	0.219
D6	0.165	0.079	0.113	0.113	0.275	0.255
D7	0.159	0.084	0.117	0.094	0.225	0.321
D8	0.161	0.105	0.123	0.110	0.267	0.234
D9	0.163	0.126	0.105	0.106	0.227	0.273
D10	0.169	0.102	0.104	0.104	0.235	0.286
D11	0.149	0.113	0.123	0.115	0.256	0.244
D12	0.118	0.100	0.145	0.096	0.192	0.349
D13	0.106	0.112	0.135	0.104	0.205	0.338
D14	0.163	0.142	0.109	0.115	0.260	0.211
D15	0.151	0.122	0.126	0.121	0.235	0.245
D16	0.163	0.101	0.126	0.142	0.232	0.237
D17	0.176	0.084	0.094	0.098	0.213	0.335
D18	0.104	0.093	0.148	0.090	0.269	0.295
D19	0.111	0.111	0.118	0.086	0.216	0.358
D20	0.105	0.090	0.135	0.117	0.168	0.385

1. Teaching
2. Consultation
3. Administration
4. Research
5. Other wakeful activities
6. Sleep

Data 11. *White-cell compositions of 30 blood cells by two different methods*

Sample no.	*Microscopic inspection*			*Image analysis*		
	G	L	M	G	L	M
S1	0.732	0.256	0.012	0.763	0.223	0.014
S2	0.664	0.280	0.056	0.681	0.262	0.057
S3	0.725	0.214	0.067	0.748	0.198	0.054
S4	0.806	0.175	0.019	0.867	0.116	0.017
S5	0.620	0.351	0.029	0.565	0.408	0.026
S6	0.856	0.113	0.031	0.885	0.086	0.029
S7	0.957	0.030	0.013	0.966	0.023	0.011
S8	0.927	0.053	0.020	0.935	0.047	0.018
S9	0.903	0.072	0.025	0.921	0.055	0.024
S10	0.936	0.055	0.009	0.943	0.048	0.009
S11	0.871	0.114	0.015	0.888	0.097	0.015
S12	0.445	0.523	0.032	0.569	0.401	0.038
S13	0.240	0.736	0.024	0.225	0.747	0.028
S14	0.475	0.472	0.053	0.577	0.370	0.054
S15	0.318	0.663	0.019	0.272	0.706	0.022
S16	0.462	0.516	0.022	0.544	0.432	0.025
S17	0.376	0.252	0.372	0.364	0.245	0.391
S18	0.440	0.240	0.320	0.496	0.180	0.324
S19	0.583	0.142	0.275	0.629	0.099	0.272
S20	0.399	0.169	0.432	0.500	0.115	0.384
S21	0.804	0.152	0.044	0.805	0.155	0.040
S22	0.655	0.263	0.082	0.659	0.247	0.094
S23	0.725	0.218	0.057	0.769	0.179	0.052
S24	0.650	0.298	0.052	0.665	0.283	0.052
S25	0.370	0.166	0.464	0.388	0.159	0.452
S26	0.175	0.802	0.023	0.262	0.709	0.028
S27	0.328	0.627	0.045	0.395	0.561	0.043
S28	0.427	0.511	0.062	0.388	0.542	0.070
S29	0.943	0.046	0.011	0.948	0.041	0.011
S30	0.860	0.108	0.032	0.886	0.086	0.027

G: granulocytes L: lymphocytes M: monocytes

Data 12. *Flesh, skin, stone proportions by volume of fruit for present and past seasons from 40 yatquat trees*

Tree no.	Present season			Past season		
	Flesh	*Skin*	*Stone*	*Flesh*	*Skin*	*Stone*
T1	0.513	0.236	0.251	0.507	0.221	0.272
T2	0.516	0.238	0.246	0.542	0.257	0.201
T3	0.515	0.231	0.254	0.535	0.222	0.243
T4	0.543	0.218	0.240	0.505	0.273	0.221
T5	0.535	0.219	0.246	0.503	0.244	0.253
T6	0.488	0.248	0.264	0.416	0.276	0.308
T7	0.531	0.219	0.250	0.429	0.253	0.319
T8	0.511	0.229	0.259	0.507	0.249	0.243
T9	0.517	0.221	0.262	0.426	0.233	0.341
T10	0.548	0.212	0.240	0.519	0.278	0.203
T11	0.542	0.226	0.231	0.571	0.210	0.219
T12	0.521	0.232	0.247	0.458	0.286	0.256
T13	0.495	0.244	0.262	0.548	0.238	0.214
T14	0.539	0.212	0.248	0.476	0.231	0.293
T15	0.561	0.203	0.235	0.572	0.222	0.206
T16	0.540	0.206	0.255	0.531	0.217	0.252
T17	0.526	0.216	0.258	0.433	0.203	0.364
T18	0.541	0.219	0.240	0.500	0.265	0.235
T19	0.514	0.235	0.251	0.469	0.295	0.236
T20	0.508	0.228	0.263	0.477	0.232	0.292
U1	0.481	0.265	0.254	0.460	0.220	0.320
U2	0.474	0.258	0.269	0.483	0.250	0.267
U3	0.500	0.250	0.250	0.499	0.267	0.234
U4	0.487	0.257	0.255	0.538	0.212	0.250
U5	0.493	0.256	0.251	0.432	0.324	0.244
U6	0.517	0.230	0.254	0.512	0.224	0.263
U7	0.500	0.247	0.245	0.512	0.244	0.243
U8	0.457	0.277	0.265	0.491	0.259	0.250
U9	0.480	0.258	0.262	0.506	0.220	0.274
U10	0.499	0.249	0.253	0.475	0.245	0.280

(Contd.)

Data 12. (*Contd.*)

Tree no.	Present season			Past season		
	Flesh	*Skin*	*Stone*	*Flesh*	*Skin*	*Stone*
U11	0.506	0.236	0.258	0.484	0.241	0.275
U12	0.519	0.225	0.256	0.556	0.222	0.222
U13	0.501	0.249	0.250	0.432	0.253	0.315
U14	0.526	0.234	0.240	0.541	0.216	0.243
U15	0.462	0.272	0.266	0.406	0.269	0.325
U16	0.479	0.277	0.245	0.506	0.219	0.275
U17	0.521	0.245	0.234	0.485	0.240	0.275
U18	0.503	0.239	0.257	0.550	0.197	0.253
U19	0.497	0.246	0.257	0.549	0.189	0.262
U20	0.484	0.258	0.257	0.602	0.190	0.208

T: treated
U: untreated

Data 13. *Serial number r, brilliance B and vorticity V of 81 girandoles*

r	B	V	r	B	V	r	B	V	r	B	V
1	6.2	0.8	21	27.2	22.9	41	15.1	15.7	61	17.4	25.8
2	10.7	8.0	22	16.6	9.7	42	14.0	11.1	62	15.4	17.8
3	13.4	17.2	23	19.0	15.3	43	10.4	16.6	63	14.3	14.0
4	7.4	8.9	24	20.6	16.7	44	15.1	9.9	64	13.1	12.2
5	20.0	10.8	25	20.4	21.7	45	14.5	13.3	65	18.9	12.6
6	8.7	9.8	26	16.3	17.7	46	28.3	16.3	66	22.2	23.7
7	16.6	15.5	27	21.6	22.0	47	16.9	10.1	67	9.3	16.6
8	18.1	12.1	28	12.6	5.3	48	22.0	17.6	68	8.5	7.7
9	12.9	9.4	29	11.3	8.7	49	15.7	19.1	69	14.4	18.0
10	14.0	7.5	30	11.9	18.2	50	9.4	16.6	70	8.3	15.8
11	13.5	8.9	31	6.5	10.4	51	9.5	10.4	71	10.3	12.3
12	12.5	19.5	32	3.1	7.7	52	17.9	27.5	72	13.1	15.8
13	13.3	5.0	33	12.8	18.0	53	7.5	12.3	73	24.8	10.3
14	10.0	10.7	34	14.2	22.3	54	12.2	17.0	74	23.9	11.4
15	15.1	6.3	35	15.6	11.4	55	9.0	8.8	75	25.2	23.3
16	12.8	13.3	36	22.2	19.5	56	11.3	13.8	76	19.2	15.0
17	15.4	10.9	37	18.5	20.2	57	20.4	21.4	77	11.3	4.2
18	10.3	13.0	38	10.7	9.9	58	15.4	21.5	78	15.8	20.9

(*Contd.*)

Data 13. (*Contd.*)

r	B	V	r	B	V	r	B	V	r	B	V
19	14.9	11.6	39	12.9	15.6	59	18.1	14.4	79	6.7	22.5
20	16.7	19.2	40	10.2	7.9	60	13.3	20.0	80	17.3	21.3
									81	17.0	22.2

Note: The 81 different mixtures form a special experimental design and have been determined in the following way. First the 3^4 possible quadruplets formed from the three values $-1, 0, 1$ are arranged in ascending order as $(-1,-1,-1,-1), (-1,-1,-1,0), (-1,-1,-1,1), (-1,-1,0,-1), \ldots, (1,1,1,0), (1,1,1,1)$. From each quadruplet, such as (z_1, z_2, z_3, z_4), a corresponding mixture, $(x_1, x_2, x_3, x_4, x_5)$, is then obtained by the computation:

$$x_i = \frac{\exp(z_i)}{\exp(z_1) + \exp(z_2) + \exp(z_3) + \exp(z_4) + 1} \quad (i = 1, 2, 3, 4),$$

$$x_5 = \frac{1}{\exp(z_1) + \exp(z_2) + \exp(z_3) + \exp(z_4) + 1}.$$

Thus the girandole corresponding to serial number $r = 4$ corresponds to $(z_1, z_2, z_3, z_4) = (-1, -1, 0, -1)$ and so is composed of a mixture with proportions $(x_1, x_2, x_3, x_4, x_5) = (0.12, 0.12, 0.32, 0.12, 0.32)$ of the five ingredients.

Data 14. *Colour–size compositions of 20 clam colonies from East Bay, arranged in 2 × 3 arrays*

Colony no.	*Colour–size composition*			*Colony no.*	*Colour–size composition*		
1	0.132	0.146	0.098	11	0.102	0.209	0.139
	0.229	0.199	0.196		0.205	0.131	0.215
2	0.149	0.135	0.159	12	0.148	0.165	0.147
	0.135	0.215	0.211		0.098	0.188	0.254
3	0.129	0.159	0.138	13	0.205	0.165	0.099
	0.174	0.180	0.221		0.132	0.193	0.207
4	0.199	0.172	0.172	14	0.084	0.177	0.154
	0.120	0.125	0.212		0.180	0.193	0.212
5	0.189	0.112	0.108	15	0.209	0.131	0.131
	0.128	0.273	0.190		0.108	0.187	0.234
6	0.196	0.136	0.151	16	0.176	0.123	0.169
	0.123	0.154	0.239		0.181	0.153	0.199
7	0.140	0.138	0.187	17	0.274	0.160	0.125
	0.193	0.145	0.197		0.114	0.119	0.208
8	0.183	0.149	0.204	18	0.185	0.153	0.109
	0.113	0.174	0.170		0.156	0.192	0.206
9	0.147	0.141	0.159	19	0.234	0.152	0.114
	0.194	0.161	0.198		0.141	0.161	0.199
10	0.155	0.094	0.157	20	0.204	0.160	0.180
	0.182	0.187	0.224		0.151	0.190	0.186

Data 15. *Colour–size compositions of 20 clam colonies from West Bay, arranged in 2 × 3 arrays*

Colony no.	Colour–size composition			Colony no.	Colour–size composition		
1	0.119	0.112	0.195	11	0.052	0.219	0.119
	0.191	0.131	0.252		0.193	0.185	0.232
2	0.075	0.146	0.166	12	0.104	0.128	0.170
	0.164	0.192	0.257		0.199	0.169	0.230
3	0.095	0.165	0.254	13	0.106	0.146	0.216
	0.114	0.115	0.255		0.169	0.131	0.232
4	0.169	0.109	0.161	14	0.130	0.112	0.224
	0.149	0.164	0.247		0.157	0.136	0.241
5	0.119	0.185	0.142	15	0.101	0.162	0.193
	0.116	0.195	0.244		0.184	0.116	0.244
6	0.136	0.111	0.158	16	0.053	0.162	0.225
	0.225	0.141	0.230		0.184	0.125	0.251
7	0.084	0.140	0.172	17	0.154	0.100	0.153
	0.175	0.162	0.267		0.217	0.145	0.231
8	0.220	0.090	0.181	18	0.093	0.126	0.197
	0.117	0.139	0.254		0.201	0.155	0.228
9	0.087	0.148	0.193	19	0.085	0.114	0.201
	0.179	0.195	0.197		0.149	0.212	0.239
10	0.095	0.122	0.197	20	0.074	0.100	0.148
	0.175	0.162	0.250		0.205	0.225	0.247

Data 16. *Serum protein compositions of blood samples of 30 patients of known disease type and of 6 new cases*

Patient no.	*Serum protein*			
	1	*2*	*3*	*4*
A1	0.348	0.197	0.201	0.254
A2	0.386	0.239	0.141	0.242
A3	0.471	0.240	0.089	0.200
A4	0.427	0.245	0.111	0.217
A5	0.346	0.230	0.204	0.219
A6	0.485	0.231	0.101	0.183
A7	0.398	0.217	0.152	0.232
A8	0.537	0.219	0.075	0.169
A9	0.316	0.213	0.212	0.260
A10	0.543	0.251	0.058	0.148
A11	0.409	0.228	0.163	0.199
A12	0.322	0.236	0.188	0.254
A13	0.372	0.229	0.163	0.236
A14	0.452	0.234	0.088	0.226
B1	0.490	0.189	0.102	0.219
B2	0.512	0.219	0.083	0.187
B3	0.429	0.180	0.110	0.273
B4	0.424	0.177	0.149	0.250
B5	0.377	0.175	0.170	0.278
B6	0.556	0.223	0.052	0.170
B7	0.264	0.206	0.202	0.328
B8	0.311	0.179	0.226	0.283
B9	0.338	0.194	0.167	0.300
B10	0.396	0.285	0.091	0.229
B11	0.438	0.244	0.082	0.236
B12	0.347	0.224	0.142	0.287
B13	0.376	0.181	0.147	0.295
B14	0.278	0.156	0.229	0.337
B15	0.333	0.190	0.158	0.319
B16	0.388	0.244	0.124	0.244

(*Contd.*)

Data 16. (*Contd.*)

Patient	*Serum protein*			
no.	*1*	*2*	*3*	*4*
New cases				
C1	0.432	0.189	0.160	0.219
C2	0.386	0.195	0.159	0.260
C3	0.417	0.233	0.187	0.310
C4	0.385	0.188	0.182	0.244
C5	0.346	0.209	0.180	0.265
C6	0.496	0.241	0.065	0.198

1. albumin
2. pre-albumin
3. globulin A
4. globulin B

Data 17. *Subjective diagnostic probabilities assigned by 15 clinicians and 15 statisticians*

Clinician no.	*Probabilities* A	B	C	*Statistician no.*	*Probabilities* A	B	C
C1	0.27	0.28	0.45	S1	0.70	0.07	0.23
C2	0.02	0.03	0.95	S2	0.19	0.16	0.65
C3	0.12	0.16	0.72	S3	0.18	0.26	0.54
C4	0.83	0.02	0.15	S4	0.02	0.02	0.96
C5	0.24	0.22	0.54	S5	0.08	0.16	0.76
C6	0.16	0.20	0.64	S6	0.14	0.18	0.68
C7	0.31	0.08	0.61	S7	0.16	0.11	0.73
C8	0.05	0.85	0.10	S8	0.04	0.06	0.90
C9	0.06	0.06	0.88	S9	0.06	0.54	0.40
C10	0.08	0.31	0.61	S10	0.12	0.22	0.66
C11	0.18	0.20	0.62	S11	0.06	0.02	0.92
C12	0.17	0.19	0.64	S12	0.16	0.04	0.80
C13	0.04	0.17	0.82	S13	0.27	0.17	0.56
C14	0.08	0.25	0.67	S14	0.21	0.51	0.28
C15	0.11	0.34	0.55	S15	0.15	0.15	0.70

A: algebritis
B: bilateral paralexia
C: calculus deficiency

Data 18. *Compositions and total pebble counts of 92 glacial tills*

Specimen no.	*Percentages by weight* A	B	C	D	*Pebble count*
S1	91.8	7.1	1.1	0	282
S2	88.9	10.1	0.5	0.5	368
S3	87.3	10.9	0.8	1.0	607
S4	84.6	13.5	0	1.9	532
S5	14.7	81.1	0.3	3.6	360
S6	37.0	36.1	4.0	22.8	470
S7	34.3	59.8	5.9	0	102
S8	21.5	75.0	0.4	3.1	544
S9	67.2	31.5	0.3	1.0	387
S10	49.9	46.9	0.7	2.7	294
S11	18.7	79.3	0.4	1.6	503
S12	11.3	88.2	0.4	0	697
S13	91.1	7.9	0.8	0.3	393
S14	81.7	17.6	0	0.8	665
S15	97.6	2.3	0	0	347
S16	3.4	95.1	0.4	1.0	791
S17	25.3	71.1	3.6	0	225
S18	56.0	26.3	2.3	15.4	175
S19	75.1	24.6	0.3	0	333
S20	15.2	80.7	1.9	2.2	269
S21	41.5	36.4	5.1	17.0	118
S22	19.5	77.9	0	2.6	154
S23	89.1	9.8	1.1	0	276
S24	93.7	5.2	1.0	0	480
S25	87.9	8.5	0	3.5	373
S26	80.2	18.4	1.4	0	369
S27	83.3	15.1	1.6	0	126
S28	82.6	16.5	0.9	0	460
S29	83.2	14.5	1.6	0.7	441
S30	64.7	34.5	0.8	0	502
S31	58.7	41.2	0	0	126
S32	66.5	28.7	0.8	4.0	376

Data 18. (*Contd.*)

Specimen no.	Percentages by weight A	B	C	D	Pebble count
S33	89.8	9.3	0.8	0	118
S34	80.5	17.8	0.7	1.0	303
S35	85.6	12.0	2.4	0	250
S36	78.4	16.7	12.0	3.8	582
S37	62.3	31.8	2.8	2.8	69
S38	72.6	22.6	3.1	1.8	226
S39	81.6	10.0	4.5	3.9	359
S40	65.3	19.9	9.9	4.9	453
S41	95.6	4.2	0.2	0	427
S42	79.0	16.8	3.0	1.2	334
S43	92.6	6.3	1.1	0	364
S44	89.5	9.8	0.5	0.1	869
S45	91.2	7.7	1.1	0	441
S46	95.3	3.6	1.1	0	615
S47	91.9	7.5	0.4	0.2	532
S48	89.4	0.7	1.0	1.9	417
S49	89.4	9.2	1.4	0	360
S50	95.2	3.6	1.2	0	580
S51	95.9	4.1	0	0	147
S52	94.0	4.8	1.2	0	500
S53	83.4	15.7	0.7	0.2	943
S54	15.7	84.3	0	0	305
S55	7.8	90.6	0.8	0.8	1151
S56	18.2	76.6	0.4	4.8	457
S57	27.5	71.4	1.1	0	637
S58	51.1	46.5	0.3	2.1	284
S59	7.0	89.4	1.8	1.3	386
S60	80.1	19.0	0	0.5	221
S61	73.1	23.6	1.4	1.9	208
S62	74.9	14.7	4.7	5.8	573
S63	79.6	17.3	0.7	2.3	565
S64	74.1	20.0	2.9	2.9	170

(*Contd.*)

Data 18. (*Contd.*)

Specimen no.	Percentage by weight A	B	C	D	Pebble count
S65	73.2	21.5	1.9	3.1	261
S66	0.1	99.9	0	0	1097
S67	77.9	21.1	0	1.0	408
S68	45.8	33.3	8.3	12.5	24
S69	26.7	39.6	2.9	14.9	890
S70	57.7	35.1	3.6	3.6	168
S71	54.5	40.9	4.5	0	22
S72	70.5	28.8	0.7	0	601
S73	11.5	87.4	1.1	0	364
S74	27.8	71.9	0.3	0	342
S75	22.4	76.9	0.6	0	867
S76	20.1	78.3	0.1	1.4	691
S77	15.8	83.5	0.6	0	462
S78	91.5	6.6	0.3	1.6	318
S79	90.0	8.2	1.3	0.4	461
S80	75.7	21.0	0.6	2.8	777
S81	85.6	13.4	1.0	0	397
S82	88.5	9.2	1.7	0.5	347
S83	71.0	25.8	1.5	1.8	744
S84	11.5	86.1	2.3	0.2	576
S85	4.5	95.0	0	0	321
S86	21.2	72.8	0.8	5.2	382
S87	19.5	78.1	0.8	1.6	645
S88	15.0	80.6	1.7	2.6	459
S89	27.0	70.2	1.3	1.5	681
S90	15.9	83.3	0.8	0	245
S91	16.9	74.3	1.2	5.9	575
S92	31.4	65.9	2.7	0	698

A: red sandstone
B: gray sandstone
C: crystalline
D: miscellaneous
Adapted from Kaiser (1962, Table 1).

Data 19. *Yat, yee, sam measurements (m) for the final jumps of the 1985 Hong Kong Pogo-Jump Championship*

Finalist	*Yat*	*Yee*	*Sam*	*Finalist*	*Yat*	*Yee*	*Sam*
Ho	4.63	8.84	7.68	Lam	4.50	8.14	6.36
	2.30	8.60	4.97		3.70	9.20	6.90
	1.92	7.63	3.77		3.42	6.86	4.98
	3.16	7.96	5.50		4.63	7.27	5.95
Ip	6.96	7.16	7.91	Mak	5.28	6.19	7.52
	8.90	6.59	6.99		6.67	5.49	7.46
	8.41	7.40	7.26		5.77	5.38	7.53
	5.69	6.79	5.35		6.35	4.93	7.99
Jao	10.00	6.15	6.79	Ng	5.78	6.94	9.32
	7.15	4.70	6.44		5.62	6.98	8.94
	13.12	4.14	8.12		4.66	5.95	8.41
	9.13	4.33	5.88		4.62	7.45	7.18
Ko	8.23	8.16	5.79				
	6.74	7.14	4.24				
	8.13	6.12	5.56				
	13.15	5.74	5.64				

Data 20. *Proportions of sand, silt and clay in sediment specimens*

Specimen no.	*Proportions by weight*		
	Sand	*Silt*	*Clay*
A1	0.36	0.60	0.04
A2	0.34	0.61	0.05
A3	0.06	0.87	0.07
A4	0.03	0.91	0.06
A5	0.08	0.87	0.05
A6	0.33	0.63	0.04
A7	0.59	0.36	0.05
A8	0.20	0.78	0.02
A9	0.48	0.51	0.01
A10	0.02	0.80	0.18
B1	0.45	0.53	0.02
B2	0.91	0.08	0.01
B3	0.69	0.25	0.06
B4	0.74	0.25	0.01
B5	0.62	0.37	0.01
B6	0.42	0.54	0.04
B7	0.46	0.51	0.03
C1	0.19	0.71	0.10
C2	0.64	0.35	0.01
C3	0.33	0.55	0.12
C4	0.21	0.74	0.05

Adapted from McCammon (1975).

Data 21. *Permeabilities of bayesite for 21 mixtures of fibres and bonding pressures.*

Experiment no.	*Proportions in mixture* A	B	C	D	*Bonding pressure* (*kg/cm*^{2})	*Permeability* (*mD*)
E1	0.30	0.30	0.30	0.10	5	399
E2	0.10	0.40	0.40	0.10	8	329
E3	0.40	0.10	0.40	0.10	8	346
E4	0.40	0.40	0.10	0.10	8	430
E5	0.50	0.20	0.20	0.10	11	315
E6	0.20	0.50	0.20	0.10	11	325
E7	0.20	0.20	0.50	0.10	11	278
E8	0.28	0.28	0.28	0.16	5	414
E9	0.12	0.36	0.36	0.16	11	263
E10	0.36	0.12	0.36	0.16	11	254
E11	0.36	0.36	0.12	0.16	11	323
E12	0.48	0.18	0.18	0.16	8	395
E13	0.18	0.48	0.18	0.16	8	335
E14	0.18	0.18	0.48	0.16	8	324
E15	0.25	0.25	0.25	0.25	5	352
E16	0.11	0.32	0.32	0.25	8	306
E17	0.32	0.11	0.32	0.25	11	242
E18	0.32	0.32	0.11	0.25	8	388
E19	0.39	0.18	0.18	0.25	11	297
E20	0.18	0.39	0.18	0.25	8	390
E21	0.18	0.18	0.39	0.25	11	246

A: short fibres
B: medium fibres
C: long fibres
D: binder

Permeability is measured in microdarcies.

Data 22. *Compositions of eight-hour shifts of 27 machine operators*

Operator no.	*Proportions of shift in activity*			
	A	B	C	D
O1	0.667	0.180	0.053	0.100
O2	0.578	0.180	0.112	0.123
O3	0.560	0.271	0.086	0.076
O4	0.490	0.316	0.091	0.103
O5	0.598	0.119	0.117	0.166
O6	0.617	0.180	0.079	0.124
O7	0.700	0.197	0.046	0.057
O8	0.577	0.266	0.080	0.077
O9	0.591	0.179	0.084	0.145
O10	0.532	0.107	0.119	0.242
O11	0.511	0.303	0.082	0.104
O12	0.625	0.219	0.082	0.074
O13	0.667	0.115	0.107	0.111
O14	0.573	0.208	0.112	0.107
O15	0.585	0.235	0.080	0.092
O16	0.558	0.245	0.100	0.096
O17	0.652	0.214	0.066	0.068
O18	0.619	0.214	0.066	0.068
O19	0.628	0.245	0.067	0.061
O20	0.596	0.228	0.090	0.086
O21	0.546	0.185	0.114	0.155
O22	0.606	0.146	0.079	0.169
O23	0.613	0.257	0.058	0.072
O24	0.680	0.173	0.060	0.088
O25	0.584	0.246	0.072	0.098
O26	0.542	0.122	0.159	0.176
O27	0.545	0.121	0.156	0.178

A: high-quality production
B: low-quality production
C: machine setting
D: machine repair

Data 24. *Measurements in degrees of the angles N, A, B for English seventeenth-century and Naqada skulls*

Skull no.	*English* N	A	B	*Skull no.*	*Naqada* N	A	B
Females							
EF1	65	75.5	39.5	NF1	68.8	74.6	36.6
EF2	64	71	45	NF2	68.1	75.4	36.5
EF3	67	74	39	NF3	69.9	76.5	33.6
EF4	67	72	41	NF4	61.1	82	36.9
EF5	70	70	40	NF5	56	91.8	32.2
EF6	63	75	42	NF6	65.9	74.9	39.2
EF7	62.5	76	41.5	NF7	73.75	73.75	32.5
EF8	66	74	40	NF8	70.2	72.8	37
EF9	65	70	45	NF9	74.3	72.7	33
EF10	64.5	73	42.5	NF10	72.2	72	35.8
EF11	62	70.5	47.5	NF11	73.2	74.8	32
EF12	62	80	38	NF12	77.7	69.8	32.5
EF13	64.5	73.5	42	NF13	72.8	69.9	37.3
EF14	65	73.5	41.5	NF14	67.5	75.4	37.1
EF15	63	74	43	NF15	62.9	80.2	36.9
EF16	68	76	36	NF16	71.9	75.9	32.2
EF17	65.5	73.5	41	NF17	71.6	73.9	34.5
EF18	66.5	72	41.5	NF18	70.3	74.4	35.3
EF19	67	74	39	NF19	73.1	68.7	38.2
EF20	68.5	68.5	43	NF20	68.8	74.9	36.3
EF21	60.5	72.5	47	NF21	65.5	79.1	35.4
EF22	57	83	40	NF22	71	74.5	34.5
Males							
EM1	62.5	77.5	40	NM1	73.1	72.2	34.7
EM2	59.5	80.5	40	NM2	64.8	75	40.2
EM3	62.5	78	39.5	NM3	74	73.6	32.4
EM4	63.5	74	42.5	NM4	59.7	87.6	32.9
EM5	65	72	43	NM5	75.4	69.4	35.2
EM6	68.5	73.5	38	NM6	67.1	77.3	35.6

(Contd.)

Data 24. (*Contd.*)

Skull no.	*English* N	A	B	*Skull no.*	*Naqada* N	A	B
EM7	66.5	73	40.5	NM7	69.8	74.6	35.6
EM8	67	73.5	39.5	NM8	71.2	77.3	31.5
EM9	66.5	74.5	39	NM9	66.6	76.6	36.8
EM10	65	70	45	NM10	68.8	76.9	34.3
EM11	69	72.5	38.5	NM11	75.2	71.9	32.9
EM12	68	71	41	NM12	71	72.2	36.8
EM13	65.5	71	43.5	NM13	61.4	80.9	37.7
EM14	66.5	73.5	40	NM14	67.6	78.2	34.2
EM15	66	71	43	NM15	67	79.1	33.9
EM16	69	67	44	NM16	73.2	71.1	35.7
EM17	67	72.5	40.5	NM17	68.8	76.8	34.4
EM18	64.5	75	40.5	NM18	71	70.3	38.7
EM19	62.5	77	40.5	EM19	80.8	64	35.2
EM20	63.5	74	42.5	EM20	69.7	72.9	37.4
EM21	66	75	39	NM21	71.8	69.3	38.9
EM22	63	76.5	40.5	NM22	66.4	77	36.6
EM23	64	75	41	NM23	64.5	78.3	37.2
EM24	68.5	71	40.5	NM24	75.3	65.8	38.9
EM25	69	72	39	NM25	71.1	71.9	37
EM26	59.5	78.5	42	NM26	65.4	77.3	37.3
EM27	66	74	40	NM27	72	71.8	36.2
EM28	69	66.5	44.5	NM28	65.5	78	36.5
EM29	59	78	43	NM29	69.6	71.8	38.6

N: nasial angle
A: alveolar angle
B: basilar angle

Adapted from Fawcett (1901) and MacDonell (1904).

Data 25. *Areal compositions by abundance of vegetation (thick, thin) and animals (dense, sparse) for 50 plots in each of regions A and B*

	Region A					Region B			
	Thick		Thin			Thick		Thin	
Plot no.	Dense	Sparse	Dense	Sparse	Plot no.	Dense	Sparse	Dense	Sparse
A1	0.533	0.277	0.072	0.118	B1	0.387	0.211	0.146	0.256
A2	0.475	0.228	0.072	0.226	B2	0.344	0.340	0.187	0.130
A3	0.517	0.204	0.145	0.139	B3	0.241	0.296	0.263	0.201
A4	0.212	0.418	0.170	0.201	B4	0.260	0.298	0.239	0.203
A5	0.385	0.279	0.126	0.210	B5	0.188	0.306	0.322	0.183
A6	0.297	0.170	0.215	0.318	B6	0.509	0.300	0.088	0.103
A7	0.197	0.226	0.303	0.275	B7	0.443	0.113	0.103	0.342
A8	0.157	0.223	0.208	0.412	B8	0.266	0.385	0.226	0.123
A9	0.273	0.178	0.346	0.203	B9	0.413	0.283	0.153	0.151
A10	0.477	0.225	0.198	0.155	B10	0.294	0.243	0.233	0.230
A11	0.253	0.326	0.175	0.245	B11	0.460	0.236	0.146	0.158
A12	0.300	0.181	0.305	0.214	B12	0.491	0.157	0.093	0.259
A13	0.182	0.445	0.133	0.241	B13	0.359	0.323	0.134	0.183
A14	0.277	0.158	0.331	0.234	B14	0.456	0.091	0.085	0.368
A15	0.624	0.300	0.018	0.059	B15	0.371	0.054	0.126	0.449
A16	0.448	0.204	0.181	0.167	B16	0.456	0.230	0.110	0.204
A17	0.617	0.304	0.028	0.056	B17	0.345	0.212	0.235	0.208
A18	0.364	0.230	0.152	0.254	B18	0.346	0.175	0.175	0.304
A19	0.259	0.184	0.238	0.319	B19	0.256	0.350	0.178	0.216
A20	0.069	0.251	0.521	0.158	B20	0.425	0.192	0.150	0.233
A21	0.196	0.425	0.118	0.261	B21	0.314	0.202	0.145	0.339
A22	0.142	0.152	0.373	0.333	B22	0.257	0.249	0.256	0.238
A23	0.637	0.171	0.080	0.113	B23	0.581	0.172	0.076	0.170
A24	0.279	0.169	0.247	0.310	B24	0.397	0.130	0.120	0.358
A25	0.328	0.450	0.063	0.159	B25	0.553	0.149	0.078	0.220
A26	0.529	0.190	0.119	0.162	B26	0.108	0.230	0.441	0.220
A27	0.464	0.359	0.083	0.094	B27	0.418	0.232	0.161	0.189
A28	0.212	0.202	0.273	0.313	B28	0.537	0.120	0.081	0.261
A29	0.120	0.235	0.324	0.321	B29	0.486	0.183	0.075	0.255
A30	0.444	0.237	0.131	0.193	B30	0.187	0.375	0.306	0.133
A31	0.299	0.167	0.318	0.215	B31	0.412	0.277	0.143	0.168
A32	0.260	0.327	0.184	0.230	B32	0.356	0.134	0.140	0.370
A33	0.432	0.266	0.103	0.199	B33	0.371	0.208	0.196	0.226
A34	0.138	0.226	0.300	0.336	B34	0.435	0.238	0.150	0.176
A35	0.482	0.295	0.099	0.124	B35	0.383	0.144	0.137	0.336

(*Contd.*)

Data 25. (*Contd.*)

	Region A					*Region B*			
	Thick		*Thin*			*Thick*		*Thin*	
Plot no.	*Dense*	*Sparse*	*Dense*	*Sparse*	*Plot no.*	*Dense*	*Sparse*	*Dense*	*Sparse*
A36	0.308	0.220	0.153	0.319	B36	0.214	0.274	0.353	0.159
A37	0.389	0.127	0.276	0.208	B37	0.357	0.195	0.197	0.252
A38	0.038	0.152	0.609	0.200	B38	0.160	0.279	0.368	0.192
A39	0.345	0.284	0.166	0.205	B39	0.395	0.227	0.203	0.177
A40	0.145	0.140	0.348	0.366	B40	0.292	0.274	0.215	0.219
A41	0.287	0.193	0.208	0.312	B41	0.477	0.236	0.132	0.156
A42	0.113	0.126	0.447	0.314	B42	0.152	0.372	0.300	0.176
A43	0.447	0.190	0.132	0.231	B43	0.167	0.217	0.331	0.286
A44	0.319	0.252	0.206	0.223	B44	0.358	0.263	0.203	0.176
A45	0.489	0.198	0.097	0.216	B45	0.246	0.350	0.253	0.151
A46	0.398	0.276	0.119	0.207	B46	0.486	0.122	0.113	0.279
A47	0.507	0.191	0.094	0.209	B47	0.341	0.120	0.155	0.384
A48	0.281	0.258	0.190	0.271	B48	0.241	0.230	0.215	0.314
A49	0.342	0.250	0.109	0.299	B49	0.373	0.171	0.114	0.343
A50	0.161	0.564	0.095	0.180	B50	0.091	0.267	0.492	0.250

Data 26. *Fossil pollen compositions from three different locations A, B, C*

Specimen no.	Proportions		
	Pinus	*Abies*	*Quercus*
A1	0.297	0.583	0.120
A2	0.184	0.484	0.331
A3	0.478	0.436	0.086
A4	0.392	0.542	0.065
A5	0.201	0.506	0.293
A6	0.308	0.527	0.164
A7	0.239	0.603	0.158
A8	0.204	0.586	0.210
A9	0.185	0.581	0.234
A10	0.170	0.521	0.309
B1	0.240	0.363	0.397
B2	0.356	0.449	0.195
B3	0.183	0.444	0.373
B4	0.347	0.504	0.149
B5	0.365	0.458	0.177
B6	0.219	0.484	0.297
B7	0.221	0.448	0.332
B8	0.338	0.475	0.187
B9	0.246	0.446	0.308
B10	0.170	0.434	0.396
C1	0.416	0.440	0.144
C2	0.341	0.479	0.180
C3	0.440	0.416	0.143
C4	0.607	0.366	0.027
C5	0.667	0.299	0.034
C6	0.545	0.362	0.094
C7	0.486	0.408	0.107
C8	0.468	0.451	0.081
C9	0.707	0.253	0.040
C10	0.494	0.396	0.111

Data 27. *Compositions of 25 specimens of lammite*

Specimen no.	*Percentages by weight* A	B	C	D	E
L1	48.7	30.6	2.8	9.5	8.5
L2	49.6	14.5	12.8	15.9	7.3
L3	40.8	11.9	20.0	9.7	17.7
L4	51.9	24.7	4.6	13.1	5.6
L5	30.1	5.1	47.6	8.4	8.8
L6	51.0	29.3	4.1	7.6	8.0
L7	48.9	22.0	8.7	14.1	6.3
L8	55.5	19.8	7.7	7.6	9.4
L9	20.8	1.6	64.4	7.5	5.7
L10	48.7	19.2	10.6	8.5	13.1
L11	45.0	39.9	1.0	8.4	5.7
L12	26.9	3.9	53.8	7.3	8.1
L13	44.8	10.1	22.9	12.6	9.5
L14	28.8	7.0	46.6	9.3	8.3
L15	53.2	19.6	7.5	12.1	7.5
L16	50.9	17.9	10.0	14.5	6.7
L17	42.8	11.9	21.9	7.9	15.5
L18	44.5	18.4	13.9	6.0	17.3
L19	50.7	30.0	2.9	10.3	6.2
L20	44.7	8.8	24.9	7.3	14.3
L21	25.2	3.2	55.8	6.3	9.4
L22	19.9	2.3	65.5	5.0	7.4
L23	52.0	20.4	7.3	11.6	8.7
L24	54.2	20.5	6.0	12.3	7.0
L25	41.6	11.1	23.1	14.2	10.0

A: albite
B: blandite
C: cornite
D: daubite
E: endite

Data 28. *Heart and total body weights of 47 female and 97 male domestic cats*

Cat no.	Weight Heart (g)	Weight Body (kg)	Cat no.	Weight Heart (g)	Weight Body (kg)	Cat no.	Weight Heart (g)	Weight Body (kg)
F1	9.6	2.3	F16	9.0	2.3	F31	8.5	2.7
F2	10.6	3.0	F17	7.6	2.1	F32	10.1	2.3
F3	9.9	2.9	F18	9.5	2.0	F33	8.7	2.1
F4	8.7	2.4	F19	10.1	2.9	F34	10.9	2.2
F5	10.1	2.3	F20	10.2	2.7	F35	7.9	2.3
F6	7.0	2.0	F21	10.1	2.6	F36	8.3	2.1
F7	11.0	2.2	F22	9.5	2.3	F37	10.1	2.9
F8	8.2	2.1	F23	8.7	2.6	F38	13.0	3.0
F9	9.0	2.3	F24	7.2	2.1	F39	8.7	2.2
F10	7.3	2.1	F25	9.8	2.1	F40	6.3	2.4
F11	8.5	2.1	F26	10.8	2.7	F41	8.8	2.4
F12	9.7	2.2	F27	9.1	2.2	F42	10.9	2.5
F13	7.4	2.0	F28	11.2	2.3	F43	9.0	2.5
F14	7.3	2.3	F29	8.1	2.1	F44	9.7	2.3
F15	7.1	2.2	F30	10.2	2.4	F45	8.4	2.3
						F46	10.1	2.6
						F47	10.6	2.3
M1	9.4	2.9	M11	10.5	2.6	M21	13.5	2.8
M2	9.3	2.4	M12	8.6	2.5	M22	10.4	2.7
M3	7.2	2.2	M13	10.0	2.8	M23	11.6	3.2
M4	11.3	2.9	M14	12.1	3.1	M24	10.6	3.0
M5	8.8	2.5	M15	13.8	3.0	M25	12.7	2.5
M6	9.9	3.1	M16	12.0	2.7	M26	15.6	3.5
M7	13.3	3.0	M17	12.0	2.8	M27	9.1	2.4
M8	12.7	2.5	M18	10.1	2.1	M28	7.6	2.2
M9	14.4	3.4	M19	11.5	3.3	M29	12.8	3.4
M10	10.0	3.0	M20	12.2	3.4	M30	8.3	2.6

(*Contd.*)

Data 28. (*Contd.*)

Cat no.	Weight: Heart (g)	Weight: Body (kg)	Cat no.	Weight: Heart (g)	Weight: Body (kg)	Cat no.	Weight: Heart (g)	Weight: Body (kg)
M31	11.2	3.4	M56	11.0	3.7	M81	12.5	2.7
M32	9.4	2.6	M57	12.4	3.0	M82	11.8	2.9
M33	8.0	2.7	M58	13.5	3.2	M83	15.0	3.6
M34	14.9	3.3	M59	14.1	3.3	M84	10.2	2.8
M35	10.7	2.2	M60	12.7	3.0	M85	11.0	2.5
M36	13.6	3.2	M61	10.1	2.9	M86	11.5	2.6
M37	9.6	2.2	M62	10.4	3.0	M87	20.5	3.9
M38	11.7	3.5	M63	7.9	2.4	M88	12.2	3.0
M39	9.3	2.5	M64	14.8	3.8	M89	9.4	2.6
M40	12.3	3.2	M65	6.5	2.0	M90	9.0	2.7
M41	13.0	3.2	M66	11.5	3.1	M91	8.8	2.5
M42	9.6	2.7	M67	9.1	2.8	M92	9.6	2.2
M43	7.7	2.6	M68	9.6	2.3	M93	13.0	3.1
M44	9.6	2.7	M69	11.6	3.0	M94	12.0	3.3
M45	6.5	2.0	M70	7.9	2.2	M95	11.1	2.7
M46	14.3	3.1	M71	12.4	3.4	M96	11.8	3.6
M47	7.3	2.4	M72	12.9	3.5	M97	11.4	2.8
M48	14.8	3.6	M73	17.2	3.5			
M49	15.7	3.5	M74	16.8	3.8			
M50	13.3	2.8	M75	8.5	2.2			
M51	9.1	2.2	M76	15.4	3.3			
M52	7.9	2.5	M77	9.8	2.7			
M53	7.9	2.4	M78	11.9	3.2			
M54	14.4	3.9	M79	10.6	2.9			
M55	12.5	3.1	M80	13.3	3.6			

Adapted from Fisher (1947, Table 1).

Data 29. *Heart and total thorax widths in X-rays of 20 normotensive (N) and 55 hypertensive (H) males*

Patient no.	Width (cm) Heart	Width (cm) Thorax	Patient no.	Width (cm) Heart	Width (cm) Thorax
N1	10.8	26.5	N11	13.4	29.0
N2	12.0	31.0	N12	13.0	25.6
N3	10.8	25.6	N13	13.0	32.9
N4	11.8	29.2	N14	13.0	28.5
N5	12.2	27.0	N15	12.9	27.5
N6	13.6	29.7	N16	14.2	29.7
N7	13.3	29.2	N17	14.4	31.0
N8	12.9	28.5	N18	13.2	34.4
N9	12.8	31.0	N19	12.6	25.0
N10	11.2	26.0	N20	13.5	28.2
H1	12.0	28.0	H21	14.3	29.2
H2	15.0	31.3	H22	13.9	22.0
H3	12.3	31.0	H23	11.5	28.8
H4	16.4	30.8	H24	10.2	22.7
H5	13.3	27.7	H25	15.4	26.3
H6	14.6	30.8	H26	11.0	26.3
H7	13.6	29.0	H27	16.7	28.8
H8	16.1	30.3	H28	16.5	27.7
H9	13.9	25.6	H29	13.0	28.0
H10	12.4	27.5	H30	11.5	25.3
H11	12.0	29.5	H31	15.1	27.6
H12	15.9	31.7	H32	15.7	30.8
H13	11.5	22.0	H33	11.4	22.2
H14	14.1	30.5	H34	14.5	26.0
H15	14.5	29.9	H35	19.8	33.6
H16	13.4	24.9	H36	15.5	27.2
H17	17.0	30.5	H37	13.2	32.8
H18	13.3	25.6	H38	15.6	26.7
H19	13.2	23.5	H39	12.1	25.5
H20	15.1	31.8	H40	13.9	31.4

(Contd.)

Data 29. (*Contd.*)

Patient no.	Width (cm)		Patient no.	Width (cm)	
	Heart	Thorax		Heart	Thorax
H41	13.6	24.8	H51	16.0	26.2
H42	16.8	34.5	H52	18.0	35.8
H43	11.8	28.3	H53	14.0	27.8
H44	14.8	28.2	H54	11.4	30.5
H45	15.3	31.2	H55	11.4	30.5
H46	14.6	27.0			
H47	11.5	24.1			
H48	11.6	24.1			
H49	13.1	28.0			
H50	13.1	25.9			

Data 30. *Household expenditures (HK$) on five commodity groups of 20 single men*

Household no.	Commodity group 1	2	3	4	5
m1	640	328	147	169	196
m2	1800	484	515	2291	912
m3	2085	445	725	8373	1732
m4	616	331	126	117	149
m5	875	368	191	290	275
m6	770	364	196	242	236
m7	990	415	284	588	420
m8	414	305	94	68	112
m9	1394	440	393	1161	636
m10	1285	374	363	785	487
m11	1102	469	243	496	388
m12	1717	452	452	1977	832
m13	1549	454	424	1345	676
m14	838	386	155	208	222
m15	845	386	211	317	280
m16	1130	394	271	490	386
m17	1765	466	524	2133	822
m18	1195	443	329	974	523
m19	2180	521	553	2781	1010
m20	1017	410	225	419	345

1. Housing, including fuel and light
2. Foodstuffs
3. Alcohol and tobacco
4. Other goods, including clothing, footwear and durable goods
5. Services, including transport and vehicles

Data 31. *Activity patterns of a statistician for 20 days*

Day no.	Proportion of day in activity					
	1	*2*	*3*	*4*	*5*	*6*
D1	0.162	0.041	0.138	0.123	0.254	0.282
D2	0.200	0.039	0.073	0.076	0.346	0.266
D3	0.201	0.082	0.115	0.146	0.194	0.261
D4	0.134	0.077	0.107	0.146	0.214	0.321
D5	0.224	0.080	0.091	0.162	0.195	0.248
D6	0.144	0.063	0.103	0.123	0.316	0.252
D7	0.125	0.054	0.137	0.102	0.312	0.270
D8	0.127	0.077	0.110	0.101	0.341	0.244
D9	0.139	0.052	0.128	0.111	0.266	0.304
D10	0.108	0.052	0.082	0.075	0.413	0.270
D11	0.187	0.091	0.113	0.116	0.264	0.228
D12	0.184	0.070	0.066	0.151	0.305	0.216
D13	0.155	0.086	0.101	0.119	0.225	0.315
D14	0.181	0.097	0.081	0.164	0.271	0.206
D15	0.224	0.096	0.101	0.142	0.203	0.234
D16	0.198	0.067	0.139	0.154	0.162	0.281
D17	0.214	0.073	0.102	0.130	0.201	0.281
D18	0.132	0.037	0.148	0.099	0.307	0.277
D19	0.167	0.073	0.127	0.122	0.266	0.245
D20	0.166	0.064	0.101	0.145	0.242	0.282

1. Teaching
2. Consultation
3. Administration
4. Research
5. Other wakeful activities
6. Sleep

Data 32. *Replicate serum protein compositions determined from eight aliquots of a blood sample*

	Serum protein			
Aliquot	A	B	C	D
1	0.133	0.261	0.215	0.390
2	0.103	0.327	0.190	0.380
3	0.124	0.316	0.186	0.375
4	0.113	0.278	0.204	0.405
5	0.128	0.336	0.196	0.339
6	0.093	0.347	0.187	0.373
7	0.106	0.351	0.180	0.363
8	0.095	0.321	0.244	0.340

A: albumin
B: pre-albumin
C: globulin A
D: globulin B

Data 33. *Serum protein compositions of seven specimens determined by three analysts*

Analyst	*Specimen*	*Serum protein* A	B	C	D
1	1	0.082	0.238	0.235	0.444
	2	0.089	0.351	0.267	0.292
	3	0.083	0.463	0.183	0.271
	4	0.150	0.290	0.225	0.335
	5	0.118	0.254	0.222	0.407
	6	0.076	0.229	0.242	0.453
	7	0.047	0.202	0.163	0.589
2	1	0.076	0.250	0.243	0.430
	2	0.093	0.279	0.291	0.336
	3	0.141	0.293	0.215	0.351
	4	0.150	0.315	0.187	0.340
	5	0.143	0.268	0.200	0.389
	6	0.104	0.243	0.190	0.463
	7	0.088	0.250	0.079	0.583
3	1	0.094	0.268	0.231	0.407
	2	0.107	0.286	0.265	0.341
	3	0.124	0.407	0.186	0.283
	4	0.126	0.271	0.263	0.341
	5	0.159	0.353	0.167	0.321
	6	0.094	0.249	0.168	0.489
	7	0.078	0.287	0.088	0.547

A: albumin
B: pre-albumin
C: globulin A
D: globulin B

Data 34. *Foraminiferal compositions at 30 different depths*

Sample no.	Proportions 1	2	3	4	Depth (m)
F1	0.74	0.19	0.03	0.04	1
F2	0.58	0.29	0.01	0.12	2
F3	0.58	0.19	0.22	0.01	3
F4	0.61	0.28	0.08	0.03	4
F5	0.82	0.13	0.02	0.03	5
F6	0.48	0.38	0.01	0.13	6
F7	0.59	0.38	0.00	0.03	7
F8	0.76	0.12	0.09	0.03	8
F9	0.81	0.12	0.04	0.03	9
F10	0.68	0.23	0.05	0.04	10
F11	0.72	0.20	0.04	0.04	11
F12	0.62	0.27	0.09	0.02	12
F13	0.45	0.25	0.29	0.01	13
F14	0.66	0.25	0.06	0.03	14
F15	0.85	0.13	0.01	0.01	15
F16	0.75	0.09	0.15	0.01	16
F17	0.69	0.25	0.00	0.06	17
F18	0.76	0.10	0.11	0.03	18
F19	0.66	0.29	0.01	0.04	19
F20	0.66	0.24	0.06	0.04	20
F21	0.50	0.46	0.00	0.40	21
F22	0.65	0.25	0.05	0.05	22
F23	0.60	0.35	0.02	0.03	23
F24	0.40	0.27	0.01	0.32	24
F25	0.60	0.10	0.30	0.00	25
F26	0.60	0.10	0.29	0.01	26
F27	0.59	0.39	0.01	0.01	27
F28	0.58	0.39	0.01	0.02	28
F29	0.61	0.34	0.02	0.03	29
F30	0.39	0.49	0.12	0.00	30

1. *Neogloboquadrina atlantica*
2. *Neogloboquadrina pachyderma*
3. *Globorotalia obesa*
4. *Globigerinoides triloba*

Data 35. *Predator–prey compositions at 25 different sites*

Site no.	Proportions P	Q	R	Site no.	Proportions P	Q	R
S1	0.123	0.649	0.227	T1	0	0.478	0.522
S2	0.218	0.557	0.225	T2	0	0.583	0.417
S3	0.072	0.787	0.142	T3	0	0.380	0.620
S4	0.103	0.703	0.195	T4	0	0.682	0.318
S5	0.158	0.450	0.392	T5	0	0.628	0.372
S6	0.257	0.372	0.370	T6	0	0.597	0.403
S7	0.129	0.439	0.433	T7	0	0.514	0.486
S8	0.120	0.531	0.340	T8	0	0.563	0.437
S9	0.096	0.677	0.227	T9	0	0.603	0.397
S10	0.131	0.666	0.203	T10	0	0.775	0.225
S11	0.241	0.581	0.178				
S12	0.134	0.600	0.266				
S13	0.184	0.579	0.236				
S14	0.053	0.688	0.259				
S15	0.106	0.673	0.221				

P: predator
Q: prey of species Q
R: prey of species R

Data 36. *Mineral compositions of 12 thin sections of granite*

Section	*Percentages*				
no.	*Quartz*	*Microcline*	*Plagioclase*	*Biotite*	*Others*
1	0.244	0.272	0.428	0.057	0.019
2	0.228	0.269	0.420	0.064	0.019
3	0.213	0.282	0.427	0.061	0.017
4	0.212	0.264	0.440	0.065	0.019
5	0.205	0.253	0.457	0.068	0.017
6	0.208	0.181	0.435	0.051	0.024
7	0.209	0.253	0.448	0.070	0.020
8	0.220	0.273	0.415	0.070	0.022
9	0.220	0.278	0.433	0.054	0.015
10	0.203	0.255	0.456	0.061	0.025
11	0.208	0.283	0.423	0.064	0.022
12	0.217	0.280	0.425	0.054	0.024

From Chayes (1971).

Data 37. *Mineral compositions of five slides as reported by five analysts*

		Percentages by volume						
Analyst	*Slide*	*Quartz*	*Mico-cline*	*Plagio-class*	*Bio-tite*	*Mus-co-vite*	*Opa-ques*	*Non-Opa-ques*
1	A	24.7	35.6	33.3	3.3	2.0	0.6	0.6
	B	26.8	35.7	32.6	3.5	0.4	0.6	0.4
	C	28.0	34.2	32.1	3.4	1.1	0.7	0.4
	D	27.8	35.0	31.5	3.3	1.0	0.9	0.5
	E	26.6	34.5	33.6	3.0	1.4	0.6	0.3
2	A	27.3	35.5	32.1	2.5	1.5	0.8	0.3
	B	27.3	35.4	31.7	3.4	1.4	0.6	0.1
	C	28.0	35.3	31.1	2.8	1.4	0.8	0.5
	D	30.1	33.8	31.5	2.6	0.9	0.7	0.2
	E	28.7	35.2	31.4	2.6	1.3	0.6	0.2
3	A	25.8	36.0	33.5	2.9	0.8	0.7	0.3
	B	25.5	33.9	33.7	4.9	0.8	1.0	0.1
	C	26.1	37.8	30.7	3.4	1.1	0.7	0.3
	D	26.2	36.0	29.5	5.7	1.3	1.1	0.1
	E	27.8	34.7	32.4	3.6	0.7	0.8	0.2
4	A	26.4	36.2	32.7	2.1	1.1	0.8	0.6
	B	26.6	36.3	31.9	3.2	0.8	0.6	0.7
	C	28.1	36.4	30.6	2.4	1.0	1.0	0.6
	D	27.1	35.9	31.6	2.7	0.9	1.1	0.6
	E	28.0	34.6	32.2	3.0	1.0	0.8	0.4
5	A	25.2	34.1	34.9	2.6	1.8	0.7	0.6
	B	28.6	34.5	31.6	2.7	1.6	0.6	0.4
	C	28.3	33.0	32.8	3.7	1.0	0.7	0.4
	D	26.3	36.1	32.3	2.9	1.1	0.8	0.5
	E	28.6	34.6	31.9	2.7	1.2	0.5	0.5

From Chayes (1956).

Data 38. *Predator–prey compositions at 25 different sites*

Site no.	Proportions P1	P2	Q	R	Site no.	Proportions P1	P2	Q	R
S1	0.395	0.095	0.311	0.199	T1	0.174	0	0.462	0.364
S2	0.223	0.205	0.308	0.264	T2	0.253	0	0.381	0.366
S3	0.170	0.317	0.247	0.266	T3	0.176	0	0.449	0.375
S4	0.172	0.256	0.357	0.215	T4	0.094	0	0.613	0.292
S5	0.443	0.192	0.157	0.208	T5	0.313	0	0.224	0.463
S6	0.269	0.219	0.327	0.185	T6	0.286	0	0.316	0.398
S7	0.287	0.299	0.213	0.201	T7	0.216	0	0.508	0.275
S8	0.323	0.199	0.277	0.201	T8	0.149	0	0.477	0.374
S9	0.305	0.095	0.323	0.277	T9	0.212	0	0.333	0.455
S10	0.209	0.299	0.291	0.200	T10	0.242	0	0.470	0.288
S11	0.279	0.241	0.217	0.263					
S12	0.426	0.125	0.195	0.254					
S13	0.427	0.079	0.383	0.111					
S14	0.372	0.141	0.190	0.296					
S15	0.090	0.489	0.265	0.157					

P1: predator of species P1
P2: predator of species P2
Q: prey of species Q
R: prey of species R

Data 39. *Microhardness of 18 glass specimens and their (Ge, Sb, Se) compositions*

Specimen no.	Percentage composition Ge	Sb	Se	Micro-hardness (kg/mm^2)
G1	20.00	5.00	75.00	4.49
G2	25.00	5.00	70.00	4.45
G3	29.17	5.00	65.83	4.38
G4	32.00	5.00	63.00	4.43
G5	35.00	5.00	60.00	4.50
G6	15.00	10.00	75.00	4.66
G7	20.00	10.00	70.00	4.57
G8	25.00	10.00	65.00	4.52
G9	30.00	10.00	60.00	4.61
G10	32.00	10.00	58.00	4.65
G11	10.00	15.00	75.00	4.74
G12	15.00	15.00	70.00	4.73
G13	20.00	15.00	65.00	4.69
G14	20.84	15.00	64.16	4.67
G15	25.00	15.00	60.00	4.71
G16	28.00	15.00	57.00	4.76
G17	16.67	20.00	63.33	4.80
G18	12.50	25.00	62.50	4.94

Adapted from Giridhar, Narasimham and Mahadevan (1981).

Data 40. *Percentages of boys in hair and eye colour categories in each of the counties of Scotland*

	Hair colour					Eye colour			
County	*Fair*	*Red*	*Medium*	*Dark*	*Jet black*	*Pure blue*	*Light*	*Medium*	*Dark*
Aberdeen	28.41	5.46	40.31	24.66	1.12	15.18	30.12	32.88	21.82
Argyll	26.93	5.05	36.94	29.18	1.90	15.16	33.15	31.26	20.43
Ayr	29.34	4.98	40.59	24.02	1.07	16.95	30.75	31.05	21.25
Banff	30.42	6.82	39.45	22.38	.93	15.44	28.37	32.91	23.28
Berwick	34.40	5.05	36.47	23.16	.92	16.13	33.79	29.59	20.49
Bute	25.93	5.42	39.41	27.97	1.27	11.87	33.64	33.98	20.51
Caithness	29.13	4.58	39.00	25.33	1.96	13.98	30.35	31.24	24.43
Clackmannan	29.29	4.05	42.63	23.79	.24	11.78	33.88	33.51	20.83
Dumbarton	28.61	4.30	39.03	26.67	1.39	13.66	34.78	29.57	21.99
Dumfries	31.44	4.76	39.53	23.08	1.19	13.67	32.27	33.18	20.88
Edinburgh	27.98	5.21	41.60	24.19	1.02	14.21	30.75	32.14	22.90
Elgin	31.72	5.08	37.59	24.05	1.56	19.16	26.21	31.90	22.73
Fife	30.36	4.72	39.48	24.59	.85	14.75	28.62	32.91	23.72
Forfar	26.02	5.21	41.89	25.48	1.40	15.36	27.79	32.07	24.78
Haddington	30.99	5.35	38.18	24.59	.89	17.50	31.58	29.50	21.42
Inverness	28.03	4.70	35.61	29.06	2.60	20.45	30.30	26.57	22.68
Kincardine	31.28	5.73	36.46	25.12	1.41	16.21	33.99	29.51	20.29
Kinross	30.23	4.94	42.40	21.48	.95	12.36	32.70	34.03	20.91
Kirkcudbright	28.63	4.79	38.87	26.40	1.31	14.26	34.37	29.80	21.57
Lanark	24.33	5.05	42.83	26.69	1.10	13.07	30.34	32.95	23.64
Linlithgow	30.60	5.29	40.18	23.13	.80	15.12	31.42	31.67	21.79
Nairn	31.67	6.21	38.10	23.40	.62	19.88	28.36	35.82	15.94
Orkney	32.59	4.95	38.23	23.04	1.19	18.26	30.49	32.82	18.43
Peebles	24.80	4.98	47.36	22.05	.81	12.20	32.01	34.35	21.44
Perth	29.29	5.14	38.40	25.47	1.70	16.04	30.01	31.11	22.84
Renfrew	24.09	4.83	42.85	26.81	1.42	13.95	29.71	33.18	23.16
Ross & Cromarty	30.63	4.97	35.84	26.49	2.07	19.75	29.28	30.77	20.20
Roxburgh	31.34	5.76	38.54	22.99	1.37	16.57	30.94	28.32	24.17
Selkirk	28.69	5.85	44.94	19.41	1.11	17.83	25.63	35.65	20.89
Shetland	33.48	5.67	36.14	23.12	1.59	25.16	24.09	27.46	23.29
Stirling	27.06	4.57	42.59	24.57	1.21	15.08	30.62	31.64	22.66
Sutherland	31.91	4.82	38.07	23.94	1.26	19.12	27.41	27.17	26.30
Wigtown	28.79	4.93	37.72	27.02	1.54	18.88	29.86	18.19	23.07

Adapted from Tocher (1908).

Author index

Subject index

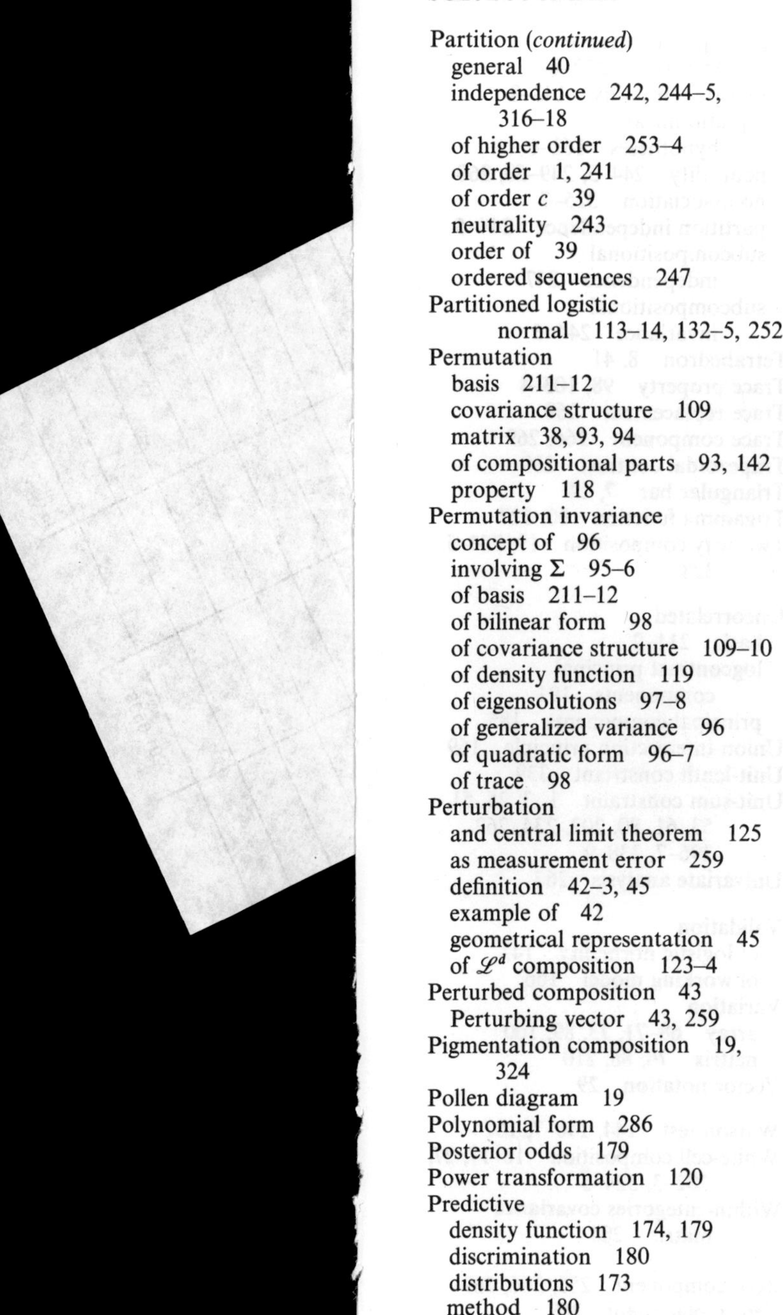

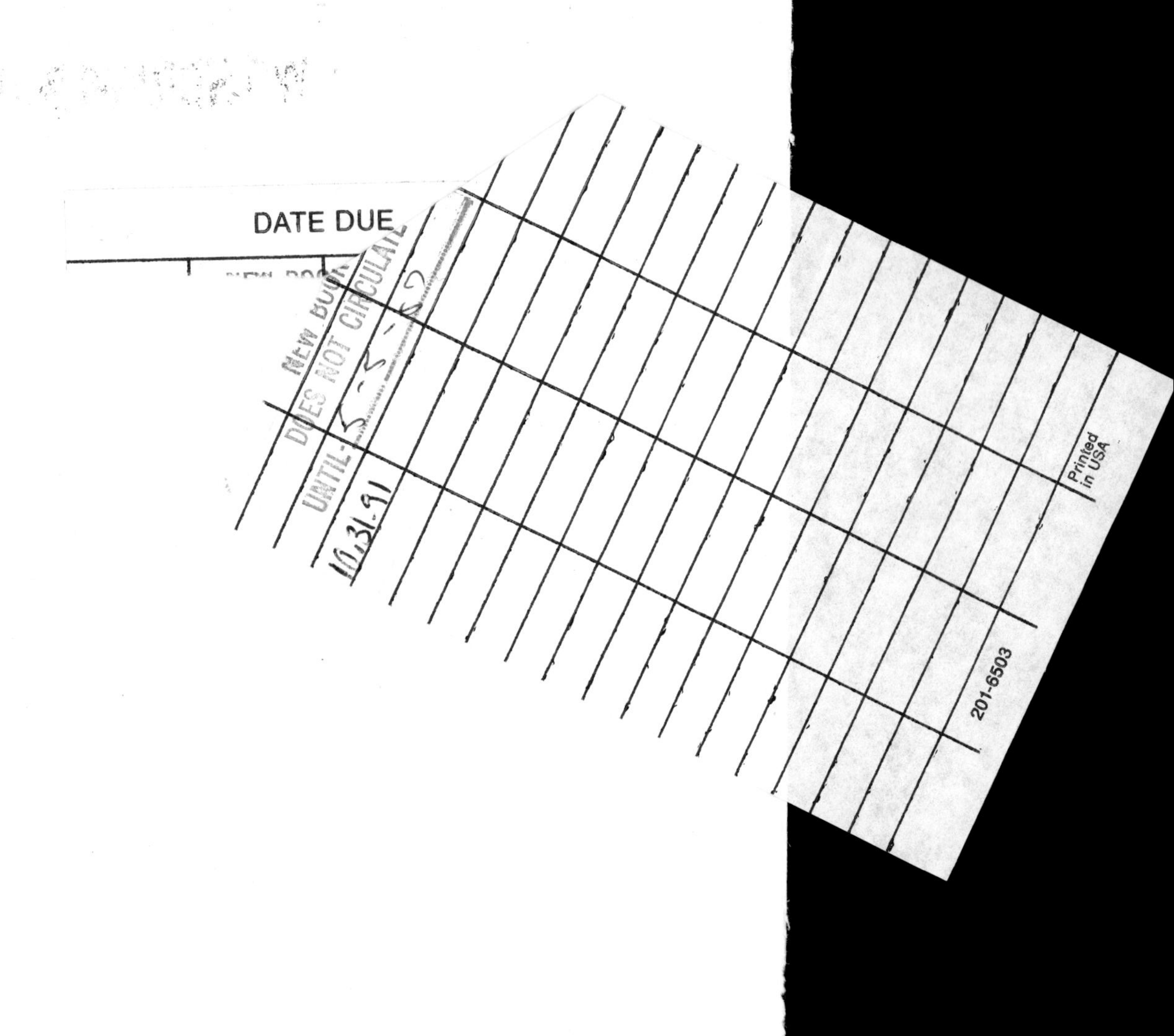
DATE DUE
NEW BOOK
DOES NOT CIRCULATE
UNTIL
Printed in USA
201-6503